S0-CAH-270

METAL-SURFACE REACTION ENERGETICS

THEORY AND APPLICATIONS TO HETEROGENEOUS CATALYSIS, CHEMISORPTION, AND SURFACE DIFFUSION

EDITOR
Evgeny Shustorovich

Evgeny Shustorovich
Corporate Research Laboratories
Eastman Kodak Company
Rochester, NY 14650-02001, U.S.A.

Library of Congress Cataloging-in-Publication Data

Metal-surface reaction energetics : theory and applications to heterogeneous catalysis, chemisorption, and surface diffusion / Evgeny Shustorovich, editor.
p. cm.
Includes bibliographical references and index.
ISBN 0-89573-776-0
1. Chemical reaction, Rate of. 2. Catalysis. 3. Surface chemistry. I. Shustorovich, Evgeny.
QD502.M48 1991
541.3'9–dc20 91-26086
CIP

British Library Cataloguing in Publication Data

Shustorovich, Evgeny
Metal-surface reaction energetics : theory and applications to heterogeneous catalysis, chemisorption, and surface diffusion.
I. Title
546.3

ISBN 3-527-27938-5

Printed in the United States of America

ISBN 0-89573-776-0 VCH Publishers
ISBN 3-527-27938-5 VCH Verlagsgesellschaft

Printing History:
10 9 8 7 6 5 4 3 2 1

Published jointly by:

VCH Publishers, Inc.
220 East 23rd Street
Suite 909
New York, NY 10010

VCH Verlagsgesellschaft mbH
P.O. Box 10 11 61
D-6940 Weinheim
Federal Republic of Germany

VCH Publishers (UK) Ltd.
8 Wellington Court
Cambridge CB1 1HZ
United Kingdom

Preface

The worldwide value of products made via catalytic technology exceeds two trillion dollars per year, and heterogeneous catalysis leads the way. This tremendous economic impact is the major thrust of intensive efforts in industry and academia to understand at the molecular level the mechanisms of chemical reactions on surfaces. The bottom line is knowledge of energetics of chemisorption, particularly the heats of adsorption and reaction activation barriers, which are notoriously difficult to accurately measure and compute. Recent years have witnessed impressive advances in quantum mechanical and phenomenological calculations of energetics of surface reactions. Remarkably, these are not "brute force" computations but, rather, heuristic calculations providing an easily comprehended view of reality. These advances have made possible a good understanding of chemical transformations occurring on the surfaces of transition metals, which are the intrinsic parts of many industrial catalysts.

This book summarizes major developments in the theory of reaction energetics on metal surfaces and its applications to chemically important phenomena. Specifically, the book focuses on ways of calculating the heats of adsorption and reaction activation barriers, and how these energetics determine the diffusion of adsorbed species and their chemical reactivity revealing in dissociation, recombination, and disproportionation.

The book comprises five chapters on these topics, written by active researchers in the field. Of course, each author chose the scope of issues and the manner of reasoning. The Editor's main task, therefore, was to make the chapters parts of *one* book, complementary in substance and coherent in conclusions. Although this is a theoretical book, it is addressed primarily to various users of theory—practitioners and students of surface science, heterogeneous catalysis, and quantum chemistry. For this reason, the authors were encouraged to emphasize the conceptual framework and logic, sometimes at the expense of technical details. The Editor expresses his gratitude to the authors for their patience in making endless "final" revisions and adjustments.

Contemporary surface science and heterogeneous catalysis deal with so diverse phenomena and make use of so diverse techniques that mutual understanding between theorists and experimentalists becomes critical for steady progress. It is hoped that this book, written by concerned practitioners, will contribute to this vital understanding.

Evgeny Shustorovich

Contents

Contributors

Roger C. Baetzold, Corporate Research Laboratories, Eastman Kodak Company, Rochester, NY 14650-02001, U.S.A.

Alexis T. Bell, Center for Advanced Materials, Materials and Chemical Sciences Division, Lawrence Berkeley Laboratory and Department of Chemical Engineering, University of California, Berkeley, CA 94720, U.S.A.

Jay B. Benziger, Department of Chemical Engineering, Princeton University, Princeton, NY 08544-5263, U.S.A.

David Halstead, Surface Science Research Centre and Department of Chemistry, University of Liverpool, P.O. Box 147, Liverpool L69 3BX, England

Stephen Holloway, Surface Science Research Centre and Department of Chemistry, University of Liverpool, P.O. Box 147, Liverpool L69 3BX, England

Evgeny Shustorovich, Corporate Research Laboratories, Eastman Kodak Company, Rochester, NY 14650-02001, U.S.A.

Per E. M. Siegbahn, Institute of Theoretical Physics, University of Stockholm, Vanadisvagen 9, S-11346 Stockholm, Sweden

Ulf Wahlgren, Institute of Theoretical Physics, University of Stockholm, Vanadisvagen 9, S-11346 Stockholm, Sweden

CHAPTER 1

Cluster Modeling of Chemisorption Energetics

Per E. M. Siegbahn and Ulf Wahlgren

I. Introduction

Ab initio quantum chemistry has undergone rapid development during the last 20 years. At the end of the 1960s high-accuracy quantum chemical results for molecules were essentially restricted to the hydrogen molecule and some other few-electron diatomic molecules. Systems of chemical interest were treated only at a low-accuracy semiempirical level. At present, gas-phase reactions can be routinely handled for up to 10 (and more) first-row atoms to a very high degree of accuracy [1]. *Ab initio* quantum chemical calculations have therefore become a more natural ingredient in studies in most areas of modern chemistry. The goal of the project covered in this chapter is to carry this potentially highly accurate approach over to the area of surface science.

From a fundamental point of view, a theoretical treatment of a metal should be based on the infinite nature of the system and thus a description of the electronic states by Bloch waves. This type of approach exists but is only in an early phase of development. Most notable is the crystal-orbital method of Pisani and co-workers [2]. For the future this method may be promising, but at present there are problems in the application of the method to metals. One problem is the basis set overcompleteness that appears when diffuse functions (which are needed) are introduced. To keep the perfect translational symmetry for chemisorption problems and still not cause too large interactions between the adsorbates, very large effective cell sizes are required, which is another problem. How electron correlation should be accurately treated is also a yet unresolved problem. When the focus is on adsorbate–adsorbate interactions rather than on metal–adsorbate interactions, another interesting method based on the infinite nature of a metal surface has been suggested by Feibelman [3]. In this method the difficult problem of determining substrate and adsorbate electronic structure at the same time is divided into two easier problems. The first problem is then to determine Bloch waves of the unperturbed substrate. The second problem is to determine the wavefunctions of the adsorption system as a whole, which are Bloch waves of the substrate scattered by the screened adsorbate-induced potential. For example, this method has been successfully applied to the study of the migration of impurity atoms in the presence of surface defects. Similar to this method in spirit are the Green's function techniques, where the Dyson equation for the adsorbate and substrate is solved starting with the solution of the unperturbed periodic system [4]. So far only a few applications have been made using this technique, and the systems studied, such as a linear chain of lithium atoms doped with hydrogen, have not been very realistic.

Another method where the infinite surface is actually included in the model is the full-potential linearized augmented plane wave (FLAPW) method of Freeman and co-workers [5]. In this method the surface is treated as a slab of a selected thickness and the electronic structure is obtained using local density methods. The FLAPW method has been used to study free surfaces, interface phenomena, and catalytic promotion and poisoning of molecular dissociation on surfaces. One limitation of the method is that the latter process can be studied only indirectly without actually dissociating the molecule. Obviously, the method is also limited by the accuracy of the local density method employed. The FLAPW method is similar in spirit to the muffin-tin Xα method, which for small molecules is known to behave in a rather unpredictive way, with sometimes excellent and sometimes very poor results. Overall, the FLAPW method appears to treat the properties of the surfaces best and, for these systems, seems to be the method of choice if the very long computational times can be afforded.

Another approach, which provides valuable new insight into the electronic structure of adsorbates on surfaces, is the effective medium approach [6]. This approach starts out with the energy of an atom in a uniform electron gas, and includes semiempirical corrections for the actual situations on a surface. Other variants of this method are the embedded atom method [7], the corrected effective medium method [8], and the embedded diatomics-in-molecules method [9]. The latter method has, for example, been shown to give a reasonably accurate representation of the potential surface for H_2 dissociation on nickel. To reach sufficient accuracy, however, the potential surface had to be heavily parameterized including more than 25 empirical parameters. The infinite crystal-orbital equations have also been solved within a purely semiempirical framework such as the extended Hückel method [10]. From a qualitative viewpoint, interesting results have been obtained using this approach for the band structure of rather complicated systems. For chemisorption problems, however, more questionable results emerge, and it is not clear whether the accuracy is high enough to be useful (see also Section VIA). For more details on these crystal orbital-type methods, the reader is referred to a recent excellent review by Sauer [11], where additional references can be found.

From a molecular chemist's point of view, the simplest and most natural model of a surface is a cluster of metal atoms. The cluster model has been used to treat chemisorption phenomena on transition metal surfaces for at least two decades. This has been done in connection with the use of both *ab initio* quantum chemical methods and local density and semiempirical methods. The most widespread semiempirical method is the atom superposition and electron delocalization (ASED) approach of Anderson [12]. This method is similar to the extended Hückel method but includes a more sophisticated atom–atom repulsion. Cluster models with up to 100 atoms have been treated using this method, and recently a crystal-orbital version of the method has also been completed. As with the extended Hückel method [10,13], the results of the ASED method depend critically on the specific parameterization, and the accuracy is therefore different from case to case. The usefulness of these methods also tends to be very dependent on the cleverness and intuition of the individual user; the general value of the methods is therefore difficult to judge. The electronic structure problem of clusters with adsorbates has also been treated by many variants of the local density approach. As a whole this type of approach must be considered as fundamentally much more rigorous than the purely semiempirical methods. Clearly, this does not mean that the results obtained using local density methods must always be closer to experimental results than the semiempirical results, but agreement in numbers is not the whole purpose of a theoretical treatment. A conceptual microscopic understanding is also important. The most advanced local density method that has been applied to cluster models of surface problems is the local spin density (LSD) method of Salahub and co-workers [14,15]. For chemisorption energies the normally quoted accuracy is about 10%, the values always being too high. This accuracy is, for example, obtained for CO chemisorption on palladium [14]. Sometimes, however, the errors using this approach can be much larger; in particular the chemisorption energies can exceed the experimental values by almost a factor of 2, as in the case of carbon chemisorption on nickel clusters [15]. In recent years the most interesting new development of the local density methods has been the incorporation of gradient terms of the density [16]. When this type of approach is applied to surface problems, great improvements in the results can be foreseen.

The pioneering work on the cluster model in connection with conventional quantum chemical methods was made by Goddard and co-workers [17]. They showed that quite reasonable results can be obtained if the transition metal atoms are treated as one-electron systems using effective core potentials (ECPs). This is of course a dramatic simplification and makes modeling of surfaces with up to a hundred atoms feasible; however, a fundamental question that has only recently been seriously addressed is how much accuracy is lost by this one-electron ECP approximation. We discuss this problem throughout this chapter but mainly in Sections II and III. If the one-electron ECP approximation is not used and the metal atoms are treated at the all-electron level, one can today perform medium-level-accuracy (Hartree–Fock) calculations with up to 5–10 transition metal atoms using conventional self-consistent field (SCF) methods. Using the direct SCF approach [18] one has been able to treat up to 34 copper atoms, but these calculations are quite expensive and, in the near future, will be interesting mainly as benchmark-type calculations. If electron correlation of the $3d$ electrons is to be included, a calculation including 4 or 5 transition metal atoms is today still a challenge, but these types of calculations will soon become more common. The best cluster model for the future, as we see it, contains at least 25 atoms of which the atoms closest to the adsorbate (up to 5 atoms) are treated at the all-electron level including 3*d* correlation. To perform this type of calculation, slightly more advanced methods and somewhat more powerful computers are needed.

Apart from the accuracy of the one-electron ECP approximation, the most fundamental question in the cluster modeling of chemisorption energetics is, how many atoms are needed to reproduce actual metal surface conditions. This question has not before been systematically investigated at a high enough level of accuracy. Strong opinions on this point exist, however. For example, it is often taken for granted that the Fermi level has to be well reproduced by a cluster if it should be reasonable as a surface model. Similar opinions exist about the reproduction of the continuous density of states. A quantitative reproduction of the Fermi level, or the work function, requires very large clusters. One can show [19] that a spherical cluster with radius 10 Å (containing up to 400 atoms) will have an ionization potential more than half an electron volt too high compared with a real surface. The results and the picture of the surface chemical bond emerging from, for example, extended Hückel calculations [10] strongly indicate that most of this error in the position of the Fermi level should enter into the chemisorption energy. A similar picture is obtained from other methods, such as the effective medium method and other embedded atom methods [6–9]. An error in the chemisorption energy of at least 10 kcal/mol would thus be expected even for a 300- to 400-atom cluster. In line with these general arguments, the results of actual cluster calculations seemed to show that the chemisorption energy is very poorly described by a finite cluster. In particular, because calculations gave results that were strongly oscillating with cluster size, it was rather firmly concluded that the cluster model was not useful for modeling chemisorption energetics. So far, attempts to remedy the poor cluster convergence by embedding methods have been moderately successful. Most notably, in the method of Whitten and Pakkanen [20], the cluster is embedded in a larger cluster treated with lower accuracy. This method is, however, still limited in accuracy by the size of the embedding cluster, which has so far been less than 100 atoms. As its usefulness for producing reliable chemisorption energies was considered doubtful, the cluster model was instead used to study more general problems of a qualitative nature, for example, characterization of the surface chemical bond as being covalent or ionic in origin. This was the situation in 1985 when our engagement in cluster modeling started. One of the first studies of our project was to systematically calculate the chemisorption energy for clusters of different size to understand the origin of the cluster oscillations in detail. It turned out that by using a quite different picture, a molecularly oriented rather than the conventional solid state-oriented picture, it was indeed possible to go beyond a qualitative understanding of the bonding [21]. In this new picture, the exact position of the Fermi level is of rather limited importance to the strength of the chemisorption bond [22]; more importantly, with this new understanding it becomes possible to correct the cluster chemisorption energies to be useful for surface predictions. These results are described in Section IV, in which we also explain why the oscillatory behavior of the reactivity as a function of cluster size, found for some transition metals in molecular beam experiments, does not directly translate into corresponding oscillations for the cluster model as it is used. There are important differences between the clusters used as theoretical models of surfaces and the actual clusters in molecular beams.

In the first half of this chapter we concentrate on the description of the cluster model and its implications for the nature of the surface chemical bond. The importance of being able to go back and perform very accurate calculations on small systems to check the model is stressed. This feature, we believe, makes the present approach unique among the different approaches available for treating surface problems, and therefore deserves to be especially emphasized. In the second half of the chapter we describe some representative results that have been obtained using this model. In Section V atomic chemisorption is discussed, exemplified by hydrogen, carbon, nitrogen, and oxygen on nickel and copper. In Section VI molecular chemisorption, in particular CH_x chemisorption, is discussed. In Section VII detailed studies are presented for H_2 and O_2 dissociation on different Ni(100) and Cu(100) sites. In that section some results for CH_4 dissociation are also discussed.

II. The Cluster Model: Use of One-Electron Effective Core Potentials

A. General Considerations

As mentioned in Introduction, the present limit for pure *ab initio* calculations on transition metal clusters is about 5–10 atoms. It is clear that one of the first steps in an unbiased investigation of clusters as models for metal surfaces has to be a study of the convergence of certain properties with cluster size. For this purpose a limit of 5–10 atoms is not enough. Within the same computational framework, there is at present only one possible and general way out of this problem—to model the transition metal atoms as one-electron systems. The work by Goddard and co-workers [17] has shown that this type of modeling, at least in a qualitative sense, can be quite reasonable. We return below to situations when this procedure might work and when it is unlikely to work.

The first general question one can ask is for which transition metals the one-electron model is expected to be best and for which metals less accurate results should be expected. The answer to this question starts out with a consideration of the ground state of a free metal atom. In the metal bulk or in a surface, transition metals (at least on the right-hand side of the periodic table) will be predominantly in an atomic state with one outermost s electron. We call this atomic state the $d^{n+1}\,s$ state. This d occupation is also found to be dominant and optimal in most complexes of these transition metals [23]. The general idea in the one-electron model of transition metals is that this state can be described as a one-electron system with s character, where all the other orbitals including the d electrons act only as a potential. For every free transition metal atom there also exist low-lying states of different types with occupations $d^n s^2$ and d^{n+2}. Unlike the $d^{n+1}\,s$ state, these latter states will not have radical character in the s shell and will therefore be less optimal for bonding.

Table 1.1. Relative Energies (eV) for Low-Lying States with Different *d*-Oribtal Occupations of Some Transition Metal Atoms

Atom	Occupation	Energy	Atom	Occupation	Energy
Fe	d^6s^2	0.00	Co	d^7s^2	0.00
	d^7s	0.87		d^8s	0.42
	d^8	4.07		d^9	3.36
Ni	d^8s^2	0.03	Cu	d^9s^2	1.49
	d^9s	0.00		$d^{10}s$	0.00
	d^{10}	1.74			
Ru	d^6s^2	0.87	Rh	d^7s^2	1.63
	d^7s	0.00		d^8s	0.00
	d^8	1.09		d^9	0.34
Pd	d^8s^2	3.38	Ag	d^9s^2	3.97
	d^9s	0.95		$d^{10}s$	0.00
	d^{10}	0.00			

It is clear that the lower the d^ns^2 and d^{n+2} states are for the free atoms, the more they will complicate the simple picture of a metal as having all the atoms in clean $d^{n+1}s$ states. The general and obvious rule is that the one-electron model of a transition metal in a surface or bulk should be best for the atoms that have $d^{n+1}s$ ground states as free atoms. The lowest-lying states for some metal atoms are given in Table 1.1, with relative energies taken from Moore's tables [24]. Of the atoms given in this table, copper and silver are expected to behave most like one-electron systems in a metal surface, but ruthenium, rhodium, and nickel might also be reasonable. Cobalt and iron have d^ns^2 ground states, whereas palladium has a d^{n+2} ground state, and for these metals it is clear that the one-electron approximation is much more questionable.

In the one-electron approximation all the effects of the *d* shell have to be approximated, and another, related property of interest in this context is the ratio between the extents of the $(n+1)s$ and *nd* orbitals, $<r_{(n+1)s}> / <r_{nd}>$ [25]. For copper this ratio is as high as 3.36. For many chemists, who might find it unusual to consider copper as approximately a one-electron system, this very large difference in radial extent of the two orbitals should be informative. For nickel the ratio between the orbital extents is 3.22, which is slightly smaller but still comfortably large to attempt a one-electron approximation. As one approaches the center of the periodic table the ratio decreases, and for chromium it is down to 2.69. For second-row atoms the situation is more critical. Even an atom like silver, which has the most favorable atomic splitting of all the atoms in Table 1.1, has a ratio of the two orbital extents of only 2.67, which is similar to that for chromium. Of the transition metals we have mentioned here, nickel and particularly copper are expected to be the least problematic to model as one-electron systems, and in this chapter we concentrate on these two atoms. Already at the onset it might be considered doubtful to pursue an approach that might work only for a few transition metals; however, transition metal surfaces like nickel, cobalt, and iron, for example, are experimentally found to be similar at a qualitative level, so that a detailed understanding of reactions on one of these metals may have more general implications than for just this metal. Also, general and important questions like the delocalized or localized nature of the surface chemical bond should be possible to address and understand from a knowledge of just a few transition metal surfaces.

It is clear that the interaction energy between a metal atom and an adsorbate is less sensitive to the details of the electronic structure of the metal atom the longer the distance is between the metal atom and the adsorbate. When an adsorbate is chemisorbed on a surface, we must therefore identify three different adsorption sites that might require different treatments. First, the shortest metal–adsorbate distance generally occurs when the adsorbate is chemisorbed at an on-top site. Typical bond distances are here the same as found between a ligand and a transition metal in molecular complexes. For nickel and copper this means distances of about 1.5 Å for a hydrogen atom and about 1.75 Å for carbon, nitrogen, and oxygen atoms. The next adsorption site to be considered is the bridge site where the bond distances are expected to be somewhat longer than at the on-top sites, and the final site of interest is the hollow site. In the hollow sites of the (100) surfaces of nickel and copper the bond distance has typically increased to 1.8 Å for hydrogen atoms and to 1.9 Å for oxygen atoms. Carbon and nitrogen atoms reconstruct the surfaces (see the discussion in Section V) and represent special cases. For a typical molecular adsorbate such as CO the bond distance to the nearest metal atom is still another 0.5 Å longer than for an oxygen atom. With the rather large differences in metal–adsorbate bond distances between the different sites, the one-electron approximation of the metal atoms is thus expected to be much better in a hollow than in an on-top site.

The first adsorption site we consider in detail here is the on-top site. Even if adsorption at this site gives rise to the shortest metal–adsorbate bond distances, this site is still the least problematic to describe. The reason for this is that the shortest metal–adsorbate bond distance concerns only one metal atom. If the one-electron approximation for this atom turns out not to be adequate, this metal atom can be described at the all-electron level instead, and the calculation is still possible to perform. The one-electron approximation for on-top chemisorption has been studied in detail by Bagus *et al.* [26]. Sequences of calculations were performed for the systems Ni_5CO and Cu_5CO with CO chemisorbed on-top. In the first set of calculations all metal atoms were described as all-electron atoms. In the second set of calculations a mixed cluster treatment was used where all metal atoms except one, the on-top atom, were described as one-electron systems. Finally, in the third set of calculations all metal atoms were described as one-electron systems. The results are quite conclusive. The mixed cluster treatment is a very accurate model, giving results close to those of the all-electron treatment for both geometry and chemisorption energy. Treating the on-top metal atom as a one-electron system, on the other hand, gives quite poor results. From these results we draw two important conclusions. First, the bond distance to the on-top metal atom is too short to allow a one-electron description of this atom, and second, the bond distance to the next nearest neighbors is large enough to safely allow this approximation to be used for these atoms. As an all-electron description of one metal atom is always affordable, the computational treatment of on-top adsorption is therefore clear: it should be of the mixed cluster type. Pushing the resources, a similar treatment of bridge chemisorption should also be possible with two metal atoms described at the all-electron level. The main question remains: Is the metal–adsorbate bond distance appearing in hollow positions large enough to allow a description where all the atoms are described at the one-electron level. From the results for on-top bonding this is not yet clear, as the bond distances from the adsorbate to the metal atoms in the hollow positions are typically in between those for the two cases discussed with respect to on-top bonding (to the on-top atom and to its nearest neighbors). The possibility of describing the hollow chemisorption in the cheapest possible way would be very important as this is the dominant chemisorption site for covalently bound atomic and molecular adsorbates. This question has therefore been emphasized from the beginning of the present project [27–29].

The present work differs from the previous work by Goddard and co-workers in two aspects. The first difference has to do with the connection between the semi-quantitative model calculations based on the one-electron ECPs and the all-electron model calculations. In the Goddard approach, all-electron calculations play a minor role and enter only at the atomic level. In our work the situation is reversed and the one-electron ECPs are used only to obtain the cluster size effect. Whereas the one-electron ECP results were the final answer in the work by Goddard and co-workers, the results obtained at this level should in the ideal use of our scheme always be corrected against all-electron results for small clusters. This difference in the two approaches is also reflected in the parameterization, where we determine the *d* orbital parameters against all-electron results for a small relevant cluster, at both the frozen and the relaxed *d*-orbital levels. In the Goddard approach all the parameters were determined at the atomic level. We left the original Goddard scheme essentially because we found that *d*-shell relaxation can affect the chemisorption energy substantially, and it is thus important to model this specific effect reasonably accurately. In our approach the one-electron ECPs serve only to model rigorous results obtained for smaller systems. The second aspect where our approach is different from the original Goddard scheme is more technical and has already been mentioned. We believe that the success of modeling the interaction with a transition metal using a one-electron ECP is strongly dependent on the distance between the adsorbate and the metal. Therefore, unlike Goddard *et al.*, we do not attempt to model on-top chemisorption using these ECPs. As mentioned in Introduction, the ultimate continuation of the present approach is to always use a mixed cluster treatment independent of the adsorption site. At present, these types of calculations are simply too expensive for hollow sites.

B. Computational Model and Details

To describe transition metal atoms as one-electron systems we use an ECP method. The basic formalism is the same as used by, for example, Bonifacic and Huzinaga [30], which means that the full nodal structure of the valence orbitals is always kept.

An important modification of the original ECP method is that we, based on work by Pettersson *et al.* [31], also describe some core orbitals explicitly as frozen orbitals. The expression for the one-electron (closed shell) Fock operator in this frozen ECP formalism is then as follows:

$$\hat{F} = \hat{h}^{\mathrm{eff}} + \sum_{c}^{\mathrm{val}} (2\hat{J}_i - \hat{K}_i)$$

where

$$\hat{h}^{\mathrm{eff}} = \hat{T} + \hat{V}^{\mathrm{eff}} + \hat{P} + \sum_{j}^{\mathrm{frozen}} (2\hat{J}_j - \hat{K}_j)$$

$$\hat{V}^{\mathrm{eff}} = \left(\frac{-Z^{\mathrm{eff}}}{r}\right)(1 + M_1 + M_2)$$

$$M_1 = \sum_{p} A_p \exp(-\alpha_p r^2)$$

$$M_2 = \sum_{q} r C_q \exp(-\gamma_q r^2)$$

$$\hat{P} = \sum_{k} |\phi_k > B_k < \phi_k|$$

Of the terms in the effective one-electron operator, $\hat{V}^{\mathrm{eff}}$ will describe a screened nuclear attraction and exchange effects, and the projection operator $\hat{P}$ will model mainly the Pauli repulsion between the core orbitals and orbitals on the same and on other centers in a molecule. The parameters that enter these expressions are B_k, A_p, α_p, C_q, and γ_q. The B_k values are usually chosen as the absolute values of the corresponding core orbital energies. A_p, α_p, C_q, and γ_q are calibrated in atomic calculations such as to reproduce the orbital energies and shapes of the valence orbitals. The outer core orbitals are expressed in the valence basis set by a least-squares fitting procedure and are kept frozen during both the parameter fitting and the molecular calculations.

As the *d* orbitals are not allowed to relax in a one-electron ECP it may appear that a requirement, which has to be fulfilled for the one-electron ECP approach to be applicable, is that the frozen *d* approximation should be valid; that is, the relaxation of the *d* orbitals should not influence the chemisorption energy appreciably. We show later that in many cases the *d*-orbital relaxation on the contrary gives substantial contributions to the chemisorption energy. It is clear that if this *d*-orbital contribution is a pure covalency effect between the *d* orbitals and the adsorbate, this effect will be rather difficult to model correctly by an ECP description. If, on the other hand, the *d* relaxation effect can be described as a polarization to avoid Pauli repulsion with orbitals on the adsorbate, such effects should not be more difficult to model by an ECP than normal repulsive effects. The essential point is that the relaxation effect must be dominated by the cluster and not by the adsorbate. From rigorous model calculations on small systems (see Section III) we have found that the major part of the *d*-orbital relaxation effect on the chemisorption energy, at least in the cases investigated, is to reduce the initial repulsive *d*-shell effect by a slight *sd* hybridization. We have therefore chosen to model this effect simply through a modification of the *d* projection operator, by "softening" it in the outer regions with the introduction of a second diffuse and slightly attractive *d* projection operator. Two different types of ECP have been developed. The first type is more accurate including a frozen 3*s* (and sometimes a frozen 3*p*) orbital and basis functions describing the 4*p* orbital. This ECP is normally used only for the metal atoms directly in contact with the adsorbate. For the other metal atoms, with larger distances to the adsorbate, a smaller ECP has been found sufficiently accurate. This second type of ECP has no frozen orbitals and no 4*p* basis functions. For both ECPs two basis functions are used to describe the 4*s* orbital. The details of the one-electron ECP used for nickel can be found in Ref. 27 and those for copper in Ref. 28.

Apart from the introduction of one-electron ECPs for the transition metal atoms, the rest of the calculations, including electron correlation, are performed at a normal *ab initio* level. In particular, we have found that the results are sensitive to the description of the adsorbate requiring large basis sets. These basis sets have included up to *f* functions on adsorbates like carbon and oxygen. Near-degeneracy effects on the energetics are surprisingly uncommon when chemisorption bonds are formed. When they occur, the general complete active space SCF (CASSCF) method [32] has been used to generate the orbitals. As will be shown later (see Section III) electron correlation has dramatic effects on the chemisorption energy. In particular, correlation of the electrons on the adsorbate and in the adsorbate–substrate bonds is necessary for even a qualitative calculation of the chemisorption energy. For this purpose the multireference contracted configuration interaction (CCI) method [33] has been used in most calculations described here. In the largest calculations basis sets with up to two hundred functions and configuration expansions with up to two million terms have been used. These calculations are obviously rather expensive, but it should be pointed out that a more qualitative treatment (still including electron correlation) for a system like Ni_5O can be performed on a medium-size computer in less than 20 min. The speed of these latter calculations has been essential for the present project as the results we will present are normally not based on one big calculation, but are the combined knowledge obtained from sometimes hundreds of rather small such calculations.

C. Improvements of the Model

The main improvement we have implemented in the above model is an attempt to model also correlation effects resulting from the *d* shell. For this purpose we have used a core polarization potential (CPP) formalism originally proposed by Müller *et al.* [34] for alkali and alkaline earth metals, which has been modified and extended for the case of copper by Pettersson and Åkeby [35]. This formalism is based on a classical picture of the interaction between the core, in the case of copper including the *d* shell, and the valence electrons, which assumes that the core and valence electrons are reasonably well separated. This latter assumption is clearly more valid for alkali metals than for copper, which means that considerable attention has to be paid to copper in terms of defining, for example, core orbital localization schemes and cutoff parameters. To represent the core–valence interaction, the core electrons in this model are described as a polarizable charge distribution with polarizability α, which leads to the classical interaction energy $-\frac{1}{2}\alpha\mathbf{f}^2$, where $\mathbf{f}$ is the field from the surrounding charges. By introducing a $-\frac{1}{2}\alpha\mathbf{f}^2$ operator into the quantum chemical calculations two rather different effects will be described: first, the static polarization, which is normally obtained at the Hartree–Fock level, and, second, a dynamical polarization of the core, which is normally obtained only at a correlated level. Calculations made by Pettersson *et al.* [36] for oxygen chemisorption on copper have shown that the latter, normally more complicated, effect is quite well reproduced by this rather simple model. The former static effect is part of the $3d$ relaxation energy, which is, on the whole, better modeled by the attractive d projection operator as described above. Results from these model calculations are presented in Section III.

A second improvement in the model we have attempted is to modify the distance dependence of the interaction between an adsorbate and a surface. In the above formalism the repulsion to the d shell is proportional to the square of the overlap, S^2. This should be the leading term in the distance dependence of the repulsion. The next term should be proportional to S^4, which can be approximately modeled by an operator proportional to $\hat{P}^2$ (one should note that the $\hat{P}$ operator described above is not an idempotent projection operator). This operator will, like the other operators, contain a parameter that can then be obtained to fit the distance behavior as obtained by all-electron model calculations. The improvements obtained by this model operator still have to be demonstrated by a number of applications and we mention this extension here mainly to point out one way in which the present model in principle can be systematically improved.

III. Energy Decomposition of the Surface Chemical Bond

One important aim of model calculations is to generate a detailed understanding of surface phenomena in terms of generally accepted concepts such as correlation effects, orbital polarization, and core–valence interaction. This is in our opinion almost as important as reproducing experimental results. It is thus necessary to compare results obtained on small clusters from both all-electron and simplified model calculations and, if necessary, to calibrate the cluster model so as to reproduce the all-electron results satisfactorily. The basic unit we decided to use for cluster model development was a five-atom cluster, arranged in a square pyramid with an oxygen atom approaching the base of the pyramid (see Fig. 1.1). This system serves as a model for oxygen chemisorption at a fourfold hollow site on a (100) surface of the metal. For the reasons given in Section IIA, the metals selected were copper and nickel. The calculations that will be described here were first carried out at the SCF level, and the technical details such as basis sets can be found in Refs. 27–29. As the

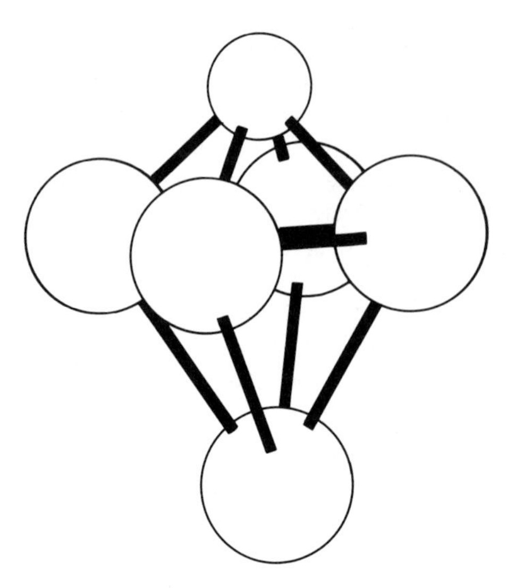

Figure 1.1. Fourfold hollow chemisorption of an atom on an $M_5(4,1)$ cluster.

aim of the calculations was to elucidate the role of *d* electrons in the chemisorption process, calculations were carried out using both frozen atomic *d* orbitals and relaxed *d* orbitals. The parameters that were determined in the calculations were oxygen binding energy and equilibrium distance above the surface. Technically the calculations were straightforward for Cu_5O, where all the *d* shells on Cu are completely filled. In Ni_5O some uncertainty is present (up to 5 kcal/mol) at the frozen *d*-orbital level because of the presence of open *d* shells. To judge the relative importance of the different effects described below it is useful to know that the experimental chemisorption energy for oxygen on Ni(100) is 115–130 kcal/mol [37] and that for oxygen on copper is 100–115 kcal/mol [38]. An effect, or an error in the description of an effect, of only a few kilocalories per mole will therefore be considered as nearly negligible.

A. Relaxation of the *d* Orbitals

The first effect we consider is the *d*-orbital relaxation energy, defined as the difference at the SCF level between the chemisorption energy obtained when the *d* orbitals are frozen in their atomic shapes and when they are allowed to completely relax in the molecular surroundings. In light of the previous use of one-electron ECPs by Goddard and co-workers, where an essentially frozen *d*-orbital description was implicitly assumed, the results of the all-electron investigations were quite surprising, particularly for the nickel system. The *d*-shell relaxation contribution to the chemisorption energy is as high as 44 kcal/mol for Ni_5O and 17 kcal/mol for Cu_5O. The total oxygen binding energy in Ni_5O is 42 kcal/mol at the SCF level, which means that, in a sense, all the binding at this level comes from 3*d*-orbital relaxation;

however, as will become clear below, the qualitative aspects of bond formation are correctly described already at the frozen *d*-orbital level.

There are two possible explanations for the large effect of *d*-orbital relaxation on the oxygen binding energy. The first explanation is that the system is trying to minimize the repulsion between the adsorbate and the *d* orbitals, in which case the origin of the relaxation is a polarization or hybridization of the *d* orbitals away from the adsorbate. This picture is consistent with the much larger tendency for *sd* hybridization found for Ni than for Cu in metal complexes. For NiCO, for example, the hybridization between 4*s* and 3*d* to minimize repulsion to the carbon lone pair is a dominant factor for the binding, and is due entirely to the presence of open *d* orbitals [23]. The second explanation for the *d*-orbital relaxation effect is that there is an attractive covalent interaction with the adsorbate. Based on the small size of the *d*-orbital coefficients in the oxygen orbitals obtained after an orbital localization, this possibility seemed rather unlikely; however, orbital coefficients may not be a certain measure of covalency effects, and to quantify the covalent contribution of the *d* orbitals to the cluster adsorbate bond a constrained space orbital variation (CSOV) [39] analysis was carried out on Cu_5O [40]. The importance of determining the magnitude of the *d* covalency effect is that the size of this contribution is expected to be connected to the reliability of the one-electron ECP description. As discussed in Section IIB, use of one-electron ECPs to model a 3*d* relaxation effect, which can be described as a slight mixing of cluster orbitals, is expected to work equally well as modeling normal repulsive effects from the 3*d* orbitals; however, using a similar approach to model a direct covalent interaction should introduce significant errors when different adsorbates are compared.

The CSOV technique is a method that provides a detailed analysis of the different contributions to a chemical bond. This is done by freezing or constraining subsets of the occupied orbitals at different stages of the calculation, sometimes in combination with constraints on the virtual space. Our CSOV analysis consisted essentially of three steps: (1) the binding energy was calculated with all *d* orbitals frozen in their atomic shapes, (2) the *d* orbitals were relaxed but the rest of the valence space (as obtained from step 1) was frozen, and (3) a fully relaxed calculation. As the bonds between the valence orbitals of the cluster and the adsorbate are formed in the first step and are not allowed to change in the second step, the energy difference between the second and the third steps provides a measure of the importance of the *d* orbitals for the covalent bonds. The results of the CSOV analysis are shown in Table 1.2. The energy difference obtained between the second and third steps is only 2 kcal/mol, which is thus our best estimate of the covalent contribution of the *d* orbitals to the chemisorption energy. In comparison to the total chemisorption energy of 100–115 kcal/mol, the covalent *d* contribution is negligible. The relaxation effect of the *d* orbitals on the binding (17 kcal/mol) is caused almost exclusively by polarization of the *d* orbitals in the cluster. The repulsion between the adsorbate and the *d* orbitals at the frozen orbital level can be rationalized in terms of Pauli exclusion effects, that is, the orthogonalization of the adsorbate orbitals against the *d* orbitals in the cluster. We should mention here that in the second step described above we

Table 1.2. CSOV Analysis of the Bonding in Cu_5O^a

Step	ΔE^b (kcal/mol)	$D_e{}^c$ (kcal/mol)	3*d* Occupancy	$Q(O)^d$
1	31.8	7.6	50.00	−0.85
2	1.6	25.4	49.78	−0.88
3	0.0	27.0	49.76	−0.84

[a]Step 1 is the frozen *d*-orbital calculation and Step 3 is the totally relaxed *d*-orbital calculation. Step 2 is a calculation where the covalent *d* interaction with the adsorbate is still not allowed (see text for further details).
[b]Difference from the relaxed 3*d* calculation.
[c]Chemisorption energy relative to $Cu_5(^4A_2)$ + $O(^3P)$
[d]Charge on oxygen (a.u.)

actually carried out several calculations using different subsets of the virtual space as well, but this gave no conflicting information.

A similar investigation on the Ni_5O cluster turned out to be much more difficult to interpret because of the presence of open *d* shells [41]. As the *d* coefficients in the localized oxygen orbitals are still quite small and the chemisorption energy of oxygen on nickel is not very different from that of oxygen on copper, a large *d* covalency effect on Ni_5O is not predicted. A similar conclusion is made based on the very small effect on the chemisorption energy obtained by increasing the *d* basis set from a single function to a description containing three functions [42]; however, the CSOV calculations indicate that the *d* covalency effect may be as large as 5–10 kcal/mol and also that the separability of the valence and the *d* orbitals does not work as well for nickel as for copper. It should be noted that the exact relationship between a covalent *d* effect of 5–10 kcal/mol and the corresponding error in the chemisorption energy is not completely clear, as this effect is included in the modeling for one particular adsorbate. Taken together, a relative error of about 5% of the total chemisorption energy from modeling both repulsive and attractive effects from the *d* shell of nickel by one-electron ECPs should not be an unreasonable error estimate for an adsorbate in a hollow position. This means an error of about 3–6 kcal/mol for hydrogen, oxygen, fluorine, chlorine, CH_3, and CH_2, which has in fact been demonstrated in explicit calculations for these adsorbates in comparisons with all-electron results. One of the main advantages of the present type of approach, compared with entirely semiempirical modeling, is that the errors are normally rather easy to understand. This means that corrections can usually be made allowing predictions that are more accurate than the actual numbers computed.

B. Effect of Valence Correlation

One of the most important results obtained in the present project is that valence correlation effects dramatically influence the chemisorption energy and that these effects are possible to evaluate using present methodology. These effects have usually been neglected in *ab initio* studies of the present type. In some calculations the effects have been inferred but not actually calculated [43]. One reason is that such calculations were simply technically too difficult and costly. The most impor-

Table 1.3. Valence Correlation Effects, Δ, on the Chemisorption Energy, D_e, of Hydrogen and Oxygen in the Fourfold Hollow Site of Cu_5 and Ni_5 (One-Electron ECP Calculations)

System	D_e (kcal/mol)		Δ
	SCF	CI	(kcal/mol)
Ni_5H	33.4	54.1	20.7
Ni_5O	42.9	105.2	62.3
Cu_5H	24.5	45.3	20.8
Cu_5O	27.6	89.0	61.4

tant achievement of *ab initio* method development, during the period 1970–1985 is that highly correlated calculations on complicated systems containing a large number of electrons became possible to perform in a nearly routine fashion.

The large effect of valence electron correlation on chemisorption energies was first demonstrated in the systems M_5X (see Fig. 1.1), where M is copper or nickel as before and X is hydrogen or oxygen [27–29]. The results of these calculations are shown in Table 1.3. By valence electrons we mean here only the *s* electrons on the cluster and the adsorbate valence electrons. The most dramatic effect was found for oxygen chemisorption. The chemisorption energy obtained at the SCF level for Ni_5O is 43 kcal/mol, and the result after valence correlation is 105 kcal/mol, an effect of 62 kcal/mol. This effect is not only large but it also brings the resulting chemisorption energy to a value resembling the experimental value of 115–130 kcal/mol [37]. In fact, increasing the basis set on the adsorbate by adding additional *d* and *f* functions increases the valence correlation contribution to the chemisorption energy by still another 8 kcal/mol, leading to a final chemisorption energy in even better agreement with experiment. The fact that such a small cluster can give results in nearly quantitative agreement with results of experiments sheds new light on the use of clusters as surface models and also on the nature of the surface chemical bond (see also Section IV). The large effect of electron correlation is sometimes understood by referring to the large correlation effect on the electron affinity of atomic oxygen (46 kcal/mol) combined with the fact that oxygen becomes negatively charged (about −1.0) on a surface; however, as the correlation effect on chemisorption energy is substantially larger than the effect on the electron affinity, and the wavefunction indicates the formation of two rather than one bond to the surface, we prefer to make a comparison to the water molecule. The correlation effect on the formation of the two covalent bonds in water is about 55 kcal/mol, which is reasonably close to the effect on the chemisorption energy of oxygen on nickel. This similarity is one reason that led us to view the bonding for oxygen on a surface as predominantly covalent in contrast to the conventional ionic picture. In the covalent picture, the negative charge on oxygen is obtained by polarization of the covalent bonds, rather than as the result of an electron jump from the cluster to the adsorbate. This bond polarization does not even have to be very significant to affect the Mulliken charge substantially.

A few additional comments should be made about valence correlation effects on chemisorption. First, the correlation effect for hydrogen is as expected much smaller than that for oxygen but it is still significant, 21 kcal/mol, which is one-third of the chemisorption energy. Another interesting result is that the correlation effects for chemisorption on nickel and copper are remarkably similar differing by 1 kcal/mol or less, which suggests that the details of the bonding in the two systems may also be very similar. So far we have discussed only valence correlation effects on the chemisorption energy. The effects on the geometry turn out to be surprisingly small. On nickel, even though the effect on chemisorption energy is increased by a factor of 2.5 for oxygen, and the frequency for motion perpendicular to the surface is as low as 300–400 cm^{-1}, the effect on distance to the surface is only 0.04 Å. For oxygen and hydrogen on copper the effects are even smaller. In general, it can be said that it is very hard to draw conclusions on surface energetics based on changes in geometries or vice versa. Even the most dramatic reconstruction of a surface can be of rather minor importance to the energetics of a catalytic reaction.

Even though this chapter is addressed mainly to readers who are not quantum chemistry experts, a few technical comments can provide some general insight into the character of the surface bond. First, for a large cluster it is not necessary to correlate all the valence electrons of the cluster. For $Ni_{21}O$ a calculation correlating only the oxygen–surface bond and the other oxygen electrons gave a result that differed by only 1.6 kcal/mol compared with a calculation where all the cluster valence electrons were correlated. It is somewhat important to be careful in localizing the electrons on the adsorbate, and when a well-defined localization cannot be achieved it may still be necessary to correlate all the valence electrons. The localized results for $Ni_{21}O$ are important not only because they lead to a drastic simplification in the calculations, but also because they indicate that the chemisorption bond is strongly localized, a fact that is also borne out by the cluster convergence study presented in the next section. From the model point of view it is also useful to know that the valence correlation effects given in Table 1.3 as calculated using the one-electron ECP model have been compared with all-electron calculations [40] for the system Cu_5O and found to be in almost perfect agreement with these results.

C. Effect of 3*d* Correlation

The final contribution to the surface chemical bond energy we consider here is the effect of 3*d* correlation. The system studied is Cu_5O, and the 3*d* correlation contribution has been computed *ab initio* through CI calculations and also by using the CPP model described briefly in Section IIC. The calculations are done at the all-electron level. Correlation of the 3*d* shells was considered only for the four copper atoms in the first layer, as earlier calculations had indicated that the chemisorption energies are insensitive to the level of treatment of the 3*d* electrons of second-layer atoms. With this restriction, this still leads to the correlation of as many as 51 electrons and is at the limit of what can be handled using present CI techniques. The details of the calculations can be found in Ref. 36. At the *ab initio* level the additional effect on chemisorption energy of correlating the 3*d* electrons was found to be 13 kcal/mol. This relatively small effect was a result that was more or less anticipated [27–29] based on the rather good agreement with experiment already obtained without correlating these electrons. A large 3*d* correlation effect would have had to be canceled by a cluster size effect because this effect is the only essential effect missing at this stage, and this seemed improbable. The simple CPP operator is quite successful in modeling the 3*d* correlation effect and the result is 12 $\pm$ 4 kcal/mol, where the error bars arise from the use of different cutoff parameters. It seems clear that the CPP technique should be very useful as a model, as this approach can be used also in conjunction with the one-electron ECP approximation of the metal atoms and can thus be used to estimate 3*d* correlation effects for quite large systems. It should be added as a drawback that the CPP technique cannot, according to our experience, be reliably used to model 3*d* relaxation effects. With the present use of cutoff parameters the estimated 3*d* relaxation effect on the chemisorption energy becomes close to zero, which is far from correct (see above). Such modeling would have been desirable but is not a severe problem in our approach, as the modeling of these relaxation effects through the attractive projection operator described above works reasonably well. In light of the poor modeling of relaxation by the CPP, the significance of previous use of the CPP operator for copper treated at the one-electron ECP level [44], where both 3*d* relaxation and correlation effects were modeled simultaneously, can be questioned. As a totally empirical approach it could of course still be useful.

D. Summary

As we have considered all the major effects that should enter the description of the adsorbate bonding in M_5X, we are now ready to summarize. The bond energy contributions for Cu_5O are given in Table 1.4, with ΔD_e representing the contribution of a specific electronic effect and D_e the sum of all the effects up to that point. We

Table 1.4. Energy Decomposition of the Surface Chemical Bond for Cu_5O (All-Electron Calculations)[a]

Effect	ΔD_e(incr.) (kcal/mol)	D_e(sum) (kcal/mol)
SCF		
3*d* frozen	7.6	7.6
3*d* covalency	1.6	9.2
3*d* relaxed	17.8	27.0
CI		
valence	61.8	88.8
3*d*	13.2	102.0
Experimental	–	100–115

[a]See text for explanations.

should point out that the numbers quoted in the table are from Ref. 36, where basis sets larger than those in some of the calculations quoted above were used, so the numbers differ slightly at some points. The first conclusion we can draw, and which has already been indicated above, is that as the final result, 102 kcal/mol, is so close to the experimental surface result, 100–115 kcal/mol, the energy decomposition given in the table should be equally valid for the actual surface chemical bond. Because we know that additional basis set extensions on oxygen are responsible for yet another 8 kcal/mol (see above) we can conclude that the correct surface value should most probably be in the upper range of the 100–115 kcal/mol found in the different experiments. Another comment worth making concerns the first entry in the table, the SCF result obtained when the 3*d* orbitals are frozen. Because this chemisorption energy is only 7.6 kcal/mol, it does not mean that the contributions to the bonding covered at this point are unimportant. On the contrary, the covalent bond formation between the oxygen orbitals and the valence orbitals of Cu_5 is probably in reality the main component of the bonding. If the orbitals of the separated fragments, Cu_5 and O, are just superimposed a negative chemisorption energy of more than 100 kcal/mol would probably result. The importance of the *sp* covalency should really be counted from this reference. It should also be added that, in spite of the large energy changes, the wavefunction changes only marginally from the frozen 3*d* level to the final result.

The results in Table 1.4 constitute a benchmark reference against which all results using the present cluster model can be referred and evaluated. If a calculation is made at a level lower than the final one in Table 1.4, the remaining effects can be estimated and discussed with reference to the results in this table. As most cluster calculations discussed below have been done using the one-electron ECP model of the metal atoms, an additional uncertainty is introduced. This uncertainty cannot be evaluated by comparison with the all-electron results in Table 1.4 because the ECP model was parameterized to fit the relaxed 3*d* result for exactly this system. From our present experience we expect that we can estimate chemisorption energies, at least in normal cases, with an uncertainty of 5–10%. This uncertainty refers to results where estimates of all the effects given in Table 1.4 are included.

IV. Cluster Convergence

The reliability of a chemisorption energy calculated using a cluster model relies on two equally important aspects of the calculation. The first aspect is that the calculation on the chosen model system be sufficiently accurate. In the preceding sections we have shown the effects that have to be accounted for and to which accuracy these effects have to be treated. The second critical aspect of the modeling is that the convergence of the property of interest with cluster size be reached, or at least understood and accounted for in some way. It should be stressed in this context that if the accuracy of the calculations is not high enough, an analysis of the cluster convergence can be quite misleading. This has been a severe problem in previous interpretations of cluster model results and is probably one reason the cluster model has been viewed with large skepticism [45a]. Another reason is that with the present solid state-oriented understanding of the nature of the surface chemical bond, it is hard to understand how a small cluster can actually be a reasonable model for calculating chemisorption energies. Normally two properties are emphasized in this type of description, both of which are very poorly modeled by a small cluster. The first of these properties is the position of the Fermi level, for which, in Introduction, we have already shown that a very slow convergence with cluster size is expected. The other property emphasized is the continuous density of states or, rather, density of one-electron levels. This property is considered important as the formation or breaking of bonds on a metal surface in the solid state picture sometimes is described as a type of resonance effect between the levels in the continuous conduction band and the adsorbate levels [6–9]. If this was a critical aspect of the surface bond description, it would clearly not be meaningful to even try to obtain chemisorption energies using the cluster model. Put bluntly, if the present solid state-oriented understanding of the nature of the surface chemical bond does not entirely miss the most important aspect, the cluster model will not work. Already, however, the quality of the results in the preceding section for the small M_5X system indicates that there is actually some other property of the cluster that governs formation of the surface chemical bond, and it therefore seems imperative to investigate the cluster model further. The obvious possibility that the M_5X system in Fig. 1.1 is just a fortuitous choice of a model has to be critically investigated by an examination of larger clusters. In this section we first show how the cluster convergence, and the nature of the surface chemical bond, for the case of hydrogen can be quite well understood. In the second subsection the relevance of describing the position of the Fermi level is investigated in simple model calculations. In the third subsection the cluster convergence of other adsorbates such as CH_x, the oxygen atom, and CO is discussed. Finally, in the last subsection we briefly discuss the cluster size oscillations of reactivities found in some cluster beam experiments.

A. Hydrogen Chemisorption

When this project started, the most common way in which the chemisorption energy was calculated using the cluster model was to restrict the wavefunction to have low spin. In this way the dangling bonds, which were considered artificial ingredients in the cluster as a model of a surface, were closed. Apart from this restriction the lowest electronic state was considered both for the cluster itself and for the cluster with the chemisorbed adsorbate [46]. Another model leading to similar results was embedding of the cluster in a larger cluster treated at a lower level of accuracy [20]. At the onset we found that releasing the low-spin restriction, otherwise keeping within the same framework as in Ref. 46, actually improved cluster convergence slightly [29]. No embedding was used. As it turns out, none of these methods is much closer than the other in treating the real origin of the cluster oscillations. As a reference we could therefore choose the results of any of these models to show how strongly the chemisorption energy oscillates using the cluster model, and here we

Table 1.5. Chemisorption of Atomic Hydrogen in Fourfold Hollow of Ni(100)[a]

Cluster	D_1 (kcal/mol)	D_2 (kcal/mol)
Ni_5(4,1)	53.1	54.1
Ni_{21}(12,9)	43.0	61.1
Ni_{21}(12,5,4)	63.0	63.0
Ni_{25}(12,9,4)	58.9	58.9
Ni_{29}(16,9,4)	53.5	58.5
Ni_{41}(16,9,16)	61.1	63.3
Ni_{50}(16,9,16,9)	61.6	61.6
$\bar{D}$	56.3	60.3
σ	6.5	3.0
Experimental	63	63

[a] D_1 is the ground-state chemisorption energy and D_2 is the bond-prepared chemisorption energy (see text). $\bar{D}$ is the average value and σ the standard deviation. The values in parentheses for the clusters denote the number of atoms per layer.

use the results of the third model [29]. In Tables 1.5 and 1.6 the ground state chemisorption energies D_1 are shown for the cases of hydrogen on Ni(100) and Ni(111). The results for Ni(111) vary strongly from cluster to cluster, whereas the results for Ni(100) happen to look much better. The result for Ni_{21}(12,9) indicates, however, that the latter is probably fortuitous.

The results for D_1 in Tables 1.5 and 1.6 certainly indicate that the cluster model is not of much use in predicting chemisorption energies and, by implication, that a solid-state oriented understanding, using some infinite model, might be necessary. Indeed, with this latter model there are severe problems in understanding the individual cluster results. One might think, for example, that the Fermi level of Ni_{21}(12,9) is particularly poorly described or that there is a very large gap in the density of one-electron levels at a critical position for this cluster. Such explanations, however, are quickly contradicted by other cluster results and they lead nowhere. Explanations based on different charge densities at the chemisorption site or explanations based on the shape, or the number of layers, of the cluster are equally unsuccessful. Instead, a molecularly oriented understanding is possible and this leads to a surprisingly simple explanation for the cluster oscillations [21].

Not much knowledge of molecular orbital theory is required to explain why the boron atom ($2s^22p^1$) forms a stronger bond to a hydrogen atom than does the beryllium atom ($2s^2$). The answer is of course that the $2p$ orbital of boron can overlap with the hydrogen $1s$ orbital and form a (covalent) bonding orbital filled by two electrons, as in BH. For BeH three electrons have to be placed in the bonding and antibonding combinations of the beryllium $2s$ and hydrogen $1s$ orbitals, which means that the antibonding orbital has to be occupied and the bonding becomes much less favorable. To reach a bonding situation similar to that in boron, beryllium needs to be excited to the $2s^12p^1$ state, but an excitation energy of 2.7 eV must be exerted with a resulting weaker bond to hydrogen. A parallel situation exists for clusters of different size. Some clusters happen to be of the boron type in their ground states so that

Table 1.6. Chemisorption of Atomic Hydrogen in Threefold Hollow of Ni(111)[a]

Cluster	D_1 (kcal/mol)	D_2 (kcal/mol)
Ni_4(3,1)	55.9	55.9
Ni_{10}(3,7)	39.4	69.5
Ni_{17}(3,7,7)	63.0	63.0
Ni_{20}(3,7,7,3)	56.1	56.1
Ni_{13}(12,1)	63.7	63.7
Ni_{19}(12,7)	27.5	65.7
Ni_{22}(12,7,3)	37.8	58.3
Ni_{25}(12,7,6)	42.4	61.7
Ni_{28}(12,7,6,3)	50.8	66.2
Ni_{40}(21,13,6)	46.5	57.2
$\bar{D}$	48.3	61.7
σ	11.1	4.5
Experimental	63	63

[a]The symbols are the same as in Table 1.5.

they already have a singly occupied orbital pointing toward hydrogen, whereas other clusters are of the beryllium type where such an orbital is missing. To form the bond to hydrogen, these latter clusters have to be excited to higher states leading to weaker chemisorption bonds. As demonstrated in Tables 1.5 and 1.6, if the chemisorption energy is calculated with respect to these states (we call them bond-prepared states), the chemisorption energy (D_2) becomes very stable from cluster to cluster and also quite close to the value obtained for actual metal surfaces. This latter result can be interpreted in several ways. The most obvious interpretation is that in an infinite surface the excitation energy to other electronic states in the conduction band is essentially zero and some of these states are bond-prepared states and can thus be reached without cost. Another, essentially equivalent interpretation is that when an adsorbate is bound to the surface some bonds in the surface may be broken. These bonds can be perfectly repaired in an infinite surface by strengthening some of the infinite number of other bonds. In a small cluster this repair is not as efficient and the cluster has to be prepared in such a way that no bonds will be broken in the chemisorption process. It turns out that a third interpretation of the results for hydrogen is the one that is easiest to combine with the results for other adsorbates such as oxygen (see below). In this interpretation the electron of hydrogen has to fit into the electronic structure of the cluster without the necessity of creating new cluster orbitals. As the hydrogen electron lies in a totally symmetric orbital, this requirement means that the free cluster needs to have a singly occupied orbital of this same symmetry, which for hydrogen is exactly the same requirement as bond preparation.

A few other points should be noted in Tables 1.5 and 1.6. First, the strength of the chemisorption bond is well described even for clusters containing as few as 10 atoms. This means that the bond and the electronic effects caused by the bonding are strongly localized close to the adsorption site. In fact, the results for the larger

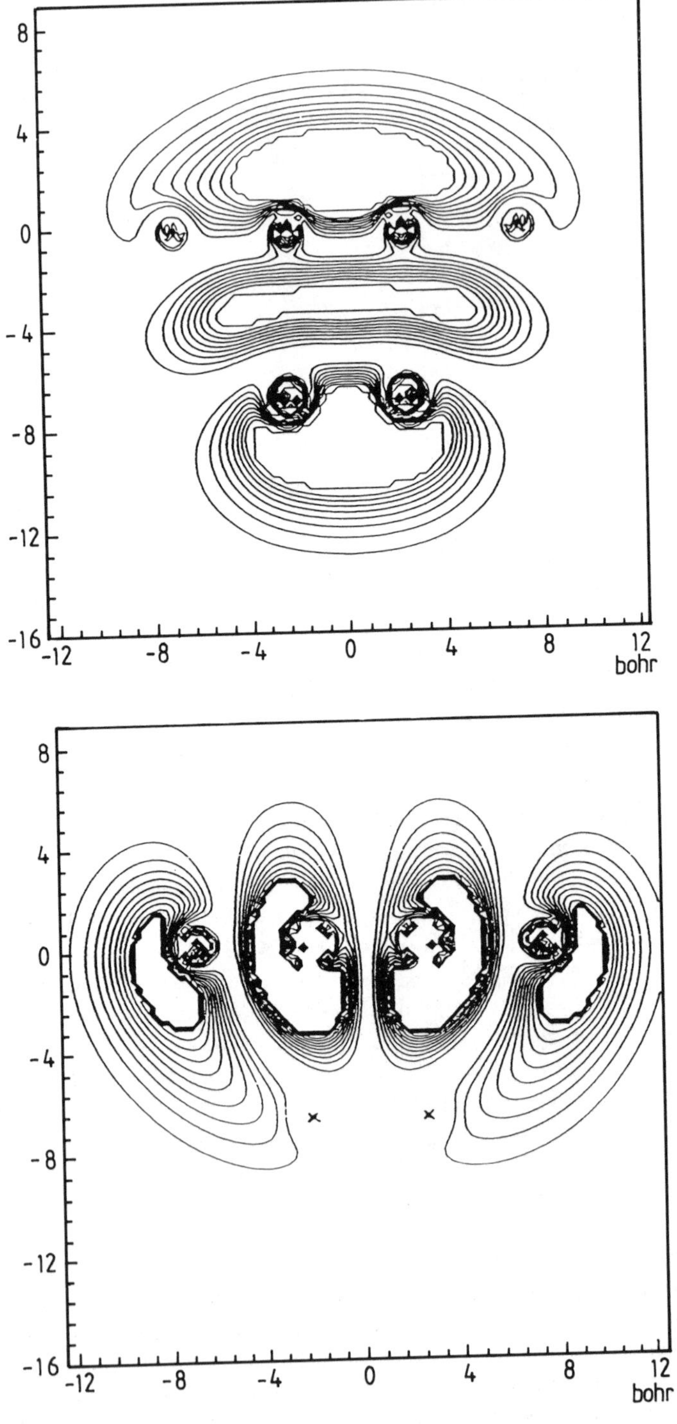

Figure 1.2. Outermost (singly occupied) orbitals of $Cu_{25}(12,9,4)$. The orbital density is plotted in a plane that goes through four of the top-layer atoms and two of the third-layer atoms. These atoms are marked with crosses (when necessary) in the figure.

clusters do not indicate that a closer agreement with surface experiments is obtained for these than for the smaller clusters. It is also difficult to understand the results in the tables if the chemisorption energy were sensitive to the exact positions of the one-electron orbital levels of the cluster (see next subsection). Also, our experience that all the bond-prepared states give reasonable chemisorption energies must indicate that the overlap with adsorbate orbitals is always sufficient and not a critical factor. A plot of the outermost orbitals of a typical cluster, shown in Fig. 1.2, confirms this experience in that these orbitals are very delocalized. A final point about the results in the tables is that the results for the clusters that have to be highly excited to become bond prepared have a slight tendency to overshoot the chemisorption energy. Therefore, if there is a choice, a cluster with a ground state that is already bond prepared should preferably be used as a model for a surface.

Probably the major reason why the simple nature of bonding between a hydrogen atom and a cluster has not been noted before is that the electronic structure of the adsorbate and the cluster appears much more complicated than it actually is. Before the adsorbate is chemisorbed, the molecular orbitals of the cluster have orbital energies in the range 5–16 eV. The outermost of these orbitals, which are completely delocalized, are the open-shell orbitals (if there are any). After chemisorption of hydrogen there is hardly any hydrogen mixing in the outermost orbitals. The hydrogen character is instead spread out among orbitals in the range 8–12 eV, that is, about 3–7 eV down into the band. This is also the picture obtained from experimental spectra. The idea of describing the chemisorption bond as a resonance effect between orbitals of similar orbital energies is therefore what most immediately comes to mind. This picture has, for example, been given by effective medium theories [6–9]. To see direct bond formation between the outermost open shell orbital of the cluster and the hydrogen orbital, localization of the orbitals is required, and this orbital localization is probably not performed until a rather clear idea of the outcome has been obtained. Finally, even if the lack of bond preparation would have been considered as at least partly responsible for the cluster oscillations, clearly many more effects are possible. Among these, the position of the Fermi level and the density of states, which do vary from cluster to cluster, were believed by theoreticians to be totally dominant. As usual in this type of science, the idea might seem trivial, but the truth is probably that without a numerical demonstration of the effect, as seen in Tables 1.5 and 1.6, the idea as such would have been nearly wasted.

B. Description of the Fermi Level

In the present type of model calculations it is very simple to test the sensitivity of the results to the position of the Fermi level and other one-electron levels. This can be done by adding electrons to or taking electrons away from the cluster. A calculation of this type has recently been done modeling atomic hydrogen adsorption on a Ni_{13} cluster [22]. In the neutral cluster the ionization potential (IP) is 4.8 eV compared with the Fermi level of a Ni(100) surface of 5.2 eV. It is common to identify the Fermi level of a cluster with the IP, but other definitions would suffice equally well for the present purpose. The main point is that the energy levels (all the outer ones) change drastically as electrons are added or taken away. If one electron is added to Ni_{13} the IP decreases by 3.7 eV to 1.1 eV, and if one electron is taken away the IP increases to 7.9 eV. When two hydrogens are added to fourfold hollow positions, the bond-prepared chemisorption energy for the neutral cluster is found to be 57.4 kcal/mol, in reasonable agreement with the experimental value which should be slightly below 63 kcal/mol [47]. For the Ni_{13}^{-} cluster the chemisorption energy is found to be 57.7 kcal/mol and for Ni_{13}^{+} it is 60.3 kcal/mol. In summary, the Fermi level changes by 6.8 eV (from 1.1 to 7.9 eV) but the chemisorption energy changes by only 0.1 eV (2.6 kcal/mol). As chemisorbed hydrogen becomes negative (by about −0.4) on the surface one might at least have expected that the chemisorption energy should decrease in the case of Ni_{13}^{+}, where electrons are much harder to obtain from the surface, but the opposite is found.

The main conclusion that can be drawn from the above results is that the chemisorption energy is very insensitive to the positions of the one-electron levels, including the Fermi level. Also, the bond-prepared chemisorption energy is in quite good agreement with experimental surface results even for rather small clusters with as few as 10 atoms. Our conclusion is that a picture of the surface chemical bond that aims at the most important aspects of this bond cannot rely on something that is present only in an infinite model of the surface. Therefore, the position of the Fermi level and a continuous density of states appear to be of less importance. Instead we emphasize the simple chemical nature of the bonding between the adsorbate and the surface. The simplest picture of this bonding is that it is covalent. An ionic picture of the bonding is also possible to use provided the electrons of the adsorbate fit into the electronic structure of the cluster. It is clear that a strongly ionic picture would indicate a high sensitivity to the charge of the cluster and the position of the levels of the cluster, which is contrary to our observations. With the picture of a covalent bond depending on the size of the overlap between the fragment orbitals, it is easy to realize that the bonding should be rather insensitive to these properties of the cluster. Another feature of chemisorption we would like to underline is the radical character of a metal surface, which is brought out by bond preparing the cluster. In fact, one extreme picture of the features of a metal surface in the context of chemisorption is where a set of nonbonding metal atoms in their radical $d^{n+1}s$ states are placed beside each other on the surface. The nonbonding character is used here to emphasize that the bonds (which obviously exist) between the surface atoms can be broken without any cost by strengthening other bonds. This picture seems very remote from the conventional band picture of the metal surface, but it should be stressed that this description is aimed at bringing out the features essential to chemisorption. For describing spectroscopic properties, other features of the surface may be stressed. We finally note that the picture of the surface chemical bond we advocate has many similarities with the bond order conservation model of Shustorovich [45], which has been successfully used to explain a large number of surface phenomena. One example of this similarity is that the bond order, which should be conserved during chemisorption in this model, is distributed over the adsorbate and the metal–adsorbate bonds but not over the metal–metal bonds. This is in line with the interpretation of bond preparation given above, namely, the bonds in the surface can be broken without any cost.

An obvious argument against the above conclusions on the surface chemical bond is that a correlation between the position of the Fermi level and the chemisorption energy can actually be found for many real metal surfaces and adsorbates. For example, the Fermi level rises as one goes from the right to the left in the periodic table for the first-row transition metals [10], as does the chemisorption energy for hydrogen [48]. To rationalize this finding an indirect connection can be used. In our covalent picture the chemisorption energy depends primarily on the overlap and thus the size of the local atomic metal orbitals in the $d^{n+1}s$ states. It is clear that a direct relation is expected between the size of these atomic metal orbitals and the position of the Fermi level. Therefore, the critical factor, namely the size of the metal orbital, will affect both the chemisorption energy and the position of the Fermi level. The correspondence between the position of the Fermi level and the chemisorption energy is thus, in our interpretation, a correlation of secondary importance and neglects the factor of main importance, the size of the orbitals in the local interaction.

C. Adsorbates Other than Hydrogen

The binding of hydrogen to a surface is the simplest possible case of chemisorption and is also the most straightforward case for our model to calculate the chemisorption energy. The only factor governing bond formation is the overlap between the hydrogen 1*s* orbital and a singly occupied orbital in the cluster. The next step in complexity is to treat chemisorption of CH_3. This case has also been studied systematically [49] and it was shown that bond preparation works equally well in this case. Recently, fluorine chemisorption has been studied as a prototype case with a strongly ionic bonding [50]. It turns out that cluster size oscillations are essentially removed also for fluorine by bond preparation and the degree of ionicity therefore does not seem to matter. To understand this, the picture where the electrons of the adsorbate should fit into the electronic structure of the cluster is quite useful. The position for the Fermi level was also in the case of fluorine found to be unimportant. Chemisorption of CH and CH_2 has recently been treated with clusters that in their ground state have three singly occupied orbitals of proper symmetry to covalently bind both these adsorbates (see Section VI). This type of bond-prepared cluster also seems to work well in the sense that results in general agreement with experiment are obtained. CH and CH_2 have also been treated by Whitten and co-workers [51] using different methods including embedding but without bond preparation, and results quite different from ours were obtained. The chemisorption energies of CH and CH_2 were thus found to be nearly the same in Ref. 51, which led to the predictions of an exothermicity for formation of CH_2 of 33 kcal/mol and an endothermicity for formation of CH of about 80 kcal/mol. Our bond preparation results are much more thermoneutral, in agreement with experiment.

Cluster bond preparation for chemisorption of the oxygen atom represents an interesting case and has recently been studied [52]. The most obvious bonding mode is formation of two covalent bonds between the cluster and oxygen, but it is not obvious, in contrast to CH_2, which directions these bonds will have. The two choices for oxygen are either two bonds parallel to the surface or one bond parallel and one bond perpendicular to the surface. It is clear that the two choices require different bond preparations. A third possible bonding mode is one that is most common; this mode is best understood from the viewpoint that the electrons of the adsorbate should fit into the electronic structure of the cluster. In this case the lone-pair electrons of oxygen in the A_1 symmetry replace a cluster orbital of the same symmetry, and the two replaced cluster electrons fill up the holes (partly covalently) of the $2p$ orbitals parallel to the surface. With the three possible bonding modes, which are essentially equivalent, it is not too surprising that the ground states of most clusters can bind oxygen quite well, which was our finding in an earlier study that did not consider bond preparation [29].

The most difficult system we have yet come across is CO. The cluster size variations of the chemisorption energy for CO have been studied in detail by Bagus *et al.* [53] for the case of copper. One reason the cluster oscillations (about ± 10 kcal/mol) appear to be particularly dramatic in this case is that the total chemisorption energy is so small, only of the order of 15–20 kcal/mol. Bagus *et al.* find that the largest chemisorption energy is obtained for clusters with high-lying outermost orbitals of the same symmetry as the CO π^* orbitals and low-lying outermost orbitals of the same symmetry as the carbon σ lone pair. These variations are easy to understand chemically in terms of the normal repulsive σ interaction and attractive π interaction, but the oscillations are not so easily reduced by bond preparation.

Quite apart from the oscillatory behavior of the chemisorption energy for CO on copper, there is an additional phenomenon we have not yet found for any other system: our calculations on the ground states of some clusters give chemisorption energies larger than those found experimentally, which clearly cannot be corrected by a bond preparation of the type described above for hydrogen. At present, the most plausible reason for this finding is that our calculations are still not accurate enough for this type of adsorbate bonding. The next step in our calculations is therefore to improve the level of correlation treatment by inclusion of the CPP operator (see Section IIC) for all copper atoms in the cluster. The other reason for the very large chemisorption energies is that the bonding of CO to copper has large contributions from other types of interactions, which are not covalent and which are unimportant for the other adsorbates we have studied. For example, it is possible that a direct electrostatic interaction between dipoles and/or induced dipoles is responsible for a large part of the interaction, and this will clearly require a different treatment than would a straightforward covalent interaction. A systematic investigation of CO on copper at a very high level of accuracy is definitely needed before these questions can be answered and it is our hope that we will be able to perform such a study in the near future.

We have here shown that the cluster oscillations have simple chemical origins and that the variations can be reduced to satisfactory levels in most cases by simple bond preparations of the cluster; however, there are some systems that require further investigations, such as CO on copper. For these systems a positive aspect is that the oscillations are quite informative about the effects that are most important for the formation of the surface chemical bond [53]. An understanding of the origin of the bonding is after all one of the main reasons the theoretical modeling is performed. One final conclusion we would like to make is that even though a full understanding of the details of the cluster oscillations for all systems has not yet been attained, the basic origin of the variations is definitely understood for most systems. This origin is intimately connected with the nature of the surface chemical bond as a very localized molecular type of phenomenon. Above all, with this understanding it is easy to see that even rather small clusters can be used as quantitatively accurate models for chemisorption on real infinite surfaces.

D. A Comment on Cluster Beam Results

A common argument against clusters as models for surfaces is that rapid oscillations in properties of clusters of different size have been found in cluster beam experiments for some metals [54]. It should first be noted that these oscillations have not been found for the metals we are focusing on here, copper and nickel. For reactions of H_2 with cobalt and iron clusters, however, individual clusters with 20 or fewer atoms show large differences. In a recent paper we discussed the case of cobalt in detail [55]; here we simply give a quick summary of the main conclusions drawn. In Table 1.1 we see that iron and cobalt are different from nickel and copper in that the ground state of the atom is not $d^{n+1}s$ but d^ns^2. As already discussed in Section IIA, for an infinite surface and thus for large clusters the $d^{n+1}s$ state is expected to dominate also for iron and cobalt. It is perhaps not surprising that for some of the smaller clusters, some of the metal atoms will still prefer to be in the d^ns^2 state. For the linear clusters Fe_3 and Co_3, for example, the central atom will be in the d^ns^2 state. In Ref. 55 we have shown that this special effect for iron and cobalt leads to a closed shell wavefunction for some of the Fe and Co clusters which makes these clusters quite unreactive with H_2. This is an interesting and potentially useful property of the clusters of these metals but does not have any impact on the use of the cluster model. As we have discussed in detail in Section II, in the cluster model the metal atoms will be assumed to be in the $d^{n+1}s$ state even for these small clusters, which will make also these clusters similar to real surfaces.

V. Atomic Chemisorption: Examples

From this section on we give examples of applications of the model described in the previous sections. The natural starting point is atomic chemisorption. The general features of atomic chemisorption have been elegantly described using the bond order conservation (BOC) model [45]. With this model it is possible to explain why practically all known atomic adsorbates prefer to chemisorb at hollow sites and the model also gives reasonable diffusion barriers for motion of the atoms on the surface (see Chapter 3). This model is, however, phenomenological, and a critical parameter is the atomic chemisorption energy which has to be taken directly from sometimes rather uncertain experiments. We feel that model calculations of the present type can sometimes be as accurate as experiment, or at least complement these, to supply more accurate input data for the BOC model. Also, to understand the chemical origin of the binding and to explain such important phenomena as surface reconstruction, quantum chemical calculations are needed. A final important reason to perform calculations on these systems is to interpret spectroscopic information which is often used to extract important surface data in an indirect way. At the end of Subsection B we give an example of where such information is obtained through the use of a Born–Haber cycle [56].

A. Oxygen, Hydrogen, and Chlorine Chemisorption

Atomic oxygen chemisorption represents a special case of the application of the present model, as $3d$ relaxation was used to directly fit a parameter in the one-electron ECP for this case (see Section II). However, a comparison between the ECP results in Table 1.3 and the all-electron results in Table 1.4 shows that the correlation effect, which was not fitted, is also well reproduced by the ECP in the case of Cu_5O. Hydrogen and chlorine chemisorption was used as the first test cases of the ECP model without parameter fitting. We have already seen from the cluster convergence study in Tables 1.5 and 1.6 that hydrogen chemisorption is well described using the present model. For Ni(100) the average value of the chemisorption energy for all the clusters studied is 60.3 kcal/mol compared with the experimental value of 63 kcal/mol. The corresponding numbers for Ni(111) are 61.7 and 63 kcal/mol, respectively.

The binding of chlorine to a copper surface is completely ionic with hardly any trace of covalency, as demonstrated by Pettersson and Bagus [57], and therefore represents a different and critical test case of the ECP model. The chemisorption energies (about 100 kcal/mol) from the all-electron SCF calculations in Ref. 57, obtained at the frozen $3d$ level, are almost perfectly reproduced by the ECP model for both Cu_5 and Cu_9 clusters. The $3d$ relaxation energies are slightly overestimated by a few kilocalories per mole, but part of this discrepancy would be reduced if a larger $3d$ basis with f polarization functions would be used in the all-electron calculations. The fairly large change in the chlorine bond distance to the surface between the two clusters of 0.3 Å is also very well reproduced.

A property of general interest is the difference in chemisorption energy for an adsorbate on two different metals. This difference is expected to be particularly well reproduced by the present type of calculations as many of the systematic errors will cancel. To obtain an estimate of this energy difference, calculations were performed for oxygen and hydrogen on a $M_{25}(12,9,4)$ cluster where M is copper or nickel [28]. The results for hydrogen give an energy difference between nickel and copper of 8 kcal/mol compared with the experimental value of 7 kcal/mol [47]. For oxygen we obtain a difference of 12 kcal/mol also in favor of nickel. In the latter case direct measurements of the two chemisorption energies have not been possible to perform because of the high desorption temperatures. Instead, an estimate of this energy difference can be obtained from core ionization experiments and the use of a Born–Haber cycle where the equivalent core approximation (ECA) is used [58]. The ECA is used in this case to identify a core-ionized nickel atom with an approximate copper atom. In this way an estimate of 19 kcal/mol is obtained for the chemisorption energy difference, which is somewhat larger than our estimate of 12 kcal/mol but is in line with our result that the chemisorption energy is smaller for copper. At this point it is probably difficult to say which of the results should be trusted most, but we feel that our theoretical estimate should be at least as accurate as the experimental one.

Diffusion barriers for motion of atoms on metal surfaces are important to know to understand the dynamics of many catalytic reactions. We have calculated this barrier for hydrogen and oxygen on Cu(100) and Ni(100). From the BOC model [45a] it follows that the diffusion barrier in this case has to be equal to the energy difference between the chemisorption energy at the fourfold hollow and that at the bridge site (see also Chapter 3). The best experimental estimate of the diffusion barrier for hydrogen on nickel is 3.5–4.0 kcal/mol [59]. This energy is thus very small and is also extremely sensitive to details in the modeling. For example, the contribution of zero-point vibration effects has been estimated to be 0.5 kcal/mol or about 15% of the barrier [9]. For a Ni_{25}(12,9,4) cluster we obtain 4.3 kcal/mol, including zero-point vibration, which thus agrees very well with the experimental estimate. For copper there is no experimental estimate but our calculations indicate a barrier even smaller than that for nickel. Our estimate for the barrier is 2 kcal/mol or less. In recent dynamic calculations, where one of the aims was to demonstrate the importance of tunneling effects on the motion of hydrogen on copper, a barrier of 12 kcal/mol was used [60]. According to our calculations, this barrier is definitely too high and will probably overestimate the relative importance of tunneling effects considerably.

For the oxygen diffusion barrier, in our model a problem of general character appears. It turns out that the ECP description is not as good for the bridge as for the hollow position, which is not unexpected in light of the shorter bond distances at the bridge site. The calculated barrier will therefore be overestimated. For oxygen on Ni(100) we obtain an ECP result of 32 kcal/mol; however, it is not impossible to obtain a better estimate. By a direct comparison between ECP and all-electron calculations for the small bridge-bonded Ni_2O system, we find that the ECP underestimates the binding by 11 kcal/mol. Our best estimate for the diffusion barrier is thus 21 kcal/mol, which is in good agreement with the estimate based on the BOC model of 18 kcal/mol [45a]. This procedure illustrates a general point about the present approach: whenever there is any doubt about the quality of the ECP results one should go back and perform comparative all-electron calculations. When bridge bonding is considered, this procedure should always be used. Similar types of calculations for oxygen on copper lead to a diffusion barrier of 25 kcal/mol. This indicates that the oxygen migration barrier should be slightly higher on copper than on nickel which is a surprising result, as the chemisorption energy is larger for nickel, but the accuracy of these estimates is probably not high enough to definitely conclude that this is so.

B. A Comparison between Carbon, Nitrogen, and Oxygen

We have already indicated that one must be very careful in drawing qualitative conclusions about the nature of the binding to a surface based on apparently large differences in metal–adsorbate distances. For example, if hydrogen or oxygen on some metal surface moves away from the hollow position by as much as 0.5 Å this does probably not mean that the binding mechanism has changed appreciably. On the contrary, such a difference may not even be visible in the electronic wavefunction. For similar reasons, it is not necessarily true that there exists a qualitative explanation why a molecule like CO prefers one site on one surface and another site on another surface. Answers to such questions require the use of extremely accurate methods. The situation is also usually the same for changes in metal–metal distances in the substrate. Even a seemingly dramatic change in these distances will often have only a marginal effect on the energetics of surface reactions. For the cluster model this means that it is usually a rather good approximation to keep the cluster model in a frozen geometry even if, for example, substantial changes in the distance between layers have been found experimentally. Also, if an incorrect adsorption site is found for a molecule on a cluster it could still mean that the bonding is qualitatively well described. Metal surfaces are thus very different in this respect from ordinary small molecules for which geometric information can often be used to draw direct mechanistic and energetic conclusions.

There are of course exceptions to the rule that differences in metal–adsorbate geometries are not energetically informative and one such exception is the difference observed for carbon and nitrogen on the one hand and oxygen on the other. As carbon and nitrogen cause reconstruction and even move into the plane of the surface [61,62], we expect that appreciable forces are present in these cases, the origin of which should be possible to understand through model calculations. Oxygen remains outside the surface by 0.86 Å [63] and hardly modifies the surface. We emphasize that we feel that this is a qualitative problem and our rather crude model calculations should therefore be appropriate.

The forces responsible for the differences in geometries between carbon and nitrogen compared with oxygen were studied in two types of geometry optimizations, using a small Ni_5 (see Fig. 1.1) cluster [64]. In the first set of calculations only the distance of the adsorbate above the four-fold hollow site of Ni_5 was optimized, with a frozen Ni_5 geometry where the Ni–Ni distances were set equal to the bulk distance. In the second set of calculations the Ni_5 geometry was allowed to relax. The fourfold hollow site was thus allowed to open and the second-layer Ni atom was allowed to move up or down. The resulting geometries and energies are given in Table 1.7. From this table we see that the geometry relaxation of the cluster gives a quite exaggerated picture of the distortions compared with what actually occurs in a real surface. This is of course expected as the restoring forces surrounding the five nickel atoms are not present; however, we feel that this is even an advantage in the present case because the reconstructing forces should now be more visible. We also note that if the cluster is not allowed to change geometry, all three adsorbates prefer to stay outside the surface at similar heights above the surface. The differ-

Table 1.7. Bond Lengths and Chemisorption Energies for Carbon, Nitrogen, and Oxygen on Ni_5 in Frozen and Relaxed Geometric Structures[a]

Adsorbate	Bond Length (Å)			Energy (kcal/mol)
	Ni_a-Ni_a	Ni_b-Ni_a	$X-Ni_a$	
		Ni_5 in a frozen geometry		
Carbon	2.49	1.76	1.18	117
Nitrogen	2.49	1.76	1.08	86
Oxygen	2.49	1.76	1.07	110
		Ni_5 in a relaxed geometry		
Carbon	2.91	2.12	−0.53	150
Nitrogen	2.94	2.06	−0.37	112
Oxygen	2.96	2.06	−0.26	116

[a]Ni_a denotes an atom in the first layer and Ni_b an atom in the second layer. The perpendicular distances are given to the first-layer plane for Ni_b and the adsorbate X, with a negative sign meaning below the surface.

ences in chemisorption energies found before and after the cluster is allowed to relax are quite illustrative of the forces present. The gain in chemisorption energy for oxygen is only 6 kcal/mol on penetration of the surface, indicative of a rather weak force. For nitrogen the corresponding energy difference increases to 26 kcal/mol and for carbon it is as high as 33 kcal/mol, indicative of rather strong forces. The expected qualitative difference between oxygen and carbon/nitrogen is thus actually seen, and it should be possible to see an origin for this difference in the wavefunction. The explanation for the stronger force for carbon and nitrogen is quite simple. Oxygen can form a maximum of two covalent bonds which are already formed above the surface. In addition to these bonds there will be an oxygen lone pair that will act repulsively toward the surface. It is this repulsive interaction for oxygen that leads to an optimal position outside the surface. For nitrogen and carbon there will not be any repulsive effects of this type to balance the attractive force resulting from the formation of bonds to the surface. In addition to the two bonds to the surface that can be formed just as for oxygen, nitrogen can use its third singly occupied $2p$ orbital to form another covalent bond to the surface. For carbon the initially empty $2p$ orbital can act as an electron acceptor and this interaction will thus also be attractive toward the surface. The larger attraction toward the surface for carbon and nitrogen than for oxygen is also clearly seen in the larger overlap populations with the surface for the former atoms.

The energies given in Table 1.7 can also be used to predict the chemisorption energies for the three different atoms; however, as the reconstruction energies for carbon and nitrogen based on the crude Ni_5 model are quite uncertain, we prefer to make a comparison with an experimental measurement, rather than use the energies in the table directly. This experiment is a core ionization measurement that was used to estimate the chemisorption energy difference between oxygen and nitrogen on Ni(100) based on a Born–Haber cycle and the ECA (see above) [56]. The result of this experiment is that nitrogen is expected to bind by 5 kcal/mol more than oxygen. In the experiment a nitrogen atom on a Ni(100) surface is core ionized and the ECA is used to identify this ionized nitrogen atom with an approximate oxygen atom. The ECA can be tested by calculations and is found to be in error by 14 kcal/mol, which is more than might have been expected. In fact, nitrogen turns out to be the worst case for the ECA. For oxygen the ECA has hardly any error. Another assumption in the experimental estimate is that nitrogen and oxygen have the same geometry on the surface. This assumption was later found experimentally to be incorrect. Nitrogen adsorbs at a height of 0.1 Å [61] and oxygen at a height of 0.86 Å [63]. The energy correction introduced by this difference in geometries was also calculated and, together with the calculated error in the ECA, leads to a corrected experimental estimate for the energy difference between nitrogen and oxygen of −6 kcal/mol. Nitrogen is thus now predicted to chemisorb more weakly than oxygen. To be able to give absolute values for the different chemisorption energies we use the calculated chemisorption energy of oxygen on Ni_{25}, which was found to be 111 kcal/mol. This calculation was done with a small oxygen basis set and without nickel 3*d* correlation. The error in this type of calculation can be estimated based on the results in Tables 1.3 and 1.4 to be about 20 kcal/mol, and our final estimate for the oxygen chemisorption energy is about 130 kcal/mol. With the result that nitrogen should bind by 6 kcal/mol less than oxygen we reach an estimate for the nitrogen chemisorption energy of approximately 125 kcal/mol. A similar estimate for carbon leads to approximately 150 kcal/mol. This is less than the 170 kcal/mol found experimentally [65] but this value is suspiciously close to the sublimation energy of graphite of 171 kcal/mol and, in our opinion, might therefore have been overestimated.

The main purpose of the preceding discussion was to illustrate a general point, namely, that the present model calculations are probably most useful in connection with and as a complement to experimental measurements that need to be analyzed and sometimes corrected. We have done a similar analysis for a valence ionization experiment with O_2 on different metal surfaces where the results were used to predict the O−O distance [66]. More recently we have analyzed the large line-broadening effects found in core ionization of CO and oxygen [67,68]. These experiments can then be used to obtain direct structural information about adsorbates on surfaces.

VI. Molecular Chemisorption: Examples

As molecular chemisorption is so often dissociative on transition metal surfaces, this section and the next section on chemical reactions on surfaces are strongly related. In this section we first discuss intermediates formed in the methanation reaction of CO and H_2 on Ni(100) and Ni(111). These intermediates are, besides the carbon atom, the three CH_x fragments ($x = 1$–3). Even though these intermediates have been known for a long time, only recently have they been directly observed. The situation is similar for the molecules discussed in the second subsection, chemisorbed H_2 and O_2. Molecular H_2 was not detected until 1986 and then only under extreme conditions on a stepped surface packed with hydrogen atoms. Molecular O_2 has not been observed on any well-defined nickel surface but has been seen on, for example, silver and platinum surfaces. We will discuss the possibility of observing molecular O_2 also on nickel. In general, it is probably true that one of the situations in which the present model calculations are most useful is for these types of short-life intermediates that are difficult to detect and observe experimentally. With the cluster model, calculations have been performed for CH_3 on Ni(111) by Schüle *et al.* [49], and for O_2 [69], H_2 [70] and CH_x [71] by Siegbahn and co-workers. The reader is referred to these papers for more detailed information.

A. Chemisorption of CH_x

The first direct demonstration of the formation of all three CH_x fragments in the methanation reaction was made using secondary ion mass spectroscopy (SIMS) on a Ni(111) surface by Kaminsky *et al.* [72]. From the dependence on temperature of the production of the CH_x fragments they concluded that the three fragments should have similar stabilities. Ceyer *et al.* [73] were the first to spectroscopically identify methyl radicals on a single-crystal surface under UHV conditions. This was also on the Ni(111) surface. Using a combination of molecular beam techniques and high-resolution electron energy loss (HREELS) spectroscopy they found an unusually low C–H stretching frequency for CH_3. They also concluded that the reason methyl radicals had not been observed before is that they are not stable to dissociation or recombination above 150 K, which is below the temperature where the methanation reaction is normally carried out.

From the preceding summary of experimental results for chemisorbed CH_3 it is clear that many basic questions remain concerning this radical. We address three of these questions here: the chemisorption energy, the adsorption site, and the origin of the observed low C–H frequency. One main point we want to emphasize by going through these calculations here is that much more certain information can be drawn from the calculations by combining many different calculations with as much experimental information as possible. If we start with the chemisorption energy we first conclude that previous knowledge of this entity, based mostly on calculations, is very uncertain. The results of different extended Hückel-type calculations differ dramatically from each other. In one study [13], CH_3 was found to be strongly unbound by 49 kcal/mol in the threefold hollow site and bound by 38 kcal/mol in the on-top site, whereas in another study [74], the preferred site was reversed, with the hollow site lower in energy by 19 kcal/mol. The experimental estimate for the chemisorption energy of CH_3 quoted by Ceyer *et al.* [73] is 75 kcal/mol and they expect an on-top adsorption site. Previous cluster calculations by Upton [75] (using the model described by Goddard *et al.* [17]) gave a chemisorption energy of 67 kcal/mol. Finally, the original version of the BOC model [45a] predicts a chemisorption energy of 32 kcal/mol (see below, however). Our results, using the recipe for bond preparing the clusters given in Section IV, are given in Table 1.8 for different clusters. As a comparison we have also added the corresponding chemisorption energy for hydrogen for each cluster in the table. We see that the calculated chemisorption energies for CH_3 range between 45 and 50 kcal/mol. As the hollow site chemisorption energies for hydrogen are between 5 and 7 kcal/mol lower than the experimental surface value (63 kcal/mol), we expect that also the calculated chemisorption energy for CH_3 should be underestimated by a similar amount at this site. By a similar comparison, no such correction is expected for the on-top site.

Table 1.8. Chemisorption Energy, D_e, for Methyl at On-Top and Threefold Hollow Sites for Different Clusters Modeling Ni(111)[a]

Cluster	D_e (kcal/mol)	
	CH_3	H
On-top chemisorption		
Ni_1	45	61
$Ni_{10}(7,3)$	47	–
Threefold hollow chemisorption		
$Ni_4(3,1)$	50	56
$Ni_{20}(3,7,7,3)$	47	56
$Ni_{22}(12,7,3)$	46	58

[a]For comparison, the hydrogen chemisorption energy for each cluster is also given (the experimental value is 63 kcal/mol).

This is reasonable as the calculations for the on-top site are performed at a slightly higher level of accuracy and include 3*d* correlation of the on-top metal atom. For the hollow site, 3*d* correlation, which should be less important than that for the on-top site, is neglected. Based solely on the calculations reported in Table 1.8, we predict the hollow site to be the preferred adsorption site with a chemisorption energy in the range 50–55 kcal/mol; however, the assignment of adsorption site is at this point rather uncertain. To make progress on this question we made calculations also of the vibrational frequencies, which could be informative as Ceyer *et al.* found an unusually low C–H frequency. It turns out that in a comparison between calculated harmonic frequencies and directly measured frequencies, which includes anharmonicity, there are large systematic errors. To avoid these systematic errors, rather than comparing absolute numbers we decided to calculate the shift in the C–H frequency with respect to a molecule for which the frequencies had been measured. The choice of molecule was $LiCH_3$ as it has a very low C–H stretching frequency. For the hollow site the calculated shift, $-108\ cm^{-1}$, is in quite nice agreement with the experimental shift of $-100\ cm^{-1}$, whereas for the on-top site a positive shift of about $+100\ cm^{-1}$ is obtained by the calculations. Comparison of the Ni–C and Li–C frequency shifts is equally conclusive. After these frequency comparisons, assignment of the hollow site as the preferred adsorption site is thus much more certain. Additional calculations, varying the geometry further, show that there is no direct attractive interaction between the hydrogen atoms and the surface, which otherwise might have been anticipated based on the low C–H frequencies found. The reason for the low values of these frequencies is instead simply a charge transfer of electrons from the surface over into the C–H bonds. As a final comment we should add, that as our predicted chemisorption energy of CH_3 is as high as 50–55 kcal/mol compared with the BOC model estimate of 32 kcal/mol [45a], this led to a slight revision of the BOC model. By separating cases of strong and weak molecular chemisorption, a BOC estimate of 48 kcal/mol is reached instead [45b], which is in much better agreement with our prediction.

Table 1.9. Chemisorption Energy, D_e, of CH_x, x = 1–3, on Ni(111) Using Bond Preparation[a]

	D_e (kcal/mol)		
Adsorbate	Bond Preparation	Embedding	BOC Model
CH	120	72	116
CH_2	88	67	83
CH_3	49	41	48

[a]For comparison, the embedding results of Whitten *et al.* [51] and the predictions by the BOC model [45b] are also given.

In similar cluster calculations we have also recently studied chemisorption of all three CH_x fragments on both Ni(100) and Ni(111). Our bond preparation results for Ni(111) are given in Table 1.9, together with results by Whitten and co-workers [51] and predictions made by the BOC model. The first conclusion that can be drawn from the results in the table is that there is very good agreement between the present bond preparation results and those predicted by the BOC model, with a difference within 5 kcal/mol for all three CH_x fragments. When the chemisorption energy for hydrogen and the gas-phase C–H bond energies are added to the CH_x chemisorption energies in Table 1.9, the reactions $CH_x \rightarrow CH_{x-1}$ + H are predicted to be almost thermoneutral for both these models, in general agreement with experimental information concerning the methanation reaction (see Ref. 72, for example). The difference between our model and the BOC model comes in when the predicted chemisorption energy of the carbon atom is also taken into account. In the BOC model the experimental value, 170 kcal/mol, is used, leading to a predicted exothermicity for the dissociation of CH as high as 37 kcal/mol. With our predicted value for the chemisorption energy of the carbon atom of 150 kcal/mol (see Section V), we predict an exothermicity of the same dissociation of only 12 kcal/mol. The experimental value is not known. If the agreement between the bond preparation results and the BOC model in Table 1.9 can be said to be quite satisfactory, this cannot be said when the results are compared with the embedding results of Whitten *et al.* The discrepancy between the bond preparation and embedding calculations is 8 kcal/mol for CH_3, which is not alarming, but increases to 21 kcal/mol for CH_2 and 48 kcal/mol for CH. The main difference between our results and the embedding results is most certainly our use of bond-prepared clusters. This is in line with the growing discrepancy between the results as one goes from CH_3 to CH_2 to CH. For CH, bond preparation is critical, requiring three singly occupied orbitals of correct nodal structure on the cluster. For CH_3 only one such orbital is required. The remaining discrepancy between our results and the embedding results is due to the use of a much larger basis set in our case (including up to f functions on carbon) and a more thorough correlation treatment. Quite generally, even a calculation on a small cluster but where a large basis and correlation treatment have been used, is in our opinion much more predictive than a large cluster calculation of medium accuracy and without bond preparation.

B. Chemisorption of Molecular H_2 and O_2

Whereas physisorbed H_2 on metal surfaces such as copper has been observed for a long time, even the existence of chemisorbed H_2 was doubted until quite recently. Mårtensson *et al.* [76] were then able to detect this species on a stepped Ni(100) surface by first adsorbing a $p(1 \times 1)$ layer of hydrogen atoms on the surface, thus blocking the dissociation. A low H–H frequency of 3210 cm^{-1} characterized the observed H_2 as chemisorbed. The gas-phase H–H frequency is 4401 cm^{-1} [77]. A similar situation existed in the chemistry of transition metal complexes. A few years ago, no molecularly bound H_2 complex had been observed, but mainly through the work of Kubas *et al.* [78], a large number of such compounds are now known. The ability to observe these compounds is important because it provides general insight into the mechanism of breaking of the H–H bond.

To study the chemisorption of molecular H_2, some cluster calculations were performed [70]. One important question, which can be answered only by calculations, is whether there is a minimum for molecular H_2 also on a flat Ni(100) surface and without packing of the surface with hydrogen atoms. If this is so, the observation of chemisorbed H_2 will be of much greater general importance. The cluster chosen for the calculations was a $Ni_{13}(9,4)$ cluster and, based on previous experience, we decided to look for chemisorbed H_2 in a parallel orientation at an on-top site. The nickel atom at this site has to be modeled at the all-electron level including 3*d* correlation. Much to our surprise we found a very deep minimum of 11 kcal/mol for chemisorbed H_2. In a rough one-dimensional search, the barrier height for dissociation into chemisorbed hydrogen atoms was found to be 9 kcal/mol counted from the molecular minimum. At the time the main interest was only to determine whether this barrier would lead to an energy point higher than the asymptote, and this is definitely not the case; however, the barrier of 9 kcal/mol is certainly an upper bound to the true barrier and a better estimate is probably about 5 kcal/mol. Zero-point vibrational effects are, rather surprisingly, not found to be very important to the size of the barrier [9], even though the H–H bond (which has a high frequency) is broken. With such a large chemisorption energy, one might still ask why molecular H_2 had not been observed earlier. The reason is found in the detailed dynamics of H_2 chemisorption. When a heavy atom, like a nickel atom, comes in contact with a nickel surface the kinetic energy is known to be immediately dissipated into the surface even though the barrier to diffusion is rather small [79]. The situation for a light system like H_2 is, however, very different as energy exchange between a heavy body and a light body is much less efficient by implication of the conservation of momentum. In fact, the energy exchange (ΔE) in a collision between a particle with kinetic energy E and mass m and a static particle with mass M is given by [80]

$$\Delta E = \frac{4mM}{(m + M)^2} E \tag{1.1}$$

This means that of the initial 11 kcal/mol of kinetic energy the hydrogen molecule obtains when it is molecularly chemisorbed, only 1.4 kcal/mol is lost in the collision with the on-top nickel atom. The remaining kinetic energy of $>$ 9 kcal/mol is thus more than enough to surmount the barrier of 5 kcal/mol for dissociation into atoms. Chemisorbed H_2 will therefore be impossible to detect without the preparations made in the above-mentioned experiment [76].

Molecular chemisorption of O_2 on a nickel surface is rather similar to H_2 chemisorption, but the energetics turn out to be even more surprising and dramatic. As already stated, chemisorbed O_2 has not yet been observed on any well-defined nickel surface. For Ag(110), where it has been observed, the chemisorption energy was found to be only 10 kcal/mol [81]. A similar chemisorption energy was found for O_2 on Pt(111) [82]. As the atomic oxygen chemisorption energy on Ag(110) is 80 kcal/mol, whereas that for Ni(100) is 130 kcal/mol, a simple scaling leads to a predicted chemisorption energy for molecular O_2 on Ni(100) of 15–20 kcal/mol. To make a more accurate estimate of this energy, cluster calculations including bond preparation were performed for O_2 chemisorbed at three different sites of Ni(100) [69]. Chemisorption at each site was modeled by different clusters centered at the site of interest. A Ni_{13}(9,4) cluster was used for the on-top site, a Ni_{20}(6,8,6) cluster for the bridge site, and a Ni_{25}(12,9,4) cluster for the fourfold hollow site (see Fig. 1.3). The resulting molecular chemisorption energies were much larger than expected. For the on-top and bridge sites this energy was 56 kcal/mol, and for the hollow site it was as large as 78 kcal/mol. It is clear that these chemisorption energies have to be rationalized in a way other than simple scaling based on the atomic values. Through an analysis of the wavefunction we found that the conventional picture of molecular bonding to transition metals, through donation and back-donation from the transition metal, is a poor picture in this case. Instead, the wavefunction shows that two normal covalent bonds have been formed, similar to the bonding in H_2O. The large chemisorption energy can now be rationalized. When an oxygen atom is chemisorbed, two covalent bonds are formed, leading to a chemisorption energy of 130 kcal/mol. This leads to an estimated energy of 65 kcal/mol per covalent bond for nickel. A similar estimate for silver is 40 kcal/mol. The gas-phase double-bonded O_2 molecule has a dissociation energy of about 110 kcal/mol, and in the chemisorption process one of the bonds is broken. This leads to an estimated loss of 55 kcal/mol. For O_2 on Ni(100) we thus expect a gain through the formation of two covalent bonds of $2 \times 65 = 130$ kcal/mol, and a loss of intramolecular bonding of 55 kcal/mol. This leads to a predicted chemisorption energy of $130 - 55 = 75$ kcal/mol, which is not far away from the results of our calculations. For Ag(110) the same procedure leads to an estimate of $80 - 55 = 25$ kcal/mol, which is larger (but not too much larger) than the experimental value. Of course, these estimates are very approximate as in the real bond formation many other effects are present. Already our calculated results show that something is different at the hollow site bonding than at the other two sites. This difference is due both to the smaller 3*d* repulsion and to the more favorable overlap of the valence orbitals at the hollow site compared with the other two sites. Because of the smaller strain on the O_2 molecule at the hollow site, the O–O bond distance is shorter and there is also some intramolecular π bonding left.

Even if the large chemisorption energy of molecular O_2 on nickel can be rationalized as above, the result may still be hard to accept as O_2 has not been seen on this

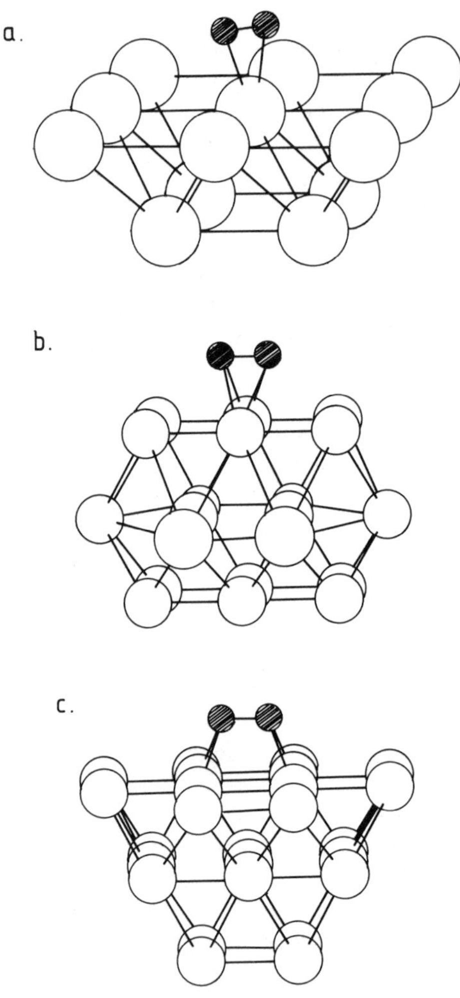

Figure 1.3. Clusters used for modeling O_2 dissociation: **(a)** Ni_{13} (9,4); **(b)** Ni_{20}(6,8,6); **(c)** Ni_{25}(12,9,4). Reprinted with permission from Ref. 69.

surface. For the bridge and the on-top site this is, however, not surprising. The O_2 chemisorption energy is large, 56 kcal/mol, but the barrier to dissociation of 8 kcal/mole is so small that the kinetic energy remaining after the collision with the surface will clearly be enough to overcome this barrier. For the hollow site the situation is less clear. The chemisorption energy here is even larger, 78 kcal/mol, and the calculated barrier to dissociation is as high as 30 kcal/mol. Again, as for H_2, this estimate of the barrier is rather approximate and a better estimate is probably about 20 kcal/mol. Equation 1.1 for the energy exchange in a collision between two particles, where one particle is the O_2 molecule and the other particle consists of four nickel atoms, predicts a loss of kinetic energy for O_2 of as much as 33 kcal/mol in the first collision. After the second collision the kinetic energy of O_2 should be 26 kcal/mol, which is barely enough to overcome the barrier of 20 kcal/mol. Situations in which there will be more than two consecutive collisions between O_2 and the surface at this site are not difficult to imagine and it might therefore be worthwhile to look for chemisorbed O_2 on this surface at low temperatures; however, it should in this context be remembered that the present picture of the dynamics of the collision between O_2 and the Ni(100) surface is oversimplified, and to make a more firm prediction of the existence of molecularly chemisorbed O_2, detailed dynamical calculations are required.

The qualitative correctness of our results for O_2 chemisorption can be further tested against detailed spectroscopic information. In recent NEXAFS and UPS experiments both the peroxo-form of O_2 (doubly bonded to the surface) and the superoxo-form of O_2 (singly bonded to the surface) have been observed but the experimental results have been difficult to interpret. In the first NEXAFS study, O_2 on Pt(111) was identified as the peroxo- form with a bond distance of 1.45 Å [83]. In a later NEXAFS experiment the results were reinterpreted, and O_2 on Pt(111) was identified as the superoxo- form with a bond distance of only 1.32 Å [84]. In this study, O_2 on Ag(110) was, on the other hand, found to adsorb in the peroxo-form with a large bond distance of 1.47 Å [84]. After these measurements there was a UPS study which, in agreement with the first NEXAFS experiment, predicted O_2 on Pt(111) to be in the peroxo- form and the bond distance was estimated by a new procedure directly from the UV spectrum to be 1.40 Å [85]. UPS studies have also been made for O_2 on Ag(110) with very conflicting results [86,87]. Under the seemingly drastic assumption that the most important features of O_2 chemisorption should be independent of metal surface, we made some comparative model calculations for O_2 on Ni_{25}(12,9,4) [66]. The qualitative conclusions based on these calculations are quite clear. The essential features of the UV spectrum measured in Ref. 85 for O_2 on Pt(111) are very well reproduced by the calculations but we found these to be characteristic of the superoxo- form of O_2, in contrast to the interpretation made in Ref. 85. When this reassignment is made, the bond distance can be optimized in the calculations leading to 1.32 Å. There is thus total agreement between the calculations and the latest NEXAFS experiment in Ref. 84 for O_2 on Pt(111). Very good agreement was also found for the UV spectrum calculated for the peroxo- form of O_2 and that measured by Barteau and Madix for O_2 on Ag(110) [86], but a comparison with several other UPS measurements on the peroxo- form of O_2 show very large differences. The additional peaks found in these latter spectra are most likely due to contaminants. The calculated O$-$O bond distance for the peroxo- form was 1.43 Å. Again, there is thus full agreement between our calculations and the interpretations and results of the latest NEXAFS experiment in Ref. 84. The overall quite good agreement between the cluster model calculations and the most reliable experiments shows that the picture and understanding of chemisorption of O_2 obtained from the calculations must be essentially correct. Similar calculations can also be used to interpret recent experiments where O_2 is photodesorbed or photodissociated on metal surfaces [88]. Such calculations are underway and have given promising results [89].

VII. Chemical Reactions: Examples

The ultimate goal of the present project is to accurately model and understand chemical reactions on transition metal surfaces. For dissociation reactions, this modeling should include not only the reactants and products but also the passage between these over possible reaction barriers. One of the main ideas in setting up the present model was therefore to construct it in such a way that the calculation of transition states for reactions should not present additional problems. For this reason, the correlation calculations are based on the open-ended CASSCF and multireference CI methods. In the first two subsections we discuss straightforward applications of the present model to H_2 and O_2 dissociation on Ni(100). For H_2 a comparison will be made with dissociation on Cu(100). In the third subsection the more complicated dissociation of CH_4 is discussed.

A. Dissociation of H_2

For H_2 dissociation on transition metal surfaces there exists a qualitative and seemingly simple problem that appears ideal for a theoretical treatment: Why does H_2 dissociate without a barrier on a nickel surface but with a barrier on a copper surface? As our previous work was concerned with simple problems in homogeneous catalysis and the same difference between nickel and copper exists there, this problem seemed to be a reasonable first question to attack in the area of heterogeneous catalysis. It was hoped that there would be a common understanding of this phenomenon in the two areas. In the case of homogeneous catalysis the difference between nickel and copper is easy to understand. When the H_2 bond is broken, two new bonds to the dissociated hydrogen atoms have to be able to form. Therefore, to break the H_2 bond, the metal atom has to have at least two singly occupied orbitals. This requirement is fulfilled for the nickel atom which has both an open 4*s* orbital and an open 3*d* orbital. In line with this, calculations show that for the simple NiH_2 system there is a very small barrier for dissociation of H_2 [90]. This is actually not more magical than the fact that the oxygen atom (^1D state) breaks the H_2 bond without a barrier. For copper, on the other hand, there is only one singly occupied orbital, the 4*s* orbital. Calculations for the system CuH_2 confirm that there is now a substantial (> 80 kcal/mol) barrier for dissociation of H_2 [91]. There is only one way these results for the simple MH_2 systems could be relevant to the heterogeneous problem—if H_2 dissociation on a surface takes place at an on-top position. To model the on-top dissociation on Ni(100), calculations were set up using a Ni_{13}(9,4) cluster as a model for the surface. The result of these calculations was that the barrier for dissociation of H_2 was essentially as small as that for the single nickel atom [92]. The 3*d* orbitals of the on-top nickel atom were shown to play exactly the same key role in breaking the H—H bond as they did in the NiH_2 system. The reason for this similarity is that cluster orbitals of appropriate symmetry do not overlap sufficiently well with the small hydrogen orbitals to overtake the role of the 3*d* orbitals (see below for O_2, however).

To study the details of the dissociation mechanism the calculations of Ref. 92 were too schematic and new, more rigorous calculations including bond preparation were

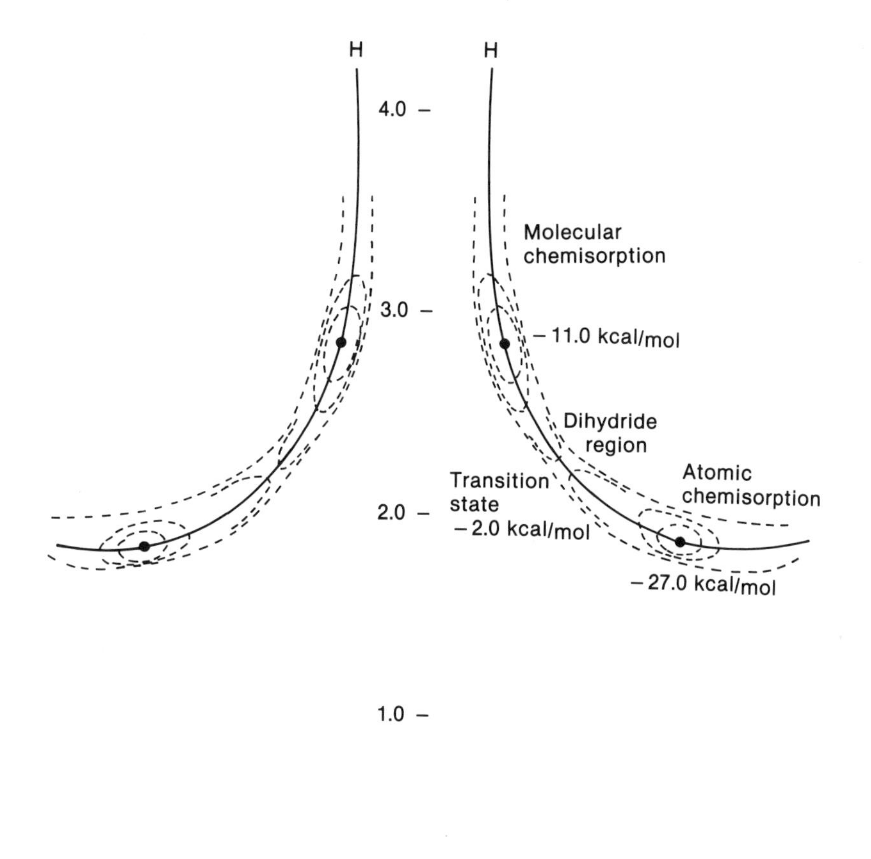

Figure 1.4. Schematic picture of the potential energy surface for the on-top dissociation of H_2 on $Ni_{13}(9,4)$. The energies given are with respect to H_2 at long distance from Ni_{13}. The solid line is the minimal energy path. Reprinted with permission from Ref. 70.

therefore undertaken [70]. In this study, similar calculations modeling Cu(100) were also performed. The results of these calculations are depicted in Fig. 1.4 for H_2 on $Ni_{13}(9,4)$ and in Fig. 1.5 for H_2 on $Cu_{13}(9,4)$. Comparison of the two figures shows the dramatic difference in dissociation behavior in the two cases. In the nickel case the H_2 molecule goes down into the fairly deep molecular chemisorption well (cf. Section VIB) and then passes smoothly over the dihydride state into the atomic chemisorption region. The energy is highest for the asymptotic gas-phase H_2 (with the possible exception of a small barrier for long distances), which means that the classical barrier height is close to zero for the dissociation reaction. In contrast, the energy for H_2 on copper increases steadily upward until a high barrier of 40 kcal/mol is passed. The minimum energy path is parallel to the surface in this region, showing

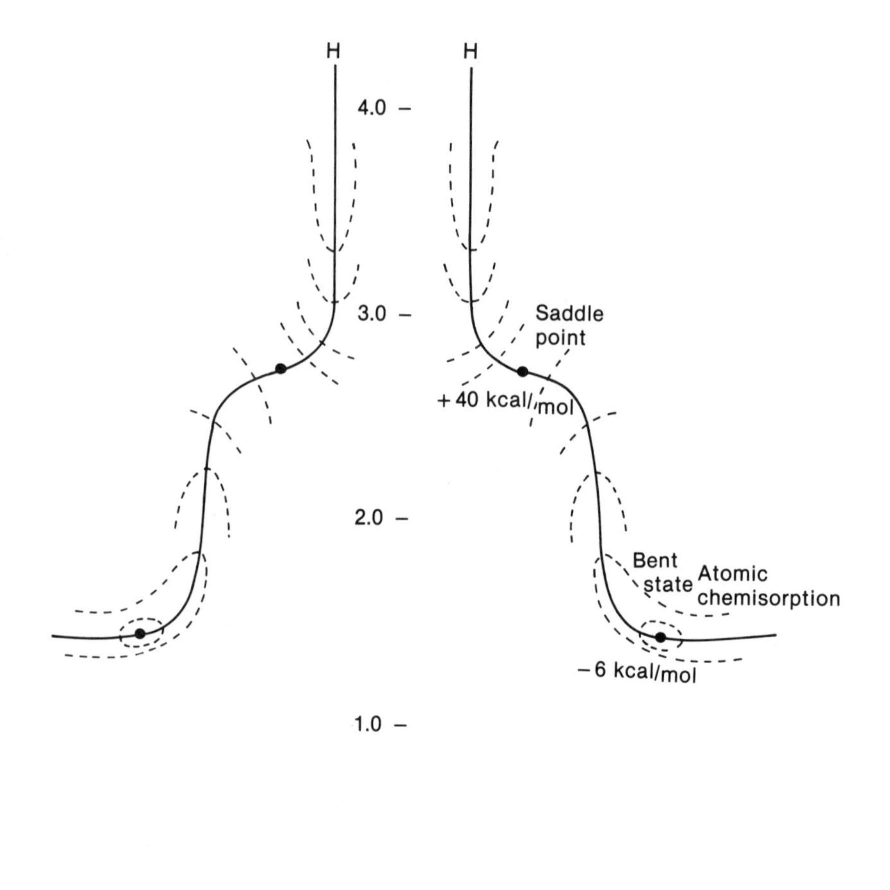

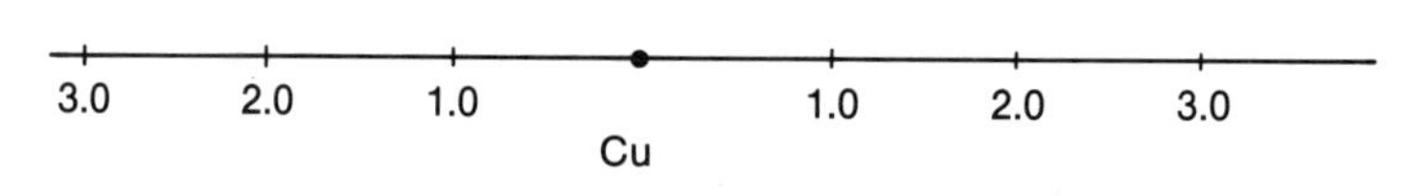

Figure 1.5. Schematic picture of the potential energy surface for the on-top dissociation of H_2 on $Cu_{13}(9,4)$. The energies given are with respect to H_2 at long distance from Cu_{13}. The solid line is the minimal energy path. Reprinted with permission from Ref. 70.

that the surface hardly participates in the final breaking of the H–H bond. For copper there is no molecular chemisorption well and no low-energy dihydride region, because the bonding in these regions requires open-shell 3*d* orbitals on the on-top metal atom. Molecular chemisorption occurs only if the repulsion between the metal 4*s* orbitals and the doubly occupied σ_g ortibal of H_2 can be reduced. This can be done by an effective s–d_σ hybridization if there is a singly occupied 3*d* orbital. The dihydride region is characterized by a direct covalent 3*d* bond to the hydrogen atoms, which obviously also requires singly occupied 3*d* orbitals. It should be noted that as the bond breaking proceeds, by moving the hydrogen atoms further out into the fourfold hollow sites, cluster orbitals start to overlap better with the hydrogen orbitals than the 3*d* orbitals do. This leads to a continuously smaller participation of the 3*d* orbitals until finally, at the atomic chemisorption sites, the 3*d* contribution to the covalent bonding is nearly negligible (see Section III).

The difference between nickel and copper surfaces is thus well demonstrated by the cluster calculations described above. There remains one problem: the difference is much larger than indicated by experiment. The nearly barrierless dissociation for H_2 on nickel is in agreement with experiment but the barrier height obtained for copper is much too high. The most often quoted experimental result for the barrier to H_2 dissociation on Cu(100) is 4 kcal/mol [93]. Most of this apparent discrepancy between theory and experiment is easy to resolve. It does not require much imagination to realize that H_2 may prefer to dissociate at a hollow site and this was actually found in additional cluster calculations that were performed. In these calculations a Cu_{25}(12,9,4) cluster was used to model the hollow site. The result for the classical barrier height at this site is 14 kcal/mol, which includes an estimate for errors in the calculations and a large (10 kcal/mol) effect of 3*d* correlation. The corresponding result for the dissociation over a bridge site is 25 kcal/mol. Agreement between theory and experiment is thus markedly improved but still cannot be claimed to be satisfactory. Based on results for nickel [9], zero-point vibrational effects are not expected to lower the barrier by more than 2 kcal/mol. A disagreement with experiment of about 8 kcal/mol thus remains. Better agreement can perhaps be obtained if the beam experiment is corrected for the possible presence of vibrationally excited H_2, as indicated in a recent interesting paper by Harris [94]. For completeness, we also give the calculated barrier heights for the dissociation of H_2 at the hollow and bridge sites of Ni(100), and they are slightly above zero and 10 kcal/mol, respectively. These results therefore indicate that there is not one (the on-top pathway described above) but two favorable dissociation channels for H_2 on Ni(100), which has actually been inferred from recent experiments [95].

B. Dissociation of O_2

The gross features of oxygen chemisorption on nickel are well known from experiments. When the Ni(100) surface is exposed to O_2, the oxygen molecules dissociate and the atoms form a $p(2\times2)$ structure with the oxygen atoms occupying fourfold hollow sites. As the oxygen exposure continues, a $c(2\times2)$ structure develops. As already discussed in Section VI, no molecularly chemisorbed O_2 has been observed on any well-defined nickel surface. The dominant theory for O_2 dissociation is that of Brundle *et al.* [96]. To explain the formation of the $p(2\times2)$ and $c(2\times2)$ structures they argue that there should be a very large repulsion between oxygen atoms at neighboring fourfold hollow sites. This repulsion favors the dissociation over the fourfold hollow site, which should dominate in the initial stage of oxygen exposure. There should also be some contribution from dissociation over on-top sites but this pathway should be less favorable. Dissociation over bridge sites should be totally excluded. They also argue that atomic oxygen migration is not likely to occur.

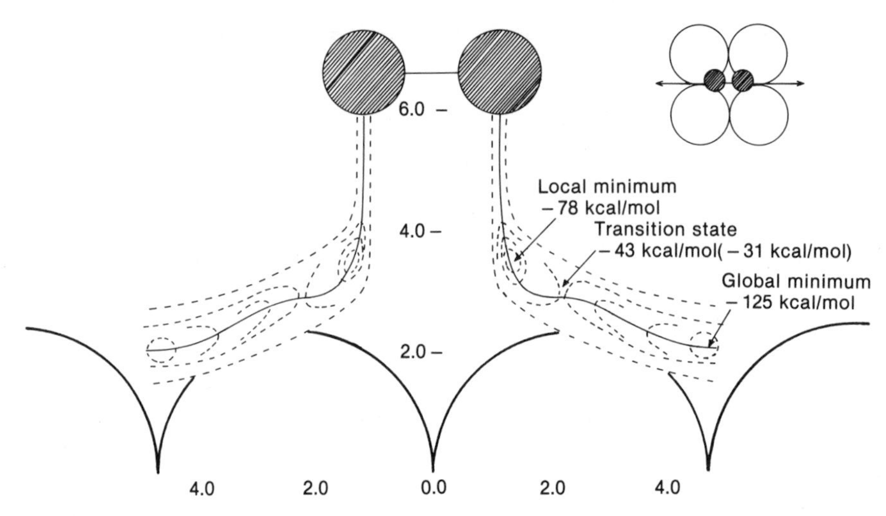

Figure 1.6. Schematic picture of the potential energy surface for the fourfold hollow dissociation of O_2 on Ni_{25}(12,9,4). The energies given are with respect to O_2 at long distance from Ni_{25}. The solid line is the minimal energy path. (The corresponding energies for the on-top dissociation and bridge dissociation are, respectively, for the local minimum −56 and −56 kcal/mol, for the transition state −50 and −48 kcal/mol, and finally, for the global minimum −98 and −125 kcal/mol.) Reprinted with permission from Ref. 69.

To test some of the assumptions made by Brundle *et al.*, cluster calculations were performed modeling the dissociation of O_2 over the three sites of interest. The clusters chosen for the calculations are shown in Fig. 1.3 and the potential energy surface for the fourfold hollow dissociation is shown in Fig. 1.6. The corresponding results for the on-top and bridge dissociation are given in the caption to Fig. 1.6. Some of the results of these calculations have already been described in Section VIA. The most important result of the calculations is perhaps that there is no barrier for dissociation at any of the three sites. One of the key assumptions in the model proposed by Brundle *et al.* [96], that the bridge dissociation should have a very high barrier, is therefore not consistent with our calculations. In Ref. 96, the conclusion that there should be a high dissociation barrier for the bridge site was reached based on the assumption of large repulsion at nearest-neighbor sites. We find that there is such a repulsion but the size of it, about 20 kcal/mol, is not decisive for the preferred dissociation. In fact, this repulsion seems to have little effect on the initial part of the dissociation curves, as judged from a comparison between the bridge and the on-top dissociations. These dissociations are almost perfectly parallel in the beginning, both having a molecular chemisorption well of 56 kcal/mol and a transition state of 48–50 kcal/mol below their asymptotes. Only at the end of the dissociation path, at the respective fourfold hollow minima, is the bridge dissociation curve less attractive than the on-top curve. This behavior indicates that the repulsion of 20 kcal/mol that we observe at nearest-neighbor sites is not due to electrostatic repulsion from two negatively charged species but, rather, is a result of local competition for bonding. This competition is not fully developed until all four bonds are formed when the two atoms are chemisorbed. Another interesting result, which was already noted in Section VIB, is that the path for the fourfold dissociation passes through a deeper molecular minimum than the other two paths. The reason for this is less 3*d* repulsion and more favorable overlaps.

As we find that there is no barrier for dissociation of O_2 over a bridge site, there must be another explanation for the fact that oxygen atoms are not found at nearest-neighbor sites. There seems to be only one possible solution to this problem: the presence of oxygen atom migration on the surface. As has been discussed above (Section VA), the oxygen diffusion barrier was estimated to be about 20 kcal/mol. When an O_2 molecule hits a bridge position the estimated loss of kinetic energy is about 40 kcal/mol (cf. Eq. 1.1). This leaves about 40 kcal/mol kinetic energy for each oxygen atom entering the fourfold hollow site after the bridge dissociation. In the collision between an oxygen atom and the fourfold hollow site, energy exchange is rather inefficient because of the large difference in masses and can be estimated from Eq. 1.1 to be only about 10 kcal/mol. The remaining 30 kcal/mol kinetic energy should be enough to overcome the barrier of 20 kcal/mol. With these simple estimates we have thus shown that oxygen migration is at least not ruled out. Clearly, detailed dynamic studies are required to definitely establish the presence of oxygen diffusion as the origin of the formation of the $p(2\times2)$ and $c(2\times2)$ structures.

In the study of H_2 dissociation described in the previous subsection, one of the most interesting results is the key role played by the 3*d* orbitals in the on-top dissociation. An interesting question for O_2 dissociation is therefore whether the same effect can be found also in this case. The experimental evidence speaks against such an effect because, in contrast to H_2, O_2 is found to dissociate quite easily on metals that do not have open *d* orbitals or have no *d* orbitals at all. To clarify the discussion of the mechanism for O_2 on-top dissociation, a schematic picture of the bond-breaking surface orbital is given in Fig. 1.7a. This orbital is antisymmetric with respect to a plane perpendicular to the O–O axis and the 4*s* orbital of the on-top atom therefore cannot participate in this surface orbital for symmetry reasons. Thus, from this atom there will be only a 3*d* contribution. As we mentioned before, the 3*d* orbital is much smaller than the 4*s* orbital (see Table 1.2) and one can easily envision situations where the overlap between the antibonding adsorbate orbital is larger with the nearest-neighbor atomic 4*s* orbitals than with the 3*d* orbital for the on-top atom in the orbital in Fig. 1.7a, even though the distance to the latter atom is much smaller. This is exactly what happens in the on-top O_2 dissociation; therefore, the 3*d* orbital will not contribute much to the bond breaking. For H_2, on the other hand, the H–H bond distance is too short for an effective overlap with the 4*s* orbitals on the nearest-neighbor atoms; the on-top 3*d* orbital will therefore play a decisive role in the bond breaking in this case.

The on-top dissociation of a general σ bond, as shown in Fig. 1.7a, is thus one situation where the metal 3*d* orbitals can play a more than marginal role in the bond breaking. It is equally clear from this figure that it is only the 3*d* orbital for the on-top atom that will be important. The overlap between the adsorbate orbitals and the 3*d* orbitals on nearest-neighbor atoms will always be much poorer than the overlap with the 4*s* orbitals on these centers because of the large difference in size between

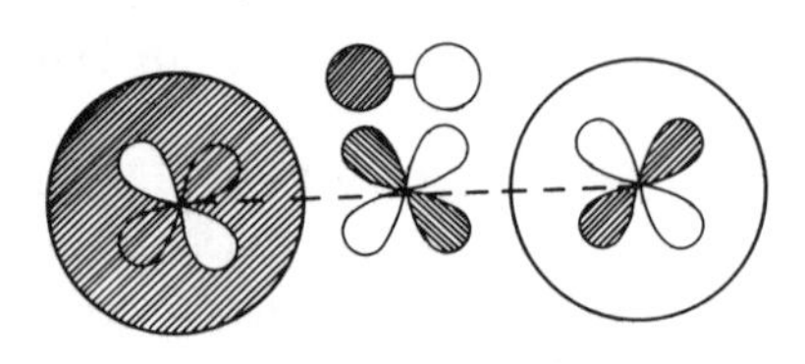

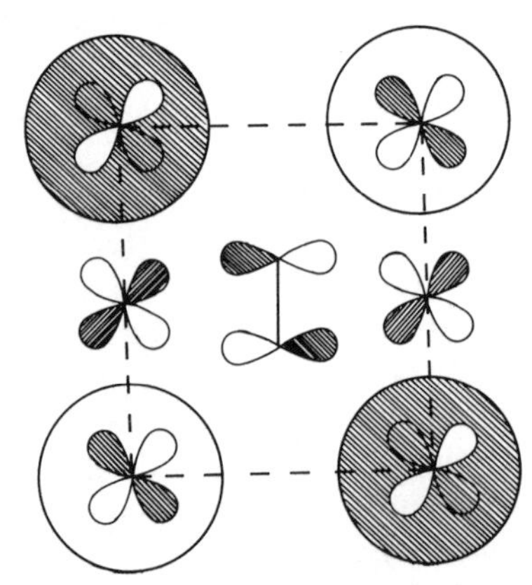

Figure 1.7. Schematic picture of bond-breaking orbitals: **(a)** a σ-bonded molecule approaching an on-top position; **(b)** a π-bonded molecule approaching a bridge position. The dashed lines connect the metal atoms in the surface and the solid lines connect the atoms of the adsorbate. Only $(n + 1)s$ orbitals (the large circles) and nd orbitals are drawn for the metal atoms. Shaded areas indicate a negative phase of the orbital.

these orbitals. It is interesting that the possible importance of the on-top $3d$ orbital follows from pure orbital symmetry arguments, in a similar fashion as symmetry arguments are used to qualitatively describe chemical reactions between organic molecules using the Woodward–Hoffmann rules [97]. Another situation in which the $3d$ orbitals can be expected to be qualitatively important is shown in Fig. 1.7b. This figure shows the dissociation of a π bond in a bridge position. For symmetry reasons, the antibonding π^* orbital of the adsorbate will not interact with the $4s$ orbitals on the bridge atoms but only with the $3d$ orbitals on these atoms. This orbital overlap will, on the other hand, be very efficient and is the main reason why the bridge site is the most plausible active site for the breaking of the very strong N_2 bond [98]. For N_2 it is also important that the repulsion to the occupied π_u orbital be minimized, which also makes the type of dissociation at the bridge site shown in Fig. 1.7b close to optimal.

C. Dissociation of CH_4

The splitting of the C–H bond in alkanes has been a topic of interest in our group for a long time. Most of these studies have been concerned with modeling the situation in homogeneous media. For this purpose, only a single metal atom with and without ligands has been used. In the most recent and accurate study the main conclusion was that the mechanism for breaking the C–H bond is very similar for all metals. The barrier for dissociation of the C–H bond in methane is in the range 20–25 kcal/mol for first-row transition metals and slightly lower, 5–15 kcal/mol, for second-row transition metals [99]. These results are not too different from known results for breaking of the C–H bond in methane for transition metal surfaces, and it is tempting to believe that the mechanism is the same. If there is a direct parallel between the homogeneous and heterogeneous situations, the bond breaking on the surface should occur at the on-top site. Model calculations [100] analogous to those for the breaking of the H_2 bond at an on-top site were therefore set up [70,92]. The calculations for CH_4 on a Ni_{13} cluster, where a full geometry optimization was performed, gave a barrier of 16 kcal/mol after correcting for basis set superposition errors. The cluster convergence was studied at the CASSCF level for two larger clusters, Ni_{18} and Ni_{26}. These clusters gave somewhat larger barriers by 7 and 5 kcal/mol, respectively. The best calculated value including a size convergence correction is thus 20 kcal/mol. This result is based on calculations of medium accuracy. A correction for deficiencies in basis sets and correlation treatment can be estimated to be at least 2 kcal/mol from an accurate calculation for the barrier for a single nickel atom [99]. Also, as there is a large loss of zero-point vibrational energy for the methane molecule, it is likely that consideration of these effects will lower the barrier further by 1–3 kcal/mol. A predicted value of 15–17 kcal/mol is thus reached for the barrier of methane dissociation. There are two different experimental results for the barrier in breaking the C–H bond in CH_4 on Ni(100). The older of these measurements gave 6 kcal/mol [101], whereas a more recent experiment gave 12 kcal/mol [102]. The first conclusion that can be drawn based on a comparison between calculated and measured values is that our predicted value is not consistent with the older measurement [101]. As a general rule in these types of calculations, the barriers are overestimated by 3–5 kcal/mol, which would bring the predicted value in even better agreement with the most recent experiment [102]. Another interesting possibility, which could explain the difference between the latest experiment and our calculations, is that tunneling is important for this reaction [103], which would tend to make the barrier appear lower in the experiments than it actually is. A final note of interest is that some preliminary measurements of dissociative chemisorption of methane on copper have recently been made yielding very large barriers (about 50 kcal/mol) [104]. This very large difference between nickel and copper strongly indicates that the dissociation on nickel should occur over on-top sites, where the difference between the metals should be largest.

One more question on the CH_4 dissociation has been examined by cluster model calculations–whether steps or kinks would drastically lower the barrier for dissociation—a classic question in surface science. To make a simple model of a kink the Ni_5 cluster (in Fig. 1.1) is turned upside down, so that the CH_4 dissociation will take place against a single atom in the first layer and with four atoms in the second layer. Calculations on this system, similar in accuracy to those described above for CH_4 on Ni_{13}, resulted in a barrier 3–5 kcal/mol (depending on the bond preparation state) higher than the one obtained for Ni_{13}. Increasing the cluster from five to nine atoms by adding four atoms in the second layer lowered the barrier height essentially back to the value found for Ni_{13}. From these results we draw the conclusion that kinks do not significantly lower the barrier height for CH_4 dissociation. At a qualitative level, the mechanism for the dissociation is exactly the same, with or without kinks in the surface, which is not entirely surprising in light of the large similarity between the single-atom and the Ni_{13} results. As our calculations do not give a relative accuracy better than 3–5 kcal/mol, a small effect on the barrier height is of course still not ruled out. Furthermore, this investigation of kinks concerns only the energetics and a large dynamic effect of kinks is therefore still possible.

VIII. Conclusions

In this chapter we summarized and exemplified the present status of cluster modeling of chemisorption energetics. The views on the adequacy of clusters as models for surfaces have varied dramatically since 1980. From 1978 to 1981 the work by Goddard and co-workers [17] on clusters modeled by one-electron ECPs showed much promise, even for quantitative calculations of chemisorption energies. Over the 5-year period following this work, several unsuccessful attempts were made to reach cluster size convergence for chemisorption of different adsorbates [46,53]. The growing knowledge that small clusters would never be able to model properly the continuous density of states of a real surface and would not obtain the correct position of the Fermi level [19] increased the skepticism for the use of the cluster model [45a]. The newest findings concerning the nature of the surface chemical bond have changed the situation again in favor of the cluster model. The most important of these new findings is that, at least for strongly bound adsorbates, the cluster size oscillations of the chemisorption energy are easy to understand and also to correct for. The use of bond-prepared clusters leads in most cases to stable chemisorption energies, even for rather small clusters of 10–20 atoms. The second finding is that these bond-prepared chemisorption energies are not only stable but also in quite good agreement with surface experiments. This means that the surface chemical bond is not as delocalized as had often been believed when, for example, embedding theories or infinite models were considered necessary. On the contrary, the surface bond must be regarded as quite localized, which is of course of utmost importance when small clusters are used as models. Finally, model calculations have shown that the position of the Fermi level is not as significant as results using other methods had indicated [6–10]. In fact, shifting the Fermi level by 7 eV, by adding and removing electrons from a cluster, changed the hydrogen chemisorption energy by less than 0.1 eV.

There are many useful aspects of the new understanding of chemisorption energetics described in this chapter, but here we mention only one. The localized nature of the chemisorption bond means that the main factor governing the strength of this bond is the overlap between adsorbate orbitals and local substrate orbitals. The availability of local orbitals of proper symmetry for efficient overlap thus becomes a key factor. It is then easy to understand that for transition metals, singly occupied $3d$ orbitals may be necessary for dissociating the H_2 molecule. The reason for this is that if the dissociation occurs at on-top positions, the bond-breaking orbital, as depicted in Fig. 1.7a, will contain only $3d$ contributions and not $4s$ contributions from the on-top atom. Similarly, the local orbital environment can be used to argue that the N_2 molecule should dissociate at bridge positions with its axis perpendicular to the line connecting the bridge atoms. As shown in Fig. 1.7b, this situation leads to the most favorable arrangement for dissociation of a π bond. The localized nature of the surface chemical bond has thus led to a possible way to discuss the mechanism of surface chemical reactions which is as simple as in the case of free molecules. This is probably one of the most important aspects of this work.

Acknowledgments

The cluster model described in this work has been developed by Itai Panas and the present authors. Discussions with Lars Pettersson and Margareta Blomberg were important to the writing of this chapter. We thank Andrew dePristo for valuable comments on the manuscript.

References

1. W. R. Hehre, L. Radom, P. R. Schleyer, and J. A. Pople, *Ab Initio Molecular Orbital Theory*, Wiley, New York, 1986.
2. C. Pisani, R. Dovesi, and C. Roetti, *Lecture Notes in Chemistry*, 48: *Hartree–Fock Ab Initio Treatment of Crystalline Systems*, Springer-Verlag, Berlin and Heidelberg, 1988.
3. P. J. Feibelman, *Annu. Rev. Phys. Chem.* **40** (1989), 261.
4. M. Seel, *Int. J. Quantum Chem.* **26** (1984), 753.
5. A. J. Freeman, C. L. Fu, and E. Wimmer, *J. Vac. Sci. Technol. A* **4** (1986), 1265; E. Wimmer, C. L. Fu, and A. J. Freeman, *Phys. Rev. Lett.* **55** (1985), 2618.
6. (a) J. K. Nørskov, and N. D. Lang, *Phys. Rev. B* **21** (1980), 2136. (b) B. I. Lundqvist, and J. K. Nørskov, in *Proceedings of the 8th International Congress on Catalysis, Berlin, 1984*, Vol. 4, p. 85; Dechema, Frankfurt-am-Main, 1984. (c) B. I. Lundqvist, J. K. Nørskov, and H. Hjelmberg, *Surf. Sci.* **80** (1979), 441. (d) B. I. Lundqvist, *Chem. Scr.* **26** (1983), 423.
7. M. S. Daw, and M. I. Baskes, *Phys. Rev. B* **29** (1984), 6443.
8. J. D. Kress, and A. E. De Pristo, *J. Chem. Phys.* **87** (1987), 4700.
9. T. N. Truong, D. G. Truhlar, and B. C. Garrett, *J. Phys. Chem.* **93** (1989), 8227.
10. (a) C. Zheng, Y. Apeloig, and R. Hoffmann, *J. Am. Chem. Soc.* **110** (1988), 749. (b) R. Hoffmann, *Solids and Surfaces: A Chemist's View of Bonding in Extended Structures*, VCH, New York, 1988.
11. J. Sauer, *Chem. Rev.* **89** (1989), 199.
12. A. B. Anderson, *Theoretical Aspects of Heterogenous Catalysis*, (J. B. Moffat, ed.), Ch. 10, Van Nostrand Reinhold, New York, 1990.
13. C. Minot, M. A. Hove, and G. A. Somorjai, *Surf. Sci.* **127** (1982), 441.

14. J. Andzelm, and D. R. Salahub, *Int. J. Quantum Chem.* **29** (1986), 1091.
15. G. Chiarello, J. Andzelm, R. Fournier, N. Russo, and D. R. Salahub, *Surf. Sci.* **202** (1988), L621.
16. A. D. Becke, in *The Challenge of d and f Electrons: Theory and Computation* (D. R. Salahub, and M. C. Zerner, eds.), ACS Symposium Series No. 394, Ch. 10, Am. Chem. Soc., Washington, D.C.
17. (a) T. H. Upton, and W. A. Goddard, *CRC Critical Reviews in Solid State and Materials Sciences*, CRC Press, Boca Raton, Fla., 1981. (b) C. F. Melius, T. H. Upton, and W. A. Goddard, *Solid State Commun.* **28** (1978), 501. (c) T. H. Upton, and W. A. Goddard, *Phys. Rev. Lett.* **42** (1979), 472.
18. J. Almlöf, K. Faegri, Jr., and K. Korsell, *J. Comput. Chem.* **3** (1982), 385; J. Almlöf, and O. Gropen, private communication.
19. D. M. Wood, *Phys. Rev. Lett.* **46** (1981), 749.
20. J. L. Whitten, and T. A. Pakkanen, *Phys. Rev. B* **21** (1990), 4357.
21. I. Panas, J. Schüle, P. Siegbahn, and U. Wahlgren, *Chem. Phys. Lett.* **149** (1988), 265.
22. I. Panas, and P. E. M. Siegbahn, *J. Chem. Phys.* **92** (1990), 4625.
23. M. Blomberg, U. Brandemark, J. Johansson, P. Siegbahn, and J. Wennerberg, *J. Chem. Phys.* **88** (1988), 4324.
24. C. E. Moore, *Atomic Energy Levels*, U. S. Department of Commerce, Nat. Bur. Stand., U.S. Government Printing Office, Washington, D.C., 1952.
25. S. P. Walch, and C. W. Bauschlicher, Jr., in *Comparisons of Ab Initio Quantum Chemistry with Experiment* (R. Bartlett, ed.), Reidel, Boston, 1985.
26. P. S. Bagus, C. W. Bauschlicher, Jr., C. J. Nelin, B. C. Laskowski, and M. Seel, *J. Chem. Phys.* **81** (1984), 3594.
27. I. Panas, P. Siegbahn, and U. Wahlgren, *Chem. Phys.* **112** (1987), 325.
28. A. Mattsson, I. Panas, P. Siegbahn, U. Wahlgren, and H. Åkeby, *Phys. Rev. B* **36** (1987), 7389.
29. I. Panas, P. Siegbahn, and U. Wahlgren, *Theor. Chim. Acta* **74** (1988), 167.
30. V. Bonifacic, and S. Huzinaga, *J. Chem. Phys.* **60** (1974), 2779.
31. L. G. M. Pettersson, U. Wahlgren, and O. Gropen, *Chem. Phys.* **80** (1983), 7.
32. B. O. Roos, P. R. Taylor, and P. E. M. Siegbahn, *Chem. Phys.* **48** (1980), 157; P. E. M. Siegbahn, J. Almlöf, A. Heiberg, and B. O. Roos, *J. Chem. Phys.* **74** (1981), 2384.
33. P. E. M. Siegbahn, *Int. J. Quantum Chem.* **23** (1983), 1869.
34. W. Müller, J. Flesch, and W. Meyer, *J. Chem. Phys.* **80** (1984), 3297.
35. L. G. M. Pettersson, and H. Åkeby, *J. Chem. Phys.* **94** (1991), 2968.
36. L. G. M. Pettersson, H. Åkeby, U. Wahlgren, and P. E. M. Siegbahn, *J. Chem. Phys.*, **93** (1990), 4954.
37. (a) W. F. Egelhoff, Jr., *Phys. Rev. B* **29** (1984), 3861. (b) D. Brennan, D. O. Hayward, and B. M. W. Trapnel, *Proc. R. Soc. London Ser. A* **256** (1960), 81.
38. (a) W. F. Egelhoff, Jr., *J. Vac. Sci. Technol. A* **3** (1985), 1305. (b) W. E. Garner, F. S. Stone, and P. F. Tiley, *Proc. R. Soc. London Ser. A* **277** (1952), 472. (c) I. G. Murgulescu, and M. I. Vass, *Rev. Roum. Chim.* **14** (1969), 1201. (d) V. E. Ostrovskii, *Russ. Chem. Rev.* **43** (1974), 921.
39. P. S. Bagus, K. Hermann, and C. W. Bauschlicher, Jr., *J. Chem. Phys.* **80** (1984), 4378.
40. U. Wahlgren, L. G. M. Pettersson, and P. Siegbahn, *J. Chem. Phys.* **90** (1989), 4613.
41. U. Wahlgren, and P. Siegbahn, in *A New Tool for the Study of Chemisorption on Transition Metals: The Quantum Chemistry Approach to Surface Reactions* (F. Ruette, ed.), Kluwer Academic, Dordrecht, 1991.
42. U. Wahlgren, P. E. M. Siegbahn, and J. Almlöf, *Theor. Chim. Acta*, in press.
43. R. L. Martin, and P. J. Hay, *Surf. Sci.* **130** (1983), L283.

44. (a) H. Stoll, P. Fuentealba, M. Dolg, J. Flad, L. von Scentpaly, and H. Preuss, *J. Chem. Phys.* **79** (1983), 5532. (b) U. Wedig, M. Dolg, H. Stoll, and H. Preuss, in *Quantum Chemistry: The Challenge of Transition Metals and Coordination Chemistry* (A. Veillard, ed.), Reidel, Dordrecht, 1986).
45. E. Shustorovich, (a) *Surf. Sci. Rep.* **6** (1986), 1; (b) *Adv. Catal.* **37** (1990), 101.
46. C. W. Bauschlicher, Jr., *Chem. Phys. Lett.* **129** (1986), 586.
47. G. Ertl, in *The Nature of the Surface Chemical Bond* (T. N. Rhodin, and G. Ertl, eds.), North-Holland, Amsterdam. 1979.
48. P. Norlander, S. Holloway, and J. K. Nørskov, *Surf. Sci.* **136** (1984), 59.
49. J. Schüle, P. Siegbahn, and U. Wahlgren, *J. Chem. Phys.* **89** (1988), 6982.
50. P. E. M. Siegbahn, L. G. M. Pettersson, and U. Wahlgren, *J. Chem. Phys.* **94** (1991), 4024.
51. H. Yang, and J. L. Whitten, (a) *J. Chem. Phys.* **89** (1988), 5329; (b) in *The Challenge of d and f Electrons: Theory and Computation* (D. R. Salahub, and M. C. Zerner, eds.), ACS Symposium Series No. 394, Am. Chem. Soc., Washington, D.C.; (c) *J. Chem. Phys.* **91** (1989), 126.
52. P. E. M. Siegbahn, and U. Wahlgren, *Int. J. Quantum Chem.*, in press.
53. P. S. Bagus, K. Hermann, and M. Seel, *J. Vac. Sci. Technol.* **18** (1981), 435.
54. (a) M. D. Morse, M. E. Geusic, J. R. Heath, and R. E. Smalley, *J. Chem. Phys.* **83** (1985), 2293. (b) M. E. Geusic, M. D. Morse, and R. E. Smalley, *J. Chem. Phys.* **82** (1985), 590.
55. I. Panas, P. Siegbahn, and U. Wahlgren, in *The Challenge of d and f Electrons: Theory and Computation* (D. R. Salahub, and M. C. Zerner, eds.), ACS Symposium Series No. 394, Am. Chem. Soc., Washington, D.C.
56. W. F. Egelhoff, Jr., *Phys. Rev. B* **29** (1984), 3861.
57. (a) L. G. M. Pettersson, and P. S. Bagus, *Phys. Rev. Lett.* **56** (1986), 500. (b) P. S. Bagus, and L. G. M. Pettersson, unpublished.
58. W. F. Egelhoff, Jr., *J. Vac. Sci. Technol. A* **3** (1985), 1305.
59. (a) S. M. George, A. M. DeSantolo, and R. B. Hall, *Surf. Sci.* **159** (1985), L425. (b) D. R. Mullins, B. Roop, S. A. Costello, and J. M. White, *Surf. Sci.* **186** (1987), 67.
60. T. N. Truong, and D. G. Truhlar, *J. Phys. Chem.* **91** (1987), 6229.
61. H. Onuferko, D. P. Woodruff, and B. W. Holland, *Surf. Sci.* **87** (1979), 357.
62. L. Wentzel, D. Arvanitis, W. Daum, H. H. Rotmund, J. Stöhr, K. Baberschke, and H. Ibach, *Phys. Rev. B* **36** (1987), 7689.
63. J. Stöhr, R. Jeager, and T. Kendeliwicz, *Phys. Rev. Lett.* **49** (1982), 142.
64. I. Panas, J. Schüle, U. Brandemark, P. Siegbahn, and U. Wahlgren, *J. Phys. Chem.* **92** (1988), 3079.
65. L. C. Isett, and J. M. Blakely, *Surf. Sci.* **47** (1975), 645.
66. I. Panas, and P. E. M. Siegbahn, *Chem. Phys. Lett.* **153** (1988), 458.
67. A. Nilsson, and N. Mårtensson, *Phys. Rev. Lett.* **63** (1989), 1483.
68. K. Børve, and P. E. M. Siegbahn, *Phys. Rev.* **B43** (1991), 9413.
69. I. Panas, P. Siegbahn, and U. Wahlgren, *J. Chem. Phys.* **90** (1989), 6791.
70. P. Siegbahn, M. Blomberg, I. Panas, and U. Wahlgren, *Theor. Chim. Acta* **75** (1989), 143.
71. P. E. M. Siegbahn, and I. Panas, *Surf. Sci.* **240** (1990), 37.
72. M. P. Kaminsky, N. Winograd, G. I. Geoffrey, and M. A. Vannice, *J. Am. Chem. Soc.* **108** (1986), 1315.
73. (a) M. B. Lee, Q. Y. Yang, S. L. Tang, and S. T. Ceyer, *J. Chem. Phys.* **85** (1986), 1693. (b) S. T. Ceyer, J. D. Beckerle, M. B. Lee, S. L. Tang, Q. Y. Yang, and M. A. Hines, *J. Vac. Sci. Technol. A* **5** (1987), 501. (c) M. B. Lee, Q. Y. Yang, and S. T. Ceyer, *J. Chem. Phys.* **87** (1987), 2724.

74. R. M. Gavin, J. Reutt, and E. L. Muetterties, *Proc. Natl. Acad. Sci. U.S.A.* **78** (1981), 3981.

75. T. H. Upton, *J. Vac. Sci. Technol.* **20** (1982), 527.

76. A.-S. Mårtensson, C. Nyberg, and S. Andersson, *Phys. Rev. Lett.* **57** (1986), 2045.

77. K. P. Huber, and G. Herzberg, *Molecular Spectra and Molecular Structure*, Vol. IV, Van Nostrand-Reinhold, New York, 1979.

78. G. J. Kubas, C. J. Unkefer, B. I. Swanson, and E. Fukushima, *J. Am. Chem. Soc.* **108** (1986), 7000.

79. A. E. dePristo, private communication.

80. L. D. Landau, and E. M. Lifshitz, *Mechanics*, Pergamon Press, Oxford, 1969.

81. (a) M. A. Barteau, and R. J. Madix, *Surf. Sci.* **97** (1980), 101. (b) C. T. Campbell, *Surf. Sci.* **157** (1985), 43.

82. (a) J. L. Gland, B. A. Sexton, and G. Fisher, *Surf. Sci.* **95** (1980), 587. (b) C. T. Campbell, G. Ertl, H. Kuipers, and J. Segner, *Surf. Sci.* **107** (1981), 220.

83. J. Stöhr, J. L. Gland, W. Eberhardt, D. Outka, R. J. Madix, F. Sette, R. J. Koestner, and U. Doebler, *Phys. Rev. Lett.* **51** (1983), 2414.

84. D. A. Outka, J. Stöhr, W. Jark, P. Stevens, J. Salomons, and R. J. Madix, *Phys. Rev. B* **35** (1987), 4119.

85. W. Eberhardt, T. Upton, S. Cramm, and L. Inococcia, *Chem. Phys. Lett.* **146** (1988), 561.

86. M. A. Barteau, and R. J. Madix, *Chem. Phys. Lett.* **97** (1983), 85.

87. K. Prabhakaran, P. Sen, and C. N. R. Rao, *Surf. Sci.* **177** (1986), L971.

88. X. Guo, L. Hanley, and J. T. Yates, Jr., *J. Chem. Phys.* **90** (1989), 5200; X.-Y. Zhu, S. R. Hatch, A. Campion, and J. M. White, *J. Chem. Phys.* **91** (1989), 5011.

89. K. Børve, I. Panas, and P. E. M. Siegbahn, to be published.

90. M. R. A. Blomberg, and P. E. M. Siegbahn, *J. Chem. Phys.* **78** (1983), 986, 5682.

91. P. E. M. Siegbahn, M. R. A. Blomberg, and C. W. Bauschlicher, Jr., *J. Chem. Phys.* **81** (1984), 1373.

92. P. E. M. Siegbahn, M. R. A. Blomberg, and C. W. Bauschlicher, Jr., *J. Chem. Phys.* **81** (1984), 2103.

93. M. Balooch, M. J. Cardillo, D. R. Miller, and R. E. Stickney, *Surf. Sci.* **46** (1974), 358.

94. J. Harris, *Surf. Sci.* **221** (1989), 335.

95. X.-Y. Zhu, M. E. Castro, and J. M. White, *J. Chem. Phys.* **90** (1989), 7442; X.-Y. Zhu, and J. M. White, *Chem. Phys. Lett.* **164** (1989), 101.

96. C. R. Brundle, J. Behm, and J. A. Barker, *J. Vac. Sci. Technol. A* **2** (1984), 1033.

97. R. B. Woodward, and R. Hoffmann, *The Conservation of Orbital Symmetry*, Academic Press, New York, 1969.

98. C. W. Bauschlicher, Jr., L. G. M. Pettersson, and P. E. M. Siegbahn, *J. Chem. Phys.* **87** (1987), 2129.

99. M. R. A. Blomberg, P. E. M. Siegbahn, U. Nagashima, and J. Wennerberg, *J. Am. Chem. Soc.*, **113** (1991), 476.

100. (a) O. Swang, K. Faegri, O. Gropen, P. Siegbahn, and U. Wahlgren, *Chem. Phys.*, in press. (b) O. Swang, Cand. Scient. Thesis.

101. T. P. Beebe, D. W. Goodman, B. D. Kay, and J. T. Yates, Jr., *J. Chem. Phys.* **87** (1987), 2305.

102. I. Chorkendorff, I. Alstrup, and S. Ullmann, *Surf. Sci.* **227** (1990), 291.

103. A. Luntz, private communication.

104. I. Chorkendorff, private communication.

CHAPTER 2

Thermochemical Methods for Reaction Energetics on Metal Surfaces

Jay B. Benziger

I. Introduction

The a priori determination of reaction energetics is a formidable task. Complete quantum mechanical treatments of energy profiles have been limited to only the simplest of systems [1]. So far effective methods to calculate reaction energetics have relied on semiempirical or phenomenological approaches. Transition state theory is the best known example of a phenomenological approach to understanding reaction energetics and predicting reaction kinetics [2–5]. The transition state or activated complex is associated with the saddle point passing from two energy minima and may be identified from quantum mechanical calculations; however, practical application of transition state theory relies on empirical approaches to estimate the change in thermodynamic properties $\Delta S^{\neq}$ and $\Delta H^{\neq}$ between the reactants and the transition state. Reaction kinetics are then found from the application of mass action kinetics with an Arrhenius rate constant, k_r,

$$k_r = \frac{kT}{h} e^{-\Delta S^{\neq}/k} e^{-\Delta H^{\neq}/kT} \tag{2.1}$$

where k is Boltzmann's constant, h is Planck's constant, and T is the absolute temperature.

Transition state theory has been enormously successful at correlating and predicting thermally activated reaction energetics. This success is the result of good simple techniques for estimating thermodynamic properties. Benson has written an excellent treatise on estimation of thermochemical data, with particular emphasis on gas-phase organic reactions [6].

In this chapter we examine phenomenological approaches for estimating reaction energetics on metal surfaces. Our approach is to estimate thermodynamic quantities that may be used to predict both reaction feasibility and reaction kinetics when coupled with transition state theory.

II. Thermodynamics of Adsorption and Reaction

The interaction of a molecule with a surface involves a sequence of three reactions: (1) adsorption, (2) surface reaction, (3) desorption. Each of these reactions may involve various elementary reaction steps. For example, adsorption may involve the formation of a weakly adsorbed "precursor" state, surface diffusion to an adsorption "site," and formation of an adsorption bond. Desorption would be expected to proceed through these events in reverse sequence. Surface reaction may involve a complex set of bond breaking and bond formation events.

From the viewpoint of the macroscopically observed reaction kinetics one is interested in identifying the rate-limiting reaction steps, in which case other reaction events are assumed to be in equilibrium. The classic example of the application of a rate-limiting step to surface reaction is "Langmuir–Hinshelwood" kinetics.

$$\mathrm{A(g)} \underset{k_{-1}}{\overset{k_1}{\Leftrightarrow}} \mathrm{A(a)} \qquad \text{(adsorption)}$$

$$\mathrm{A(a)} \underset{k_{-2}}{\overset{k_2}{\Leftrightarrow}} \mathrm{B(a)} \qquad \text{(surface reaction)}$$

$$\mathrm{B(a)} \underset{k_{-3}}{\overset{k_3}{\Leftrightarrow}} \mathrm{B(g)} \qquad \text{(desorption)}$$

In the above expressions and throughout the text, gas-phase species are designated with a (g) and adsorbed species with an (a). The observed kinetics follow different rate laws dependent on which of the reaction steps is rate limiting [7].

Adsorption equilibrium may be described by a simple lattice gas model; the equilibrium constant for the reaction is then given by [7,8]

$$\frac{\theta_A}{(1-\theta_A)P_A} = e^{\Delta S_{ad}/k} e^{-\Delta H_{ad}/kT} \tag{2.2}$$

where θ_A is the fraction of adsorption sites occupied at pressure P_A, and ΔS_{ad} and ΔH_{ad} are the entropy and enthalpy of adsorption. Adsorbate interactions and surface heterogeneity have been neglected in this simple treatment. This chapter considers well-defined crystal orientations where heterogeneity is not relevant. Adsorbate interactions will be considered to affect only the adsorption enthalpy (the Bragg–Williams approximation [8]). The entropy of adsorption is dominated by the loss of translation motion in going from a freely translating gas molecule to an adsorbed molecule. In the absence of specific knowledge about the adsorbed species the entropy changes associated with rotational and vibrational motion may be neglected relative to the change associated with the loss of translational entropy. The entropy of adsorption may be approximated by [8,9]

$$\begin{aligned} \Delta S_{ad} &= -k \ln\left[\left(\frac{2\pi mkT}{h^2}\right)^{3/2} \frac{kTe^{5/2}}{P_o}\right] \\ &= -k\left\{18.1 + \ln\left[\left(\frac{T}{300}\right)^{5/2}\left(\frac{m}{28}\right)^{3/2}\right]\right\} \end{aligned} \tag{2.3}$$

Table 2.1. Adsorption Entropies (cal/mol-K) for CO and H_2 on Metal Surfaces

Surface	Adsorbate	Adsorbate Coverage	ΔS_{ad} Eq. 2.3	ΔS_{ad} Experiment	Reference
Ni(100)	CO	0.2	−36	−27	*a*
Ni(100)	CO	0.5	−36	−37	*a*
Ni(111)	CO	0.2	−36	−26	*b,c*
Ni(111)	CO	0.4	−36	−33	*b,c*
Ru(0001)	CO	0.2	−36	−26	*d–f*
Ru(0001)	CO	0.5	−36	−39	*d–f*
Ni(100)	H_2	0.5	−28	−22	*g*
Ni(100)	H_2	0.2	−28	−20	*g*

[a] J. C. Tracy, *J. Chem. Phys.* **56** (1972), 2736.
[b] K. Christman, O. Schober, and G. Ertl, *J. Chem. Phys.* **60** (1974), 4719.
[c] H. Ibach, W. Erley, and H. Wagner, *Surf. Sci.* **92** (1980), 29.
[d] T. E. Madey, and D. Menzel, *Jpn. J. Appl. Phys.* (1974), Suppl. 2, 229.
[e] J. C. Tracy, and P. W. Palmberg, *Suf. Sci.* **14** (1969), 274.
[f] H. Pfnür, P. Feulner, H. A. Engelhardt, and D. Menzel, *Chem. Phys. Lett.* **59** (1978), 481.
[g] K. Christman, O. Schober, G. Ertl, and M. Newmann, *J. Chem. Phys.* **60** (1974), 4528.

where m is the molecular weight of the molecule and P_0 is the standard state pressure of 1 atm.

Table 2.1 summarizes the data for adsorption entropies for CO and H_2 on several metal surfaces obtained from isosteric measurements. Equation 2.3 generally underestimates the adsorption entropy, with the largest error occurring at low adsorbate coverages. The discrepancy comes from neglecting the configurational entropy of the adsorbed layer and the motion of the adsorbed molecules. At high adsorbate coverages adatom interactions result in ordering, reducing the configurational entropy of the adsorbed layer, so the total loss of all translational entropy on adsorption is a better approximation. The neglect of rotational and vibrational contributions to the entropy of adsorbed species can result in Eq. 2.3 underestimating the entropy change for highly ordered adlayers, such as found for half-monolayer coverage of CO on Ru(0001) [12] .

As the entropy associated with adsorbed species is generally small, the entropies of surface reactions may frequently be neglected. The surface reaction equilibrium may be determined by simply estimating the enthalpy change associated with the reaction obtained from bond energy considerations.

The enthalpy of adsorption is the net change in bond energies associated with adsorption. We assume the bonding may be described by additive pairwise interactions between atoms. For atomic adsorption this is the sum of the bond energies between the adatom (A) and the surface metal atoms (M_i):

$$\Delta H_{ad}(\mathrm{A}) = \sum_{\substack{\text{surface}\\ \text{metal}\\ \text{atoms}}} D(\mathrm{M}_i - \mathrm{A}) + \sum_{\substack{\text{metal}\\ \text{atoms}\\ i \neq j}} \Delta D(\mathrm{M}_i - M_j) \tag{2.4}$$

In Eq. 2.4 the quantity $D(\mathrm{M}_i - \mathrm{A})$ refers to the two-center bond energy between metal atom i and the adsorbate atom A. The second term in Eq. (2.4) accounts for the change in metal–metal bonding resulting from reorganization and relaxation of the metal surface on adsorption. In the case of molecular adsorption the weakening of bonds in the molecule, as well as the bonds formed with the surface, must be accounted for. When a diatomic AB molecule adsorbs, the heat of adsorption involves the formation of M–A and M–B bonds and the weakening of the A–B bond.

$$\Delta H_{ad}(AB) = D(M(A-B)) - D(A-B) + \sum_{\substack{\text{surface}\\ \text{metal}\\ \text{atoms}}} [D(M_i - A) + D(M_i - B)] + \sum_{\substack{\text{metal}\\ \text{atoms}\\ i \neq j}} \Delta D(M_i - M_j) \tag{2.5}$$

The first two terms in Eq. 2.5 are the difference in the two-center A–B bond energy in the adsorbed molecule, $D(M(A-B))$, and the bond energy of the free gas-phase molecule, $D(A-B)$. The sum over surface atoms is the bond energies associated with M–A and M–B bonding, and the last term in Eq. 2.5 is the relaxation energy of the metal surface resulting from adsorption. This formalism may be extended to any polyatomic molecule adsorbing on a metal surface, where one would sum over all pairwise interactions. Difficulty in estimating bond energies for the metal relaxation frequently leads to a frozen lattice approximation, where the surface is an abrupt termination of the bulk crystal with no relaxation. In general we assume a frozen lattice and neglect changes in the metal–metal bonding.

The thermodynamic analysis of adsorption and reaction equilibria may be used to estimate kinetics of reactions, choosing one state as the transition state. Only in the case of desorption, where the transition state lies between an adsorbed species and a gas molecule, will there be a significant entropy of activation. Thus, the energetics associated with the bonding of the adsorbed species will generally dominate the rate of surface reaction.

The basis for our phenomenological approach to surface reactivity is thermochemical in nature. Based on hypotheses for reactions at metal surfaces, conceivable reaction pathways and intermediates are constructed. The thermodynamic state variables of entropy and enthalpy are approximated from simple techniques such as bond additivity and bond order conservation. In the following sections we describe these two approaches to estimate adsorption bond energies, and apply these to examine a variety of problems associated with reactions at metal surfaces.

III. Estimation of Surface Bond Energies

Phenomenological approaches to determining reaction energetics require values of the bond energies. Ideally one would like to have thermodynamic measurements of heats of adsorption, such as isosteric heats of adsorption. The heat of adsorption can be determined from the variation of pressure with temperature at constant surface coverage by application of the Clausius–Clapeyron equation [10]. Unfortunately, experiments to obtain thermodynamic data are difficult and there are a limited

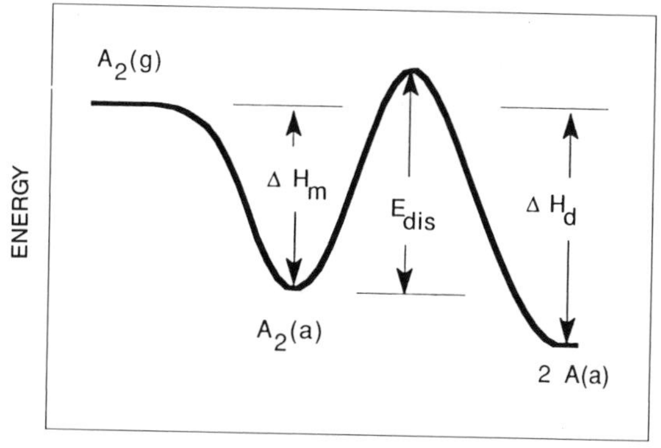

Figure 2.1. Reaction energetics for dissociative adsorption. One-dimensional reaction trajectory of A_2 dissociation via a molecularly adsorbed intermediate, $A_2(g) \rightarrow A_2(a) \rightarrow 2A(a)$. ΔH_m and ΔH_d are the adsorption enthalpies of molecular A_2 and dissociated A_2, respectively. E_{dis} is the barrier for dissociation of adsorbed A_2.

amount of compiled data. The activation energy for desorption, E_d, is an experimental quantity that is easily obtained from temperature-programmed desorption (TPD) [14]. The activation energy for desorption can be estimated from the temperature of the maximum desorption rate, T_p, the heating rate, β, and assumed values of preexponential factors, ν [14]. For a first-order desorption the activation energy for desorption was shown by Redhead [14] to be given by

$$E_d = kT_p \left\{ \ln \frac{\nu T_p}{\beta} - 3.64 \right\} \tag{2.6}$$

A preexponential factor of 10^{13} s^{-1} is assumed in most instances to obtain the activation energy for desorption [7,14]. For second-order reactions (e.g., atomic recombination) Eq. 2.6 may still be used with an effective first-order preexponential factor. The latter is the product of the true second-order preexponential factor multiplied by the surface coverage. The effective first-order preexponential factor for second-order reactions is also approximately 10^{13} s^{-1} [7,14]. Variations in preexponential factors of two or three orders of magnitude will introduce errors in the activation energy of desorption of 2–3 kcal/mol [14]. The majority of adsorption bond energies reported have been determined from TPD, generally using an approximation similar to Eq. 2.6.

The enthalpy for dissociative adsorption of diatomics, ΔH_d is equal to the activation energy for desorption only when there is no activation barrier for dissociation, E_{dis}. In the simple one-dimensional reaction trajectory, shown in Figure 2.1, the activation energy for desorption is greater than the dissociative adsorption enthalpy by the difference in energy between the molecular adsorption enthalpy, ΔH_m, and the barrier for dissociation. Existence of activation barriers for dissociation is suggested by low sticking probabilities for dissociative adsorption or molecular desorption preceding dissociation in thermal desorption. When the reaction probability is low one must distinguish whether the reaction occurs at defects and/or if the transition state is highly constrained. Precise concentrations of surface defects and the types of defects are seldom known; however, the fraction of surface defects to surface atoms at single crystals is expected to exceed 10^{-4}. (In particular, this concentration of step and kink density will be produced by misalignment of 0.1° during cutting of a single-crystal surface.) Gibson and Dubois have recently suggested that steps were the sites of methanol decomposition of Pt(111) crystal surfaces and they estimated the step concentration was $>10^{-2}$ [15]. Molecular dynamics simulations have suggested that surface mobility is high during adsorption so adsorbates sample large parts of the surface, including defects at low concentrations [16]. In the absence of strong evidence otherwise, reaction probabilities that have preexponential factors smaller than 10^{-4} times normal are suspected of occurring at defects (normal preexponential factors are $\nu_1 = 10^{13}\ s^{-1}$ for a first-order reaction and $\nu_2 = 10^{-2}\ cm^2$/atom-s for a second-order reaction).

The reasonableness of reaction kinetics as estimates for thermodynamic quantities may also be evaluated from thermodynamic consistency which requires that adsorption and desorption rates be equal at equilibrium [2,7]. At equilibrium for dissociative adsorption, $A_{2,g} \Leftrightarrow 2\ A_{ad}$, the sticking probability, s, and the preexponential factor, ν_α, for desorption with order α must satisfy the relationship with the adsorption entropy given in

$$k \ln \frac{s(\theta,T)}{\sqrt{2\pi mkT}\nu_\alpha N^\alpha} = -\Delta S_{ad} \leq S^0_{A_2} \tag{2.7}$$

In Eq. 2.7 N is the surface coverage and S^0 is the absolute entropy of A_2. Equation 2.7 indicates that for a one-dimensional reaction trajectory, low sticking probabilities for adsorption should be accompanied by low preexponential factors for desorption. When thermodynamic consistency is violated, reaction trajectories for adsorption and desorption are different and kinetic measurements do not provide proper estimates of thermodynamic quantities.

Table 2.2 summarizes metal–hydrogen adsorption energies on metal single-crystal surfaces. The majority of the values reported in Table 2.2 are activation energies for desorption determined from TPD. Isosteric heats of adsorption for dihydrogen adsorption on metals have been reported on Ni(100) [13], and agree with the activation energies for desorption presented in Table 2.2. Tabulations of dissociative sticking probabilities and preexponential factors for associative desorption for dihydrogen [17] show that the values do not differ from expected values ($s \sim 1$ and $\nu_2 \sim 10^{-1}$) by more than a factor of 100 for the transition metals with the exception of the group IB metals [18]. This indicates that the activation energies for associative desorption of dihydrogen should be good estimates of the adsorption enthalpies on all transition metals but not on the IB metals (Cu,Ag,Au).

Activation barriers for desorption of dioxygen and dinitrogen are also available for a number of surfaces and are frequently used as estimates of the adsorption enthalpy. In the case of dioxygen adsorption, tabulations of dissociative sticking probabilities and preexponential factors for associative desorption [17] suggest that the

Table 2.2. Metal–Hydrogen Adsorption Bond Energies (kcal/mol)

Surface	D(MH)[dd]	Reference	Surface	D(MH)[dd]	Reference
Fe (110)	65	*a*	Rh (111)	61	*n*
(100)	63	*a,b*	(110)	61	*o*
(111)	62	*a*	Pd (100)	64	*p*
Co (0001)	60	*c*	(111)	62	*q*
(1010)	60	*c,d*	(110)	64	*q*
Ni (100)	63	*e,f*	Ag (111)	52	*r,s*
(110)	62	*e*	W (100)	68	*t*
(111)	63	*e*	(110)	68	*u*
Cu (100)	<60	*g*	W (111)	70	*v*
(110)	<58	*g*	(112)	69	*w*
(111)	<60	*g*	Ir (110)	63	*x*
(311)	56	*h*	(111)	58	*y*
Nb (100)	65	*i*	Pt (100)	59	*z*
Mo (100)	66	*j*	(110)	58	*aa*
(110)	65	*k*	(111)	59	*bb*
Ru (0001)	66	*l*	(111)	61	*cc*
	61	*m*			

[a] F. Boszo, G. Ertl, M. Grunze, and M. Weiss, *Appl. Surf. Sci.* **1** (1977), 103.
[b] J. B. Benziger, and R. J. Madix, *Surf. Sci.* **94** (1980), 119.
[c] K. Christmann, *Surf. Sci. Rep.* **9** (1988), 1.
[d] M. E. Bridge, C. M. Comrie, and R. M. Lambert, *J. Catal.* **58** (1979), 28.
[e] J. Lapujoulade, and K. Neil, *Surf. Sci.* **35** (1973), 288.
[f] S. J. Johnson, and R. J. Madix, *Surf. Sci.* **108** (1981), 77.
[g] G. Anger, A. Winkler, and K. D. Rendulic, *Surf. Sci.* **220** (1989), 1.
[h] J. Pritchard, T. Catterick, and A. K. Gupta, *Surf. Sci.* **53** (1975), 1.
[i] D. I. Hagen, and F. F. Donaldson, *Surf. Sci.* **45** (1974), 61.
[j] H. R. Ham, and L. D. Schmidt, *J. Phys. Chem.* **75** (1971), 227.
[k] M. Mahning, and L. D. Schmidt, *Z. Phys. Chem.* **80** (1972), 71.
[l] P. Fuelner, and D. Menzel, *Surf. Sci.* **154** (1985), 465.
[m] H. Shimizu, K. Christmann, and G. Ertl, *J. Catal.* **61** (1980), 412.
[n] J. T. Yates, Jr., P. A. Thiel, and W. H. Weinberg, *Surf. Sci.* **84** (1979), 427.
[o] M. Ehsasi, and K. Christmann, *Surf. Sci.* **194** (1988), 172.
[p] R. J. Behm, K. Christmann, and G. Ertl, *Surf. Sci.* **99** (1980), 320.
[q] M. Conrad, G. Ertl, and E. E. Latta, *Surf. Sci.* **41** (1974), 435.
[r] Ph. Avouris, D. Schmeisser, and J. E. Demuth, *Phys. Rev. Lett.* **48** (1982), 199.
[s] X.-L. Zhou, J. M. White, and B. E. Koel, *Surf. Sci.* **218** (1989), 201.
[t] P. W. Tamm, and L. D. Schmidt, *J. Chem. Phys.* **51** (1969), 5352.
[u] P. W. Tamm, and L. D. Schmidt, *J. Chem. Phys.* **52** (1970), 1150.
[v] P. W. Tamm, and L. D. Schmidt, *J. Chem. Phys.* **54** (1971), 4775.
[w] R. R. Rye, B. D. Banford, and P. G. Cartier, *J. Chem. Phys.* **59** (1973), 1693.
[x] D. E. Ibbotson, T. S. Wittrig, and W. H. Weinberg, *Surf. Sci.* **110** (1981), 294.
[y] J. R. Engstrom, W. Tsai, and W. H. Weinberg, *J. Chem. Phys.* **87** (1987), 3104.
[z] F. Netzer, and G. Kneringer, *Surf. Sci.* **51** (1975), 4775.
[aa] R. W. McCabe, and L. D. Schmidt, in *Proceedings, 7th International Vacuum Congress*, Vol. 2, 120, 1977.
[bb] M. Salmeron, R. J. Gale, and G. A. Somorjai, *J. Chem. Phys.* **70** (1979), 2807.
[cc] K. Christman, G. Ertl, and T. Pignet, *Surf. Sci.* **54** (1976), 365.
[dd] Calculated from $D(\mathrm{MH}) = \frac{1}{2}[D(\mathrm{H_2}) - \Delta H_{\mathrm{ad}}(\mathrm{H_2})]$, assuming $\Delta H_{\mathrm{ad}}(\mathrm{H_2}) = E_{\mathrm{d}}$, where E_{d}, the activation energy for desorption, is given by Eq. (2.6).

adsorption enthalpy will be approximately equal to the activation energy for desorption on all metals except Ag and Au where the low sticking probabilities suggest significant barriers for dissociation. For example, measurements for adsorption and desorption of O_2 on Ag(110) [19] show that the sticking probability is low, 2×10^{-4}, whereas the preexponential factor for desorption is abnormally high, first-order desorption with $\nu = 10^{15}\ s^{-1}$. The thermodynamic inconsistency of these results suggests that reaction trajectories for adsorption and desorption are not the same, and the activation energy for desorption is not a reliable estimate of the adsorption enthalpy.

The dissociative adsorption of dinitrogen on group IVB, VB, and VIB metals appears to proceed with minimal activation. On metals such as Mo and W, dinitrogen dissociates before desorbing following low-temperature adsorption [20,21]. In contrast, dinitrogen adsorbed on Fe single-crystal surfaces below 200 K desorbs without dissociation [22]. Very low dissociative sticking probabilities of dinitrogen on Fe surfaces were also observed [22]. These results suggest that dissociative adsorption of dinitrogen on Fe surfaces is activated. The barrier for molecular desorption is estimated from Eq. 2.6 to be 8–10 kcal/mol, assuming a preexponential factor of $10^{13}\ s^{-1}$. The activation barrier for dinitrogen dissociation on Fe surfaces is therefore estimated to be greater than 8–10 kcal/mol. Ertl has suggested the barrier for dissociation to be less than 5 kcal/mol based on atomic nitrogen surface concentrations after exposure of an iron crystal to dinitrogen gas at different temperatures [23]; however, the very low sticking probabilities reported makes it impossible to decide whether the dissociative adsorption occurs on the well-defined surface or the dissociation occurs at defects, and the apparent barrier results from diffusion away from the defects. Molecular desorption also precedes dissociation of dinitrogen on Ni, Pd, Pt, and the group IB metals, and the dissociative adsorption probabilities on group VIII and IB metals are vanishingly small, indicative of significant barriers for dissociation. Hence, activation energies for dinitrogen desorption will be reasonable estimates only for the adsorption enthalpies of nitrogen on group VIB and earlier transition metals.

Adsorption isotherms for carbon are available only for nickel surfaces [24], where equilibrium between carbon in a solid solution and carbon segregated to the nickel surface has been measured. Adsorption enthalpies for sulfur, phosphorus, halogens, and other atoms are virtually nonexistent.

The dearth of information on adsorption enthalpies leads us to search for alternative approaches to estimate adsorption bond energies to supplement our limited knowledge. It is common to view surface structures as arising from a cut through a three-dimensional crystal. As a first-order approximation, the bonding at the surface is assumed to resemble that occurring in a bulk crystal. This assumption will form the basis for estimation of surface bond energies.

A. Adsorption Bond Energies from Thermodynamics of Bulk Compounds

The simplest approach to estimate adsorption bond energies is to assume bonding at the surface is similar to bonding in bulk compounds. It had been previously proposed that the adsorption enthalpy $\Delta H_{ad}(A)$ was approximately equal to the normalized enthalpy of formation of the bulk compound $\Delta H_f(M_xA_y)$ [9].

$$\Delta H_{ad}(A) = \frac{1}{y}\,\Delta H_f(M_xA_y) \tag{2.8}$$

The rationale behind this idea was that for a series of metal oxides the heat of formation of the metal oxide per oxygen atom was approximately constant and was close to the experimental values for the binding energies of oxygen on metal surfaces [9,25–27]. Metal–adsorbate bond energies were estimated from the reaction of the adsorbate atom with the metal.

$$D(MA) = \left\{ -\frac{1}{y}\,\Delta H_f(M_xA_y) + \Delta H_f(A(g)) \right\} \tag{2.9}$$

$D(MA)$ is the total bond energy of the A adatom on the metal surface. We distinguish $D(MA)$ from $D(M-A)$: the former is the total bond energy of the adsorbate with the metal and the latter refers to the two-center M–A bond.

Equations 2.8 and 2.9 neglect both disruption of metal–metal bonding and long-range electrostatic effects in ionic crystals. Including the disruption of metal–metal bonding would increase $D(MA)$ by adding in some fraction of the heat of sublimation to the expression in Eq. 2.9. The ionic bonding contribution would tend to decrease $D(MA)$ by normalizing the bond energy by a lattice sum (Madelung sum) [28]. Tables 2.3 and 2.4 compare metal–oxygen and metal–nitrogen bond energies determined from bulk thermodynamic data using Eq. 2.9 with experimental values obtained from temperature-programmed desorption. The correlation between the estimated values and the experimental quantities is remarkable for the metal–oxygen bonds. The excellent correlation between the binding energies of adsorbed oxygen and bulk oxides suggests that the magnitude of the long-range ionic interactions in metal oxides and the disruption of the metal–metal bonding are comparable.

The data for metal–nitrogen bond energies show that estimates from bulk nitrides are consistently less than the values obtained from thermal desorption. The difference between the two values is greatest when dissociative adsorption is activated (e.g., on Cu). In the case where there appears to be an activation energy for dissociation, the metal–nitrogen bond energy based on thermal desorption is an upper limit on the bond energy and the estimate from the bulk thermodynamic data is a lower limit.

Equation 2.9 is useful for estimating metal–carbon and metal–nitrogen bond energies when heats of adsorption are not available. Extensive tabulations of enthalpies of formation of metal carbides, nitrides, and oxides are available from which unknown adsorption bond energies may be estimated. The adsorption bond energies are expected to roughly correlate with the bond energies in bulk compounds: enthalpies of formation will provide a lower limit to the adsorption enthalpy. When the two quantities differ substantially, one should review the data for the adsorption bond energy and the thermodynamic data for a possible error (perhaps caused by impurities or defects), or determine if there is a barrier to dissociation. Of course, direct measurements of adsorption bond energies are preferred whenever reliable data are available.

Even when using bulk thermodynamic data to estimate adsorption energies one finds gaps in our knowledge because the thermodynamic properties of many metal

Table 2.3. Metal–Oxygen Bond Energies (kcal/mol)

Metal	Metal Oxide	$\Delta H_f(M_xO_y)$	D(MO) Eq. 2.9	Surface	D(MO)[aa] Experimental	Reference
Ni	NiO	−57.3	116.8	Ni(111)	112.1	*a*
Cu	CuO	−37.6	97.1	Cu foil	103	*b*
	Cu_2O	−40.3	99.8	Cu(110)	89	*c*
Mo	MoO_2	−140.8	129.9	Mo(100)	118	*d*
	MoO_3	−178.0	118.8			
Ru	RuO_2	−72.9	96.0	Ru(101)	95.4	*e*
				Ru(001)	99.5	*f*
Rh	Rh_2O_3	−82.0	86.8	Rh(111)	87.5	*g*
Pd	PdO	−20.4	79.7	Pd(110)	99.4	*h*
				Pd(110)	83.9	*i*
				Pd(100)	84.0	*j*
				Pd(111)	87	*k*
Ag	Ag_2O	−7.4	66.9	Ag(110)	79.5	*l*
				Ag(110)	76.0	*m*
				Ag(110)	77.5	*n*
				Ag(110)	80.2	*o*
				Ag(111)	77.0	*p*
W	WO_2	−138.8	128.9	W(110)	126	*q*
	WO_3	−201.5	126.7	W(110)	125	*r*
				W(110)	139	*d*
				W(100)	136	*d*
				W(100)	125	*s*
Ir	IrO_2	−65.5	92.3	Ir(111)	92.0	*t*
				Ir(110)	91.5	*u*
Pt	Pt_3O_4	−38.9	69.2	Pt(111)	85	*v*
				Pt(111)	85	*w*
				Pt(111)	83.5	*x*
				Pt(112)	84.3	*x*
				Pt(100)	82	*y*
Au				Au(110)	78.9	*z*

[a] J. B. Benziger, and R. E. Preston, *Surf. Sci.* **141** (1984), 576.
[b] E. Gianmello, B. Fubini, P. Lauro, and A. Bossi, *J. Catal.* **87** (1984), 443.
[c] G. Ertl, *Surf. Sci.* **7** (1967), 309.
[d] N. P. Vas'ko, Yu. G. Ptushinskii, and B. A. Chuikhov, *Surf. Sci.* **14** (1968), 448.
[e] P. D. Reed, C. M. Comrie, and R. M. Lambert, *Surf. Sci.* **64** (1977), 603.
[f] T. E. Madey, H. A. Engelhardt, and D. Menzel, *Surf. Sci.* **48** (1975), 304.
[g] P. A. Thiel, J. T. Yates, Jr., and W. H. Weinberg, *Surf. Sci.* **82** (1979), 22.
[h] M. Milun, P. Pervan, M. Vajic, and K. Wandelt, *Surf. Sci.* **211/212** (1989), 887.
[i] J.-W. He, and P. R. Norton, *Surf. Sci.* **204** (1988), 26.
[j] S.-L. Chang, P. A. Thiel, and J. W. Evans, *Surf. Sci.* **205** (1988), 117.
[k] H. Conrad, G. Ertl, J. Kuppers, and E. E. Latta, *Surf. Sci.* **65** (1977), 245.
[l] H. A. Engelhardt, and D. Menzel, *Surf. Sci.* **57** (1976), 591.
[m] H. Albers, W. J. J. VanDerWal, O. L. J. Gijzeman, and G. A. Bootsma, *Surf. Sci.* **77** (1979), 1.
[n] C. Backx, C. P. M. DeGroot, and P. Biloen, *Surf. Sci.* **104** (1981), 300.
[o] M. Bowker, M. A. Barteau, and R. J. Madix, *Surf. Sci.* **92** (1980), 528.
[p] G. Rovida, F. Pratesi, M. Maglietta, and E. Ferroni, *Surf. Sci.* **43** (1974), 230.
[q] C. Kohrt, and R. Gomer, *J. Chem. Phys.* **52** (1970), 3283.
[r] T. Engel, H. Niehus, and E. Bauer, *Surf. Sci.* **52** (1975), 237.
[s] M. Bacal, J. L. Desplat, and T. Alleau, *J. Vac. Sci. Technol.* **9** (1972), 851.
[t] V. P. Ivanov, G. K. Boreskov, V. I. Sauchenko, W. F. Egelhoff, Jr., and W. H. Weinberg, *Surf. Sci.* **61** (1976), 207.
[u] J. L. Taylor, D. E. Ibbotson, and W. H. Weinberg, *Surf. Sci.* **79** (1979), 349.
[v] C. T. Campbell, G. Ertl, H. Kuipers, and J. Segner, *Surf. Sci.* **107** (1981), 220.
[w] J. L. Gland, *Surf. Sci.* **93** (1980), 487.
[x] A. Winkler, X. Guo, H. R. Siddiqui, P. L. Hagans, and J. T. Yates, Jr., *Surf. Sci.* **201** (1988), 419.
[y] G. Kneringer, and F. P. Netzer, *Surf. Sci.* **49** (1975), 125.
[z] N. D. S. Canning, D. Outka, and R. J. Madix, *Surf. Sci.* **141** (1984), 240.
[aa] Calculated from $D(MO) = \frac{1}{2}[D(O_2)-\Delta H_{ad}(O_2)]$, assuming $\Delta H_{ad}(O_2) = E_d$, where E_d, the activation energy for desorption, is given by Eq. (2.6).

Table 2.4. Metal–Nitrogen Bond Energies (kcal/mol)

Metal	Metal Nitride	$\Delta H_f(M_xN_y)$	D (MN) Eq. (2.9)	Surface	D (MN)[m] Experimental	Reference
Fe	Fe_4N	−2.5	115.5	Fe(100)	142	*a*
				Fe(110)	141	*b*
				Fe(111)	138	*a*
Ni	Ni_3N	+1.9	111.1	Ni(111)	138	*c*
				Ni(110)	137	*d*
Cu	Cu_3N	+17.7	95.3	Cu(110)	135	*e*
Mo	Mo_2N	−19.5	132.5	Mo(100)	156	*f*
				Mo(110)	153	*g*
W	W_2N	−23.9	136.3	W(100)	151	*h*
				W(100)	152	*i*
				W(111)	151	*h*
				W(110)	153	*j*
Re				Re(0001)	144	*k*
Pt				Pt Foil	121	*l*

[a]F. Boszo, G. Ertl, M. Grunze, and M. Weiss, *J. Catal.* **49** (1977), 18.
[b]F. Boszo, G. Ertl, and M. Weiss, *J. Catal.* **50** (1977), 519.
[c]J. B. Benziger, and R. E. Preston, *Surf. Sci.* **141** (1984), 576.
[d]Y. Kuwahara, M. Fujisawa, M. Onchi, and M. Nishijima, *Surf. Sci.* **207** (1988), 17.
[e]D. Heskett, A. Baddorf, and E. W. Plummer, *Surf. Sci.* **195** (1988), 94.
[f]H. R. Han, and L. D. Schmidt, *J. Chem. Phys.* **75** (1971), 227.
[g]J. Lapujoulade, and K. S. Neil, *C.R. Acad Sci. Ser. C* **274** (1972), 2125.
[h]T. A. Delchar, and G. Ehrlich, *J. Chem. Phys.* **42** (1965), 2686.
[i]L. R. Clavenna, and L. D. Schmidt, *Surf. Sci.* **22** (1970), 365.
[j]P. W. Tamm, and L. D. Schmidt, *Surf. Sci.* **26** (1971), 286.
[k]G. Hasse, and M. Asscher, *Surf. Sci.* **191** (1987), 75.
[l]J. J. Vajo, W. Tsai, and W. H. Weinberg, *J. Phys. Chem.* **89** (1985), 3243.
[m]Calculated from $D(MN) = \frac{1}{2}[D(N_2)\text{-}\Delta H_{ad}(N_2)]$, assuming $\Delta H_{ad}(N_2) = E_d$, where E_d, the activation energy for desorption, is given by Eq. (2.6).

carbides are unknown. To obtain estimates of bond energies where no data exist we rely on extrapolated periodic trends in bond energies. Shown in Figure 2.2 are the periodic trends in bond energies for metal–oxygen, metal–carbon, and metal–nitrogen bonds based on Eq. 2.9. Metal-oxygen bonds show the same general trend for all three rows of the transition metals; the metal–oxygen bond energy decreases rapidly from group IV metals to group VII metals, and then shows only a modest decrease from the group VII metals to the group IB metals. In Figure 2.2b the bond energies for metal carbides, nitrides, and oxides all show the same general trend but the total variation of bond energy for the metal–nitrogen bond was much greater than that for the metal–carbon or metal–oxygen bond. These trends were extrapolated to predict metal–carbon and metal–nitrogen bond energies for group VIII and IB metals in rows 5 and 6. Table 2.5 summarizes averaged bond energies for metal carbon, nitrogen, and oxygen adsorption bonds. The plain entries are averaged from thermal desorption experiments to give a value corresponding to the most dense surface, for example, fcc (111), hcp (0001), bcc (110). Subjective judgment was used to weight the experimental data. Entries in parentheses are averaged values based on known thermodynamic data of bulk metal compounds. The entries in square brackets are estimated from extrapolated periodic trends. The extrapolated values

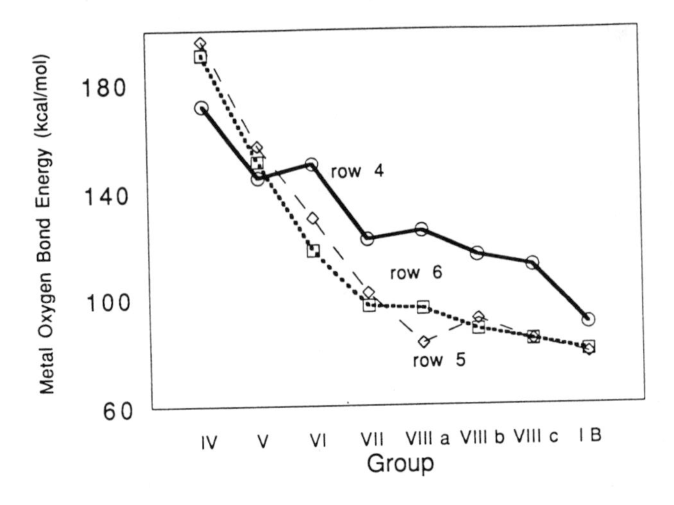

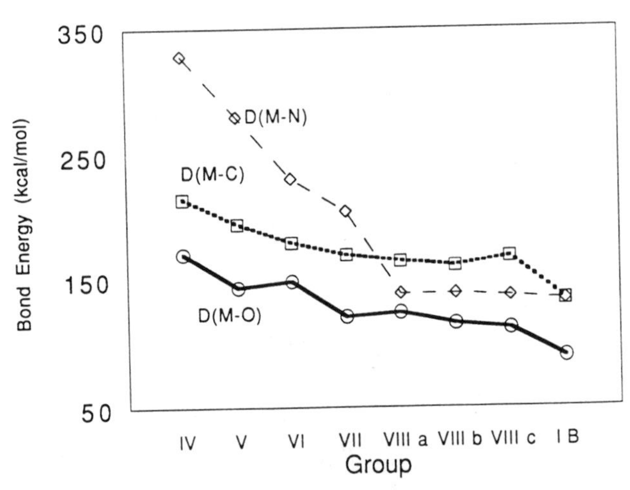

Figure 2.2. Periodic trends in bond energies for metal carbides, nitrides and oxides. **(a)** Metal–oxygen bond energies for the fourth-row (solid line), fifth-row (dashed line), and sixth-row (dotted line) transition metals calculated from Eq. 2.9. **(b)** Metal–carbon (dotted line), metal–nitrogen (dashed line), and metal–oxygen (solid line) bond energies for the fourth-row transition metals calculated from Eq. 2.9.

Table 2.5. Atomic Adsorption Bond Energies (kcal/mol)[a]

	Group							
	IVB	VB	VIB	VIIB	VIII	VIII	VIII	IB
Element	Ti	V	Cr	Mn	Fe	Co	Ni	Cu
D(MC)	(216)	(196)	(181)	(171)	(166)	(162)	169	[135]
D(MN)	(329)	(281)	(232)	(206)	140	(167)	138	135
D(MO)	(172)	(145)	(150)	(122)	(125)	(116)	112	90
Element	Zr	Nb	Mo	Tc	Ru	Rh	Pd	Ag
D(MC)	(220)	(205)	(183)	[165]	[140]	(130]	[130]	[70]
D′(MN)	(200)	(169)	155	[125]	[120]	[120]	[120]	(39)
D(MO)	(191)	(151)	118	(97)	96	88	84	80
Element	Hf	Ta	W	Re	Os	Ir	Pt	Au
D(MC)	(232)	(294)	(260)	[180]	[145]	[130]	[130]	[50]
D(MN)	(201)	(178)	(137)	144	[120]	[120]	[120]	[35]
D(MO)	(196)	(157)	130	(102)	(83)	92	84	79

[a]Plain values are averaged values from experimentally determined adsorption energies.
Values in parentheses are estimated from enthalpies of formation of bulk compounds. Values in brackets are extrapolated values from periodic trends in bond energies.

were chosen so the periodic trends for metal–carbon and metal–nitrogen bond energies in rows 5 and 6 showed a variation similar to that of the metals in row 4. The combined experimental values and estimated values listed in Table 2.5 will be used to evaluate thermodynamic driving forces for surface reactions.

B. Bond Order Conservation–Bond Energy Approach to Adsorbate Bond Energies

Approximating the atomic adsorption bond energy using Eq. 2.9 is a simple method. The limitation to the technique is that it lacks any structural specificity, so differences in the bond energy resulting from substrate structure cannot be addressed. Structural effects have been accounted for in a simple fashion using bond energy/bond order approaches [29–33]. Shustorovich proposed a bond order conservation (BOC) approach based on the Morse potential (MP) [29,33]. The two-center bond order χ is described by

$$\chi = e^{-(r-r_o)/a} \tag{2.10}$$

where r_0 is the equilibrium bond distance and a is related to the force constant for the bond. The energy of a chemical bond is represented by a Morse potential

$$Q(\chi) = Q_{AB}(2\chi - \chi^2) \tag{2.11}$$

where Q_{AB} is the equilibrium bond energy of an A–B bond. To make calculations for coordination of an adsorbate atom A to a metal surface in an n-fold coordination site (M_nA coordination) Shustorovich made four assumptions [29]:

1. Each two-center bonding interaction is described by the Morse potential (Eqs. 2.10 and 2.11).
2. The two-center bonding interactions, MA, are additive.
3. The total bond order of M_nA is conserved and normalized to unity.
4. Only nearest-neighbor bonding is considered.

Changes in M–M bonding are not explicitly accounted for in this treatment. Relaxation in metal–metal bonding can be implicitly included in the definition of the M–A bond energy. Postulates 2 and 3 require that the sum of the adsorbate–metal bond orders between nearest neighbors be normalized.

$$\sum_{i=1}^{n} \chi_{A,i} = 1 \tag{2.12}$$

From this set of postulates it follows that the bond energy for atomic adsorption in an n-fold coordination site is [29]

$$Q_n = Q_{MA}(2-1/n) \tag{2.13}$$

where Q_{MA} is the two-center M–A bond energy for an A atom coordinated to a single metal atom (on-top coordination) in the adsorption-adjusted metal lattice. At low coverages atomic adsorption is predicted to occur at the highest coordination sites, which agrees well with experimental data [34]. Furthermore, if differences in M–M bonding for different surfaces can be neglected, the adsorption bond strength is greatest on surfaces with the highest coordination sites available. This model projection accounts for the observation that adsorption is preferred at surface defects such as steps and kinks where higher coordination to the metal atoms is possible [34].

To be quantitative the BOC-MP formalism requires structural and thermodynamic data to evaluate the three parameters r_o, a, and Q_{AB}. As discussed above, thermodynamic data for Q_{AB} are limited and the combination of thermodynamic and structural data is rare. To make the BOC-MP approach independent of the structural parameters r_o and a, Shustorovich has considered only stationary points, minima and maxima, of the potential energy surfaces, in which case only thermodynamic data of Q_{MA} and Q_{AB} are required [29,33]. Values of Q_{MA} may be obtained from atomic adsorption energies, $D(MA) = Q_n$, by use of Eq. 2.13. Averaged values of atomic adsorption energies are listed in Table 2.5. Where specific adsorption site data are unavailable, a threefold site will be assumed to obtain an estimate of Q_{MA}.

The BOC-MP postulates require some clarification regarding their applicability. The Morse potential is frequently used to represent the potential for diatomic molecules where the bond energy, Q, depends only on distance, r. As metals, both simple and transition, normally have densely packed lattices [28], a spherical potential function may be expected to be a reasonable approximation. Slater has successfully used the Morse potential to correlate physical properties of metals to structural properties [35]. For atoms the assumption of spherical symmetry is a good approximation; it is expected that the BOC-MP approach will work well for atomic adsorption. For molecules, the potential functions may have strong angular dependences. Thus, describing the interaction between a molecule and a metal surface by the Morse potential may result in misleading predictions for some molecules. The Morse potential description of multicenter bonding is a good approximation when repulsive terms for chemisorption bonding are negligible, particularly on metal surfaces, as a result of the absence of a band gap [31g]. The two-center pairwise additivity of bonding is chosen for simplicity and because adatom bonding to a metal surface is known to be a localized phenomenon [36]. Van Santen has pointed out that conservation of bond order to unity is equivalent to normalization of the weighting functions within the valence bond theory [37].

Creation of a metal surface results in empty positions above the surface atoms with unsaturated metal valences. These unsaturated positions permit adsorption bonding without requiring significant disruption of the metal surface. The experimental value of the adsorption bond energy from which the Morse potential energy parameter, Q_{MA} in Eq. 2.11, is determined includes the effect of metal–metal reorganization caused by adsorption at low adsorbate coverage. It further assumes that no further changes in metal–metal bonding occur as the adsorbate coverage is increased. At low coverages each metal atom is coordinated to a single adsorbate atom, and the adsorption energy is given by expression 2.13. Above a critical coverage, larger than $1/n$ for an M_n-A unit mesh, metal atoms must bond to more than a single adatom, say as A_m-M, where m adsorbate atoms are bonded to a single metal atom. The BOC model 1 is applied to both the adatom and the metal atom.

$$\sum_{i=1}^{n} \chi_{Ai} = 1 \qquad (2.14a)$$

$$\sum_{j=1}^{m} \chi_{Mj} = 1 \qquad (2.14b)$$

Obviously, the energies of the same two-center bond, M–A, in the M_n-A mesh and the A_m-M mesh are identical which gives the bond energy per adsorbate atom, Q_n^m.

$$Q_n^m = Q_{MA}\,(2-1/n)(2-1/m)/m \qquad (2.15)$$

The bonding of a metal atom to multiple adsorbates, A_m-M, reduces the two-center bond energy by a factor of $(2-1/m)/m$ in this formalism [29]. This treatment predicts qualitatively that adsorption bond energies decrease with increasing adsorbate coverage and these effects occur in stepwise fashion at well-defined coverages. This result is in agreement with experimental results [10–13,38] and the predictions of other theoretical models [39,40]. Quantitatively, the BOC-MP model predictions are exaggerated for bond energy changes with coverage. The BOC-MP model predicts a large change in the adsorption bond energy per adsorbate as the coverage increases above critical coverages, which should lead to distinct sets of ordered adsorbate structures. On a fcc(100) surface the differential adsorption enthalpy decreases by 25% when adsorbate coverage increases above the $p(2\times2)$ structure,

Table 2.6. Atomic Adsorption Bond Energies on Nickel (kcal/mol)

System	Adsorption Site	Q			Reference
		Eq. 2.9	Eq. 2.15	Experiment	
Ni(100)–$p(2\times2)$C	Fourfold	156	$169^{c,e}$	169	*a*
Ni(111)–$p(2\times2)$C	Threefold	156	$160^{c,e}$	160	*a*
Ni(s)–(7(100)×(133)), $\theta<1/3$	Sixfold	156	$176^{c,e}$	179	*a*
Ni(100)–$p(2\times2)$O	Fourfold	116	$112^{d,e}$	105	*b*
Ni(100)–$c(2\times2)$O	Fourfold	116	$84^{d,f}$	101	*b*

[a] L. C. Isett, and J. M. Blakeley, *Surf. Sci.* **47** (1975), 645.
[b] H. J. Grabke, and H. Viefhaus, *Surf. Sci.* **112** (1981), L779.
[c] Assuming a metal–carbon two-center bond energy Q_{MC}=96 kcal/mol.
[d] Assuming a metal–oxygen two-center bond energy Q_{MO}=64 kcal/mol.
[e] $m=1$ in Eq. (2.15).
[f] $m=2$ in Eq. (2.15).

which indicates that the $p(2\times2)$ should always be formed at low temperatures by adatoms on a fcc(100) surface. Most adsorbates form ordered $p(2\times2)$ LEED structures on (100) crystal surfaces as predicted from the model [34,41,42]; however, stable structures are formed at higher coverages that would be thermodynamically unfavorable based on Eq. 2.15. The adsorption bond energies predicted by Eq. 2.15 are compared with experimental data for atomic oxygen adsorbed on Ni(100) in Table 2.6. The experimental results show a small decrease, ~4 kcal/mol, in the bond energy with coverage above one-fourth monolayer, whereas Eq. 2.15 predicts a larger change, ~28 kcal/mol. These discrepancies result from BOC to unity applied to the A_m-M mesh. This assumption neglects changes in the metal–metal bonds in the BOC condition applied to the metal atom, which exaggerates the magnitude of the decrease in the M–A bond energy.

Weinberg and Merrill (WM) have also used BOC to model adsorption bonding, where bond order was conserved to atomic valence [43]. For their bond energy/bond order correlation they used a simple power law variation of bond energy with bond order,

$$Q(\phi) = Q_0\phi^p, \qquad p \geq 1 \tag{2.16}$$

where Q_0 is the bond energy of a single bond and ϕ is the bond order in the Lewis–Pauling sense (i.e., the number of electron-pair bonds). Conceptually the WM bond energy/bond order correlation is different from that proposed by Shustorovich as it treats bond order in the traditional sense of chemical valence. The bond energy/bond order correlation was adopted from gas-phase reaction kinetics [44]. Unfortunately, the simple power law correlation given in Eq. 2.16 has only an attractive term and does not represent the potential function well. Equation 2.16 predicts the bond energy increases more than linearly with bond order. In the traditional Lewis–Pauling concept of bond order, the bond energy increases less than linearly with bond order (e.g., $Q_{C-C} > \frac{1}{2}\,Q_{C=C} > \frac{1}{3}\,Q_{C\equiv C}$). The Morse potential as introduced by Shustorovich [29,30], coupled with BOC, predicts that bond energy increases less than linearly with bond order which is more representative of actual systems.

Historically, the Lewis–Pauling concept of bond order has proved to be exceptionally helpful in understanding chemical bonding and structure. When considering adsorption of diatomic or polyatomic molecules, it would be desirable to use the Lewis–Pauling bond order to understand how adsorption to a metal may be modifying the bonding in a molecule. To this end we introduce the concept of bond order conservation to atomic valence coupled with the Morse potential. The ratio of bond order to the atomic valence is given by the same exponential function used for the Morse potential.

$$\phi / V = e^{-(r-r_0)/a} \tag{2.17}$$

In Eq. 2.17, ϕ refers to the bond order in the Lewis–Pauling sense, and V is the atomic valence in the traditional chemical sense, that is, the number of unpaired valence electrons (e.g., $V_H=1$, $V_C=4$, $V_N=3$, $V_O=2$, $V_{Cl}=1$). The parameters r_0 and a now refer to a bond where the bond order is equal to the atomic valence. We define the bond energy based on the central bonding atom, such as the C atom in methane or the O atom in water, to be correlated with the bond order through a Morse potential, given by Eq. 2.18, where Q_{AB} is the two-center bond energy.

$$Q(\phi) = Q_{AB}\left\{2\left(\frac{\phi}{V_A}\right) - \left(\frac{\phi}{V_A}\right)^2\right\} \tag{2.18}$$

Equations 2.17 and 2.18 are helpful in allowing us to recognize the difference in bonding between different atoms. When applied to homonuclear diatomic molecules and diatomic molecules where both atoms have the same valence (e.g., HCl) there is no difference from Eqs. 2.11 and 2.12 and the application and meaning of the quantities r_0, a, and Q_{AB} are the same as Shustorovich has proposed. Bond order conservation to atomic valence offers some advantages of handling energy partitioning in polyatomic molecules or fragments.

Normalized BOC for metals does not explicitly treat the bonding capacity of the surface metal atoms. The problem is to identify what is the bonding capacity of a metal surface. In a covalent crystal, such as silicon, the bonding capacity of the surface is dependent on both the surface structure and the adsorbate. The bonding capacity of the surface atoms is determined by their unsaturated valence, which is equal to the number of A–A bonds broken to form the surface. The bonding capacity of a metal surface may be defined using the effective atomic number rule, which gives a coordination number that is generally obeyed. The metal valence, V_M, may be defined as one half the number of electrons required to make a closed shell [45]; values are given in Table 2.7. These empirically derived values attempt to quantify metal bonding capacity in a valence bond framework. The values are similar to those

Table 2.7. Coordination Valence, V_M, of Metals

Group	IVB	VB	VIB	VIIB	VIIIa	VIIIb	VIIIc	IB
V_M	4	5	6	5.5	5	4.5	4	3.5

Table 2.8. Coordination Unsaturation of Surface Metal Atoms

Crystal Face	f_s[a]
fcc (100)	1/3
fcc (111)	1/4
fcc (110)	1/2
bcc (100)	1/2
bcc (110)	1/4
bcc (211)	1/2
hcp (0001)	1/4

[a] $f_s = V_{M,ex}/V_M$.

proposed by Pauling based on physical and magnetic properties [46,47]. The values given above may also be found from the coordination in transition metal carbonyls, assuming a bond order of unity for each M–CO and M–M bond.

At a metal surface the excess metal valence, $V_{M,ex}$, characterizes the bonding capacity of a surface metal atom. The excess metal valence is given by the metal valence, V_M, multiplied by the fraction, f_s, of M–M bonds broken to create the surface. The values of f_s for some common crystal surfaces are given in Table 2.8. Adsorbate bonding will be assumed to occur in an n-fold coordination site with no disruption of M–M bonding as long as the sum of the metal–adsorbate bond orders does not exceed the excess metal valence.

$$\sum_{j=1}^{m} \phi_{Mj} \leq V_{M,ex} \tag{2.19}$$

In an n-fold coordination site, M_n–A bonding, the two-center metal–adsorbate bond order is the adsorbate valence divided by n. The valence of the adatom is the valence in the traditional sense of chemical bonding. This approximate concept of metal valence allows us to describe differences in bonding capacity of metal surfaces for different adatoms.

As long as Eq. 2.19 is satisfied, the bonding capacity of the surface metal atoms is not exceeded and adsorption occurs without disruption of M–M bonds, while the adatom valence will be saturated. Within this approximation, the adsorption bond energy is given by the BOC-MP formalism as expressed in Eq. 2.13. If the adsorbate bonding requires Eq. 2.19 to be violated, the adsorption occurs if either (1) the valence of the adsorbate atom is unsaturated, or (2) the surface must restructure by reducing the M–M bonding to accommodate the M–A bonding. In case (1), the adsorption bond energy is given by

$$Q_n^i(M_n - A) = \left(\frac{nV_{M,ex}}{mV_A}\right) Q_{MA}\,(2 - 1/n) \tag{2.20}$$

The quantity $nV_{M,ex}$ represents the fraction of the valence on atom A that can be satisfied in the adsorption bond. Consistently, $Q_n^i < Q_n$ as Eq. 2.20 is applied only when Eq. 2.19 is violated.

The alternative bonding possibility (case 2) requires reduction of M–M bonding, which gives the adsorption bond energy

$$Q_n^{ii} = Q_{\mathrm{MA}}\left(2 - \frac{1}{n}\right) - Q_{\mathrm{M}}\left(\frac{V_{\mathrm{A}}}{n} - \frac{V_{\mathrm{M,ex}}}{m}\right) \tag{2.21}$$

The quantity Q_{M} is the metal–metal bond energy derived from heat of sublimation data.

$$Q_{\mathrm{M}} = \Delta H_{\mathrm{f}}(\mathrm{M(g)}) \tag{2.22}$$

The second term in Eq. 2.21 is to account for the decrease in the M–M bonding required to accommodate bonding of the adsorbate atom.

The BOC to atomic valence predicts that the differential heats of adsorption for adatoms will change in stepwise fashion at critical coverages when the incremental change in coverage requires Eq. 2.19 to be violated. For oxygen adsorption the critical coverage of oxygen occurs at 0.5 monolayer, corresponding to the $c(2\times2)$ structure. Below 0.5 monolayer the adsorption bond is predicted to be constant with the experimental value of 105 kcal/mol. The differential adsorption energy decreases to 70 kcal/mol for coverages above 0.5 monolayer, according to Eq. 2.20. The decrease in adsorption bond energy results in the $c(2\times2)$ structure being stable relative to increased oxygen adsorption under UHV conditions at room temperature. (The differential adsorption enthalpy of oxygen above 0.5 monolayer is calculated to be -12 kcal/mol. Based on Eqs. 2.2 and 2.3 the temperature must be below 220 K at a pressure of 10^{-} Torr for the oxygen coverage to exceed 0.5 monolayer.) In contrast, BOC to unity as expressed in Eq. 2.15 predicts that the differential adsorption energy of oxygen decreases from 105 to 79 kcal/mol above 0.25 monolayer, further decreasing to 46 kcal/mol above 0.5 monolayer. Based on Eq. 2.2 and 2.3 these values predict the $c(2\times2)$ structure is unstable in UHV above 550 K and coverages above 0.5 monolayer could never be achieved at pressures below 10^4 Torr (>10 atm). Experimental results show oxygen adsorbed on Ni(100) forms a $p(2\times2)$ structure at 0.25 monolayer. The adsorption bond energy decreases from 105 to 101 kcal/mol, from 0.25 to 0.5 monolayer (see Table 2.6) [48–50]. Oxygen coverage in UHV at 300 K saturates at 0.5 monolayer with a $c(2\times2)$; at 200 K, oxygen coverages exceeding 0.5 monolayer have been observed [48].

Equation 2.19 predicts that carbon and nitrogen adsorption have critical coverages of 0.25 monolayer on Ni(100) corresponding to the $p(2\times2)$ structure. Above this coverage Eq. 2.20 predicts that the differential adsorption bond energy for carbon decreases by 56 kcal/mol. The adsorption equilibrium coverages determined from Eq. 2.2 indicate that carbon coverages above 0.25 monolayer should be thermodynamically unstable relative to graphite. BOC to unity, Eq. 2.15, predicts that the differential bond energy for carbon above 0.25 monolayer decreases by 42 kcal/mol, also suggesting that carbon coverages above 0.25 monolayer on Ni(100) are unstable relative to graphite. Isett and Blakely found that for carbon coverages greater than the $p(2\times2)$ structure on Ni(100), adsorbed carbon atoms were no longer stable; rather, graphite precipitated onto the nickel surface [24], supporting the BOC-MP models. The differential adsorption energy for nitrogen above the $p(2\times2)$ structure, 0.25 monolayer, decreases by 15 kcal/mol, as calculated from Eq. 2.19, and by 35 kcal/mol, as calculated from Eq. 2.15. The experimental results show that in UHV at 300 K nitrogen forms a $p(2\times2)$ adlayer [49,50]. At low temperatures a $c(2\times2)$ nitrogen adlayer has been prepared on Ni(100) [51]. The decrease in the nitrogen adsorption bond energy from the $p(2\times2)$ to the $c(2\times2)$ layer was estimated from thermal desorption to be 18 kcal/mol [51], which is in good agreement with the predicted value based on the BOC to atomic valence model.

The BOC to atomic valence approach introduced here finds that the enthalpy changes with coverage depend on the adsorbate valence as well as the adsorption bond energy. This differs from the BOC to unity approach proposed by Shustorovich, where relative changes of the adsorption bond energy as a function of coverage were predicted to be the same for all adsorbates.

The BOC-MP analysis has allowed us to describe energetic differences between adsorption sites and surfaces based on the two-center M–A bond energy. It is still necessary to identify the two-center bond energy Q_{MA}. Shustorovich defines Q_{MA} from Eq. 2.13, which makes Q_{MA} an experimental quantity. Experimental heats of adsorption, $Q_n=D(MA)$, are available for hydrogen and frequently for oxygen, so that values of Q_{MA} may be obtained from the application of Eq. 2.13, equating the adsorption enthalpy to the atomic bond energy for the appropriate bonding configuration. The relevant experimental values of adsorption bond energies are given in Tables 2.2, 2.3, and 2.4 for hydrogen, oxygen, and nitrogen. For the adsorption of carbon, nitrogen, and oxygen where heats of adsorption are not available, estimates obtained from bulk compounds, given in Table 2.5, are used to obtain the two-center bond energies. The two-center bond energy in the BOC to atomic valence approach is the same as that defined by Shustorovich. Equations 2.9 and 2.13 define the values of $D(MA)$ for A in the bulk compound M_xA_y and A adsorbed as M_nA. Clearly, the coordination of the A atom is different in these two cases. The utility of Eq. 2.9 is to provide an estimate of the adsorption bond energy in the absence of experimental values.

IV. Enthalpies of Formation of Adsorbed Molecular Species

To characterize the energetics of chemical reactions on metal surfaces it is necessary to have estimates of the enthalpies of formation of adsorbed molecules and molecular fragments. Of particular interest are those species that have been encountered and identified in surface science studies. Experimentally there is little known about the enthalpies of formation of surface species except in a few cases where reversible adsorption occurs and adsorption isotherms have been measured. Three approaches to estimate adsorption enthalpies are discussed: a simple bond additivity approach and two approaches based on the Morse potential (one based on BOC to unity and one based on BOC to atomic valence).

The bond additivity approach for estimating enthalpies of formation follows a simple bond energy/bond order approach. Adsorption of a molecule may be envisioned as the chemical reaction between a molecular fragment, radical ZX•, with a metal surface forming the M–X bonds.

$$ZX\bullet + M(\text{surface}) \rightarrow M{-}XZ$$

The enthalpy of formation of M–XZ, which represents adsorbed species ZX, is equal to the enthalpy of formation of gas species ZX• minus the energy of forming the M–X bond (by convention, reactions that produce energy have negative enthalpies).

$$\Delta H_f(ZX_{ad}) = \Delta H_f(ZX\bullet) - \frac{\phi_{M-X}}{V_X} D(MX) \tag{2.23}$$

$D(MX)$ is the energy of the multicenter M–X bond, approximated as the adsorption bond energy (given by the values in Tables 2.2–2.5). The factor ϕ_{M-X}/V_x is the fraction of the atomic valence, V_x, of atom X, associated with bonding to the surface, where ϕ_{M-X} is the bond order to the surface in the Lewis–Pauling sense.

Bond additivity approaches of this type are common in organic chemistry. Bond energies of a wide variety of single, double, and triple bonds have been established experimentally that show that an X–Y bond energy is approximately constant, independent of the structure of the rest of the molecule. More detailed correlations, such as that compiled by Benson [6], have shown how the X–Y bond strength varies with other moieties bonding to X and Y. A shortcoming of the approach described in Eq. 2.23 is the linear approximation of bond energy with bond order for M–X bonding. This generally results in a low estimate of the adsorption bond energy as the bond energy generally increases less than linearly with Lewis–Pauling bond order (e.g., $D_{C-C} > \frac{1}{2} D_{C=C} > \frac{1}{3} D_{C\equiv C}$).

Shustorovich has suggested a technique for extending the BOC-MP approach to molecular adsorption [33]. Within this approach, Shustorovich treated all molecules as basically diatomic (see Eqs. 5.1–5.17 in Chapter 5). An alternative approach based on the Morse potential and BOC to atomic valence is presented below. This approach partitions energy in polyatomic molecules based on bond order in the Lewis–Pauling sense. In this section we consider various classes of adsorbed species and the three methods of approximating their enthalpies of formation.

A. Homonuclear Diatomics

The homonuclear diatomics H_2, N_2, O_2, Cl_2, and so forth, in valence bond theory are closed-shell systems. These molecules physically adsorb to solid surfaces because of dispersion forces. The adsorption enthalpies for molecular adsorption of these species may be described by a dispersion energy of the diatomic interacting with a free electron metal [52]. These physical adsorption forces are similar to the intermolecular forces in liquids, and the enthalpy for physisorption is comparable to the heat of vaporization [53].

In addition to the physical adsorption, there will be chemisorption. Adsorption occurs with the weakening of the A—A bond and the formation of M—A bonds. The simple bond additivity approach uses a linear correlation of bond energy with bond order. Such a model does not predict stable molecular chemisorption; simple bond additivity predicts either no stabilization for molecular adsorption or dissociation of the molecule.

The BOC-MP model predicts stable molecular bonding for diatomics. The two extremes in bonding configurations of A_2 are the molecule bonded linearly (l) with its axis normal to the surface and the molecule bonded bridged (b) with its axis parallel to the surface symmetrically bonded to two n-coordination sites. Shustorovich applied the condition of normalized bond order to the A—A′ bond and the M—A and M—A′ bonds [29].

$$\sum_{i=1}^{n} (\chi_{Ai} + \chi_{A'i}) + \chi_{AA'} = 1 \tag{2.24}$$

The alternative approach that applies conservation of valence about each of the A atoms is given by the set of equations

$$\sum_{i=1}^{n} \frac{\phi_{Ai}}{V_A} + \frac{\phi_{AA'}}{V_A} = 1 \tag{2.25a}$$

$$\sum_{i=1}^{n} \frac{\phi_{A'i}}{V_A} + \frac{\phi_{AA'}}{V_A} \leq 1 \tag{2.25b}$$

The inequality in Eq. 2.25b accounts for situations where the metal–adsorbate distances restrict the M—A bond orders. When A_2 is linearly adsorbed to a metal surface with its molecular axis normal to the surface the M—A′ bond distances for the A′ atom away from the surface are much greater than the M—A bond distances for the A atom near the surface. Hence, only the M—A interactions are considered in the adsorption bonding and M—A′ interactions may be neglected, which results in the A′ atom having unsaturated valence. The difference in the BOC conditions stated in Eqs. 2.24 and 2.25 is manifested in the case of bridge bonding. Equation 2.25 allows for a smooth transition from an adsorbed diatomic molecule to two adsorbed atoms. Equation 2.24 describes the M_n—AA′ bonding from the gas-phase AA′ up to a transition state for AA′ dissociation.

Using the Morse potential with the BOC to unity, the adsorption bond energy for A_2 adsorbed linearly in an n-coordination site is given by

$$\Delta H_{ad} = -\frac{nQ_{MA}^2}{Q_{MA} + nQ_{AA}} \tag{2.26}$$

In arriving at Eq. 2.26 it was assumed that in the linear configuration the M—A′ distance is large so that the bonding interaction between the A′ atom and the surface can be neglected. Q_{MA} is the adsorption bond energy for atomic adsorption in an on-top site previously defined. The quantity Q_{AA} refers to the total bond energy of the A_2 molecule.

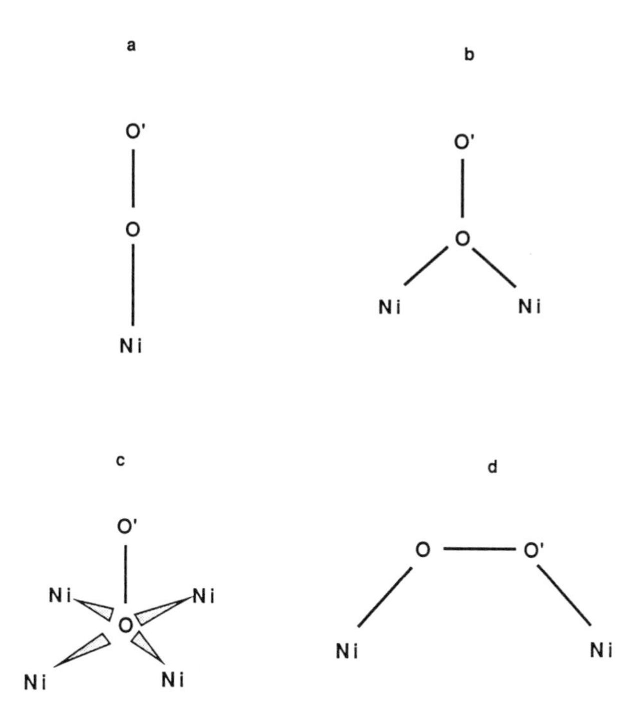

Figure 2.3. Configurations of O_2 adsorbed on Ni(100): **(a)** linear adsorption, on-top site; **(b)** linear adsorption, twofold bridge site; **(c)** linear adsorption, fourfold hollow site; **(d)** parallel adsorption, symmetric bridge bonding to two Ni atoms.

The magnitude of the error in neglecting the next-nearest-neighbor M−A bonding interactions may be estimated by considering a specific system for which experimental data are available. Dynamical LEED studies [54], vibrational spectroscopy [55], and thermal desorption [56] have been carried on for oxygen adsorption on Ni providing values for r_0, a, and Q_{NiO} [29]. Constants for the Morse potential of O_2 have previously been tabulated [35]. Figure 2.3 shows three linear bonding configurations of O_2 on Ni(100). The bond lengths and the bond energies based on the BOC-MP approach and neglecting the next-nearest-neighbor interactions are tabulated in Table 2.9. The error in neglecting the MA′ interactions is only 2 kcal/mol for

Table 2.9. Bond Distances (Å) and Energies (kcal/mol) for O_2 Adsorbed to Ni(100)[a]

Bonding Site[b]	$-\Delta H_{ad}$	r_{OO}	Q_{OO}	r_{MO}	Q_{MO}	$r_{MO'}$	$Q_{MO'}$
On-top	−22[c]	1.36	104	1.83	37	3.19	2
Twofold bridge	−27[c]	1.40	98	2.06	24	3.04	3
Fourfold hollow	−30[c]	1.44	92	2.31	14	2.93	4
Symmetric bridge	−27[d]	1.45	90	2.02	28	2.77	6

[a]Morse potential constants r^0_{MA} = 1.36 Å, Q_{MA} = 64 kcal/mol, a_{MA} = 0.45 Å [29]; and r^0_{OO} = 1.2 Å, Q_{OO} = 119 kcal/mol, a_{OO} = 0.37 Å [35].
[b]See Figure 2.3
[c]Equation (2.26).
[d]Equation (2.29).

on-top site, corresponding to a 10% error in the adsorption enthalpy. For the O_2 sitting in the fourfold site the error in the adsorption bond energy (accounting for the four Ni–O′ bonds) is 16 kcal/mol, or nearly 50% of the adsorption enthalpy.

Shustorovich has dismissed the use of Eq. 2.26 to treat linear adsorption of A_2 because of the error involved in neglecting the bonding interactions with the atom further away from the surface [33]. The discussion presented above suggests that the error is not significant for adsorption on an on-top site, but becomes quite significant when considering adsorption into higher coordination sites. Shustorovich has also discussed treating the M–A′ interaction with some alternative potential function to account for donor and acceptor species [29]. These interactions attempt to account for electron redistribution resulting from interactions of filled and empty molecular levels with the metal. This modification requires detailed information about molecular orbitals of the adsorbate that is seldom available, and it represents an extension beyond the BOC-MP framework.

Using the valence conservation as expressed in Eq. 2.25 one also arrives at Eq. 2.26 when neglecting the M–A′ interactions. Again these interactions are neglected as we assume only nearest-neighbor bonding interactions are important and the A′ atom away from the surface has no nearest-neighbor metal atoms.

The BOC-MP models predict that the highest symmetry site is the preferred adsorption site for linearly bonded A_2; however, the variation of adsorption energy with coordination is much weaker than in the case of adatom adsorption. For nitrogen adsorption on the first-row group VIII metals, where $Q_{NN} = 226$ kcal/mol and $Q_{MN} \sim 90$ kcal/mol, the adsorption enthalpy is approximately 25 kcal/mol, with the variation between an on-top site and a fourfold site being 6 kcal/mol. Shustorovich pointed out that the variation in adsorption enthalpy with coordination should be even smaller because of interactions of filled (donor) or empty (acceptor) molecular orbitals with the metal [29,31g]. These interactions are expected to be comparable to the variations in adsorption enthalpy predicted by the BOC-MP correlation and could reverse the predicted trends of preferred coordination [29,31].

The total energy of A_2 adsorbing in the symmetric configuration, Q_b, in the BOC to atomic valence approximation is given by conserving the bond order on both A atoms.

$$
\begin{aligned}
Q_b = {} & 2n\, Q_{MA}\left(2\left(\frac{\phi_{MA}}{V_A}\right) - \left(\frac{\phi_{MA}}{V_A}\right)^2\right) \\
& + Q_{AA}\left(2\left(\frac{V_A - n\phi_{MA}}{V_A}\right) - \left(\frac{V_A - n\phi_{MA}}{V_A}\right)^2\right)
\end{aligned}
\tag{2.27}
$$

The first term in Eq. 2.27 represents the total bond energy of each A atom interacting with n nearest-neighbor metal atoms in an n-fold coordination site. The second term is the AA bond energy that is reduced as a result of M–A bonding. At the limit of no interaction with the surface, $\phi_{MA} = 0$ and $Q_{TOT} = Q_{AA}$, which is the total bond energy of the A_2 molecule. At the other extreme of two adsorbed atoms, $\phi_{MA} = V_A/n$ and $Q_{TOT} = 2Q_{MA}(2-1/n)$. Implicitly assumed in Eq. 2.27 is that the bonding capacity of the metal atoms is not exhausted as stated in Eq. 2.19. The bonding capacity of the surface atoms is important when considering barriers for dissociation as will be discussed in Section VI. The A–A bond lengths are generally much less than the metal lattice spacing so the symmetric bonding is physically limited to two on-top sites (it is possible to have both A atoms sit in adjacent M_2A sites but that possibility is not considered here). The adsorption enthalpy of bridged bonded A_2 is

$$\Delta H^1_{\mathrm{ad,b}} = \frac{-4Q^2_{\mathrm{MA}}}{2Q_{\mathrm{MA}} + Q_{\mathrm{AA}}}. \qquad (2.28)$$

Based on Eq. 2.28 bridge bonding is always preferred to linear bonding. Alternatively, one could use Shustorovich's approximation of total bond order conserved to unity (Eq. 2.24) from which the adsorption enthalpy for bridge-bonded A_2 is

$$\Delta H^2_{\mathrm{ad,b}} = \frac{-2Q^2_{\mathrm{MA}}}{Q_{\mathrm{MA}} + 2Q_{\mathrm{AA}}} \qquad (2.29)$$

Equations 2.28 and 2.29 reflect the different assumptions about bond order conservation.

Shustorovich considered only the case of bridge-bonded A_2 adsorption, suggesting that the approximations for linear bonded species were not adequately justified [29,33]. For the symmetrically bonded A_2 molecule, Shustorovich pointed out that the spacing between the next-nearest-neighbor metal atom and A atom was such that those interactions should be considered. Figure 2.3d shows the symmetric bridge-bonding configuration of O_2 on Ni. Bond lengths and bond energies for the symmetrically bonded A_2 molecule, neglecting the next-nearest-neighbor interactions, are given in Table 2.9. The next-nearest-neighbor, MA′, interactions are about 15% of the nearest-neighbor, MA, interactions so neglecting the MA′ interactions will introduce a significant error. For convenience, Shustorovich assumed that both A atoms bonded to both metal atoms equally. This reduced the M–A bond energy resulting from the coverage effect discussed previously (Eq. 2.15). With these assumptions the heat of adsorption for symmetrically bonded A_2 derived by Shustorovich is given by

$$\Delta H^3_{\mathrm{ad,b}} = \frac{-\frac{9}{2}Q^2_{\mathrm{MA}}}{3Q_{\mathrm{MA}} + 8Q_{\mathrm{AA}}} \qquad (2.30)$$

The three expressions shown above display the differences in adsorption energy depending on the way BOC is interpreted.

Table 2.10 compares experimental results for molecular desorption of N_2 and O_2 from metals with calculated values using Eq. 2.26 for linear bonding and Eqs. 2.28, 2.29, and 2.30 for bridge bonding. For linear bonding on-top adsorption has been assumed; higher coordination will result in larger adsorption enthalpies. The experimental values refer to high coverages of N_2 and O_2. At low coverages the molecules dissociate before desorbing. In the low-coverage limit the heats of adsorption are expected to be more exothermic.

Table 2.10. Molecular Adsorption Enthalpies of N_2 and O_2 (kcal/mol)

	$-\Delta H_{ad}$					
System	Eq. 2.26 ($n=1$)	Eq. 2.28	Eq. 2.29	Eq. 2.30	Experiment	Reference
O_2/Ag (110)	13	41	15	9	10	*a*
O_2/Pt (111)	15	47	18	11	8.4	*b*
O_2/Pt (112)	15	47	18	11	7.7–13.1	*b*
O_2/Pt (110)	15	47	18	11	10.8, 12.1	*c*
N_2/Re (0001)	33	101	40	23	8	*d,e*
O_2/Pt(111)–($\sqrt{3}$x$\sqrt{3}$) R 30°O		14				
N_2/Re(0001)–($\sqrt{3}$x$\sqrt{3}$) R30°N		38				

[a]C. Backx, C. P. M. DeGroot, and P. Biloen, *Surf. Sci.* **104** (1981), 300.
[b]A. Winkler, X. Guo, H. R. Siddiqui, P. L. Hagans, and J. T. Yates, Jr., *Surf. Sci.* **201** (1988), 419.
[c]J. Fusy, and R. Ducros, *Surf. Sci.* **214** (1989), 337.
[d]G. Hasse, and M. Asscher, *Surf. Sci.* **191** (1987), 75.
[e]J.-W. He, and D. W. Goodman, *Surf. Sci.* **218** (1989), 211.

Shustorovich's expression for bridged bonding, Eq. 2.30, agrees well with the experimental results, though the experimental results refer to high coverages where the reduction of the MA interactions caused by coverage effects may be important. Shustorovich's approximation overestimates the MA′ bonding at low coverages but his inclusion of the reduction in Q_{MA} resulting from a coverage effect makes a correction that brings the calculated values in line with the experimental results. Equation 2.28 predicts adsorption enthalpies that seem too large. This probably results from the assumption of spherical potentials for the interaction of the diatomic molecule with the metal. The molecular orbitals of N_2 and O_2 have large angular variations and hence the interaction potential with the metal is expected to show angular variations. In the case of N_2 the bonding orbitals are expected to be nearly colinear with the N_2 axis. Linear bonding is probably more reasonable for N_2 than the symmetric bridge bonding. Additionally, the implicit constraint of not exceeding the metal bonding capacity is almost always violated for N_2 adsorption and frequently violated for O_2 adsorption. The effect of metal bonding capacity or atomic coverage may be included in Eq. 2.27. The bond order ϕ_{MA} is set equal to the excess metal valence less the MA bond order associated with atomic adsorption ($V_{m,ex}-V_A/n$). This reduces the predicted adsorption enthalpy. The effect of one-third monolayer coverage of O on Pt(111) and of N on Re(0001) has been factored into the adsorption enthalpies of O_2 on Pt(111) and of N_2 on Re(0001). Equation 2.27 was used with $\phi_{MA}=V_{m,ex}-V_A/3$. When accounting for the adsorbed atomic species the molecular adsorption enthalpies predicted by the BOC to atomic valence model more closely agree with experiment. The bond energies approximated by Eq. 2.28 represent an upper limit to the adsorption enthalpy within the BOC to atomic valence model. Equation 2.29 gives reasonable values for bridge-bonded diatomics.

It is believed this results from the limitations imposed on the M–A bond orders when bond order is conserved to unity, which can compensate for the nonspherical potentials and limited bonding capacity of the surface metal atoms.

B. Heteronuclear Diatomics

Adsorption of heteronuclear diatomic molecules, particularly NO and CO, is of great interest because they are important reactants for many reactions catalyzed by metals. Carbon monoxide and nitric oxide are unusual molecules from the viewpoint of atomic valence as both have atoms with nonstandard valence configurations in the Lewis–Pauling sense of electron-pair bonds. The unsaturated atomic valence configurations in NO and CO allow them to form strong chemical bonds to metal surfaces with relatively small perturbations to the molecular bonding.

Linear coordination of CO to a metal atom may be viewed as bonding to the carbon to satisfy its normal bonding coordination of 4. Bonding through the oxygen is unreasonable as it requires the oxygen atom to make four electron-pair bonds. The adsorption bond of CO to the metal can be viewed in the Lewis–Pauling sense as

:C:::O: + M → M:C:::O:

CO bonding as shown applies to adsorption on-top a metal atom with the CO axis normal to the surface, where interactions between the oxygen atom and the surface are negligible. A lower limit to the adsorption enthalpy is given by

$$\Delta H_{\mathrm{ad}}(\mathrm{CO}) = -\tfrac{1}{4} D(\mathrm{CM}) \tag{2.31}$$

Equation 2.31 overestimates the bond energies as it fails to account for any relaxation of the carbon–oxygen bond. Adsorption enthalpies for nitric oxide can be formulated in a fashion similar to that for carbon monoxide allowing for relaxation of the NO bond order from $2\frac{1}{2}$ to 2.

:N::·O:+M → M:N̈::Ö

$$\Delta H_{\mathrm{ad}}(\mathrm{NO}) = -\tfrac{1}{3} D(\mathrm{NM}) + \{D(\mathrm{NO}) - D(\mathrm{N{=}O})\} \tag{2.32}$$

The N=O bond energy is estimated from various nitrosyl compounds to be 100 kcal/mol [45,46].

Better estimates of the adsorption enthalpies of carbon monoxide and nitric oxide may be obtained from BOC following the same formalism used for homonuclear diatomics. The valence unsaturation suggests that only linear bound species are important, with the atom having the unsaturated valence bound to the metal. Shustorovich reasoned that linear adsorption of CO and NO is preferred for ending transition metals and concluded that the carbon or nitrogen was bound to the surface because the M–C and M–N bonds were stronger than the M–O bonds [29]. The adsorption enthalpy for carbon monoxide or nitric oxide adsorbing in an n-fold coordination site is

$$\Delta H_{\mathrm{ad}}(\mathrm{XO}) = -\frac{nQ_{\mathrm{MX}}^2}{Q_{\mathrm{MX}} + nQ_{\mathrm{XO}}} \tag{2.33}$$

where X = N or C, Q_{XO} is the dissociation energy of the appropriate oxide, and Q_{MX} is the two-center MX bond energy. Equation 2.33 had previously been derived by Shustorovich for linear bonding of an XO molecule with bond order conserved to unity, and neglecting the metal–oxygen interactions, it is identical in form to Eq. 2.29. Equation 2.33 is also valid for BOC to atomic valence with the same approximations. The XO bond order for linear coordination of CO or NO is

$$\frac{\phi_{XO}}{V_X} = \frac{Q_{MX} + n(Q_{XO} - Q_{MX})}{Q_{MX} + nQ_{XO}} \tag{2.34}$$

Adsorption enthalpies for bridge-bonded CO and NO may also be determined within the BOC-MP framework. As the bonding is not symmetric, the expressions are more complicated than Eqs. 2.28–2.30. For expressions of the adsorption bond energy of bridge-bonded species based on BOC to unity, the reader is referred to Shustorovich [33].

Equations 2.33 and 2.34 predict that the bond energy to the surface increases with M–X bond strength and the XO bond order decreases with both M–X bond strength and coordination number. The vibrational stretching frequency is generally assumed to increase with the XO bond order [57]. The decrease in bond order with increased coordination is consistent with experimental observations of vibrational spectra [57]. Experimental data do not show any correlation between vibrational frequency and adsorption enthalpy. For example, the adsorption enthalpy of CO on Pt and Ni are almost the same, yet the CO stretching frequency is nearly 100 cm^{-1} higher on Pt than on Ni [58,59]. In contrast, the CO stretching frequencies for CO adsorbed on Cu and Pt are almost the same but the adsorption enthalpy on Cu is only half that on Pt [59,60].

Equation 2.33 predicts that CO and NO adsorption is always favored at the highest coordination site. This is a consequence of assuming spherical symmetry in the bonding potential and the fact that n bonds of order $1/n$ are stronger than one bond of order 1. As previously discussed, Shustorovich has suggested that interactions between adsorbate donor or acceptor orbitals and the metal orbitals can alter the adsorption energy of a bonding site in a nonmonotonic fashion [29,33]. These interactions will also cause variations in XO bond order, so that correlations of vibrational frequency are not generally valid [33].

Some BOC-MP model predictions of adsorption enthalpies of CO on metals are shown in Table 2.11. The comparison of the adsorption enthalpies with experimental data is quite favorable. The agreement with experiment is not very good for CO adsorption on Pd and Pt. The adsorption enthalpies for CO suggest the Pd–C and Pt–C bonds are stronger than indicated from periodic trends shown by bulk compounds. It is tempting to use larger values for the M–C bond energies with Pd and Pt; however, spectroscopic evidence has shown that adsorbed carbon forms carbide-type structures on Ni, Fe, and group VI metals [24,61], whereas the interactions of Pd and Pt with carbon show phase separation [62]. These findings indicate that the Pd–C and Pt–C bonding interactions are probably weaker than on those metals where alloying is observed. More careful examination of the adsorption of carbon on group VIII and IB metals is called for to improve the quantitative analysis of surface reactions.

Table 2.11. Adsorption Enthalpies of Carbon Monoxide (kcal/mol)

Surface	$-\Delta H_{ad}$			Reference
	Eq. 2.31	Eq. 2.33 (n=1)	Experiment	
Fe(100)	41	26	25	*a*
Co(0001)	40	27	25	*b*
Ni(100)	42	27	26	*c*
			30	*d*
Ni(110)	42	27	32	*e*
		30 (n=2)	25	*f*
Ni(111)	42	25	26	*g*
Cu(100)	34	18	16	*h*
Cu(311)	34	18	15	*i*
Ru(001)	35	21	30	*j*
Ru(1010)	35	21	28	*k*
Rh(111)	32	18	30	*l*
Pd(100)	31	16	36	*m,n*
Pd(111)	31	17	34	*o*
Ag(111)	17	5	6	*p*
Ir(110)	32	18	37	*q*
Ir(111)	32	18	39	*r*
Ir(111)	32	18	44	*s*
Pt(100)	30	15	32	*t*
Pt(111)	30	16	30	*u*
			33	*v*

[a] J. B. Benziger, and R. J. Madix, *Surf. Sci.* **94** (1980), 119.
[b] M. E. Bridge, C. M. Comrie, and R. M. Lambert, *Surf. Sci.* **67** (1977), 393.
[c] S. W. Johnson, and R. J. Madix, *Surf. Sci.* **108** (1981), 77.
[d] J. C. Tracy, *J. Chem. Phys.* **56** (1972), 2736.
[e] C. R. Helms, and R. J. Madix, *Surf. Sci.* **52** (1975), 677.
[f] M. M. Madden, J. Kuppers, and G. Ertl, *J. Chem. Phys.* **58** (1973), 3401.
[g] K. Christmann, O. Schober, and G. Ertl, *J. Chem. Phys.* **60** (1974), 4719.
[h] J. C. Tracy, *J. Chem. Phys.* **56** (1972), 2748.
[i] H. Papp, and J. Pritchard, *Surf. Sci.* **53** (1975), 371.
[j] T. E. Madey, and D. Menzel, *Jpn. J. Appl. Phys.* (1974), Suppl. 2, 229.
[k] P. D. Reed, C. M. Comrie, and R. M. Lambert, *Surf. Sci.* **59** (1976), 33.
[l] D. G. Chastner, B. A. Sexton, and G. A. Somorjai, *Surf. Sci.* **71** (1978), 59.
[m] J. C. Tracy, and P. W. Palmberg, *J. Chem. Phys.* **51** (1969), 4852.
[n] H. Conrad, G. Ertl, and E. E. Latta, *Surf. Sci.* **43** (1974), 462.
[o] G. Ertl, and J. Koch, *Z. Naturforsch.* **25A** (1970), 1906.
[p] G. McElhiney, H. Papp, and J. Pritchard, *Surf. Sci.* **54** (1976), 617.
[q] J. E. Demuth, and T. N. Rhodin, *Surf. Sci.* **45** (1974), 249.
[r] C. M. Comrie, and W. H. Weinberg, *J. Chem. Phys.* **64** (1976), 250.
[s] D. I. Hagen, B. E. Nieuwenhuys, G. Rovida, and G. A. Somorjai, *Surf. Sci.* **59** (1976), 177.
[t] M. A. Barteau, E. I. Ko, and R. J. Madix, *Surf. Sci.* **102** (1981), 99.
[u] R. W. McCabe, and Z. D. Schmidt, *Surf. Sci.* **66** (1977), 101.
[v] C. T. Campbell, G. Ertl, H. Kuppers, and J. Segner, *Surf. Sci.* **107** (1981), 207.

C. Polyatomic Molecular Adsorption

Adsorption of polyatomic molecules introduces complexities that require additional assumptions to estimate adsorption enthalpies. A lower limit to the adsorption enthalpy may be obtained from the dispersion forces between the molecule and a free electron metal. The dispersion energies are slightly greater than the energy of sublimation. The enthalpy of vaporization may be used as an upper limit to the adsorption enthalpy of molecules.

The BOC methodology provides a way to account for the bond relaxation that occurs during adsorption. Shustorovich approached the adsorption of polyatomic molecules by treating them as pseudodiatomic species [32,33]. A molecule $A{-}B_z$, where B can be any molecular fragment, was treated as a diatomic where the bond energy, Q_{AB_z}, is the energy required to dissociate AB_z into A and zB fragments.

$$Q_{AB_z} = -\Delta H_f(AB_z) + \Delta H_f(A) + z\Delta H_f(B) \tag{2.35}$$

Adsorption through the A atom with bond order conserved to unity requires

$$\sum_{i=1}^{n} \chi_{Ai} + \chi_{A-B_z} = 1. \tag{2.36}$$

The bond energy was found assuming a Morse potential for a pseudodiatomic. The resulting adsorption energy is found to be

$$\Delta H_{ad}(AB_z) = -\frac{nQ_{MA}^2}{Q_{MA} + nQ_{AB_z}} \tag{2.37}$$

The effective constants for the Morse potential refer to the dissociation of the molecule into fragments. This is not a simple reaction that depends only on distance between two centers, but involves a multidimensional reaction trajectory. In Eq. 2.37, Q_{MA} is the two-center M–A bond energy. Table 2.12 lists heats of formation of some common molecular fragments for use in Eq. 2.35. This model predicts that the adsorption enthalpy decreases as the strength of A–B bonds decreases. It also predicts that adsorption is preferred in the highest coordination sites; however, the variation of adsorption enthalpy with coordination is generally small and bonding effects not considered in the BOC-MP formalism may alter the preferred adsorption site [29,33].

Bonding of symmetric polyatomic molecules such as ethene and ethyne, H_xCCH_x, was treated assuming the CH_x group to be a pseudoatom so the bridge-bonded species has an adsorption enthalpy given by Eq. 2.29 or 2.30, with Q_{AA} replaced by the total bond energy for each carbon atom.

$$Q_{H_xCCH_x} = -\Delta H_f(H_xCCH_x) + \Delta H_f(CH_x) + x\Delta H_f(H) + \Delta H_f(C) \tag{2.38}$$

The alternative BOC-MP approach conserves atomic valence. For the AB_z molecule bonding to the surface through the A atom, the BOC condition is

$$\sum_{i=1}^{n} \phi_{Ai} + \sum_{j=1}^{z} \phi_{A-B_j} = V_A. \tag{2.39}$$

Equation 2.39 assumes that the bonds in the molecular fragments B are unaltered, which is equivalent to assuming that only the bonds that are nearest neighbors to the adsorption site are perturbed by adsorption. The total bond energy of the adsorbed

Table 2.12. Heats of Formation of Molecules and Radicals (kcal/mol)[a]

Species	ΔH_f
CH_4	−17.9
CH_3	34
CH_2	86
CH	142
C	171
C_2H_4	12.5
C_2H_6	−20.2
$HC{=}CH_2$	69
$C{-}CH_3$	114
CH_3OH	−48
CH_3O	3.5
CH_2O	−27
CO	−26.4
HCO	9
HCOOH	−90.5
HCOO	−36
H	52
OH	9.4
O	59.6
NH_2	49
NH	90
N	113
CN	101

[a]S. W. Benson, *Thermochemical Kinetics*, Wiley, New York, 1976.

molecule is given by the sum of three terms: (1) the bond energies associated with the molecular fragments, (2) the bond energies of the A−B bonds; and (3) the energy of the M−A bonds to n_M metal atoms. These three terms are shown explicitly below.

$$Q_{TOT} = \sum_{i=1}^{z} Q_{Bi} + \sum_{j=1}^{z} Q_{A-B_j}\left\{2\left(\frac{\phi_{A-Bj}}{V_A}\right) - \left(\frac{\phi_{A\text{-}Bj}}{V_A}\right)^2\right\} + n_M Q_{MA}\left\{2\left(\frac{\phi_{MA}}{V_A}\right) - \left(\frac{\phi_{MA}}{V_A}\right)^2\right\} \quad (2.40)$$

In Eq. 2.40 it was assumed that all the two-center bonds in the molecule may be described by Morse potentials. The A−B bond energies are obtained from the application of Eq. 2.40 to free gas-phase molecules. Table 2.13 summarizes bond energies, Q_{AB}, for common bonds.

The adsorption enthalpy of AB_z is the difference in the total bond energy between the free molecule ($\phi_{MA}=0$) and the adsorbed molecule. If all B groups are the same the adsorption enthalpy reduces to Eq. 2.37. When there are different groups bonded to the A atom, Eq. 2.39 indicates that the adsorption bond energy is maximized by reduction in bond order of the weakest A−B bond. This result is consistent with experimental observations for methanol adsorption on Cu, Ag, and Pt

Table 2.13. Morse Potential Bond Energies, Q_{AB} (kcal/mol)

Bond A–B	Basis Molecule	$Q_{TOT}{}^{a}$ Eq. 2.40	$Q_{TOT}{}^{b}$ Experimental	Q_{AB}
C–H	CH_4	$\frac{7}{4} Q_{CH}$	397	225
C–C	Diamond	$\frac{7}{8} Q_{CC}$	171	196
C–O	CO_2	$\frac{3}{2} Q_{CO}$	383	255
O–H	H_2O	$\frac{3}{2} Q_{OH}$	219	146
O–O	O_2	Q_{OO}	119	119
N–N	N_2	Q_{NN}	226	226
N–H	NH_3	$\frac{5}{3} Q_{NH}$	278	167
C–N	HCN	$\frac{7}{16} Q_{CH} + \frac{15}{16} Q_{CN}$	335	253

[a]Total bond energy around atom A in the basis molecule.
[b]Energy required to dissociate the basis molecule into atomic constituents.

where vibrational spectra suggest that the weaker O–H bond is perturbed more than the H_3C–O bond [63].

When considering bridge bonding of symmetric polyatomic molecules, B_zA–AB_z, in the BOC to atomic valence formalism valence is conserved about both A atoms.

$$\sum_{i=1}^{z} \phi_{A-Bj} + \phi_{A-A} + \phi_{MA} = V_A \tag{2.41}$$

The total bond energy is partitioned as

$$\begin{aligned} Q_{TOT} = 2\Bigg\{ & \sum_{i=1}^{z} Q_{Bi} + \sum_{j=1}^{z} Q_{A-Bj}\left(2\left(\frac{\phi_{A-Bj}}{V_A}\right) - \left(\frac{\phi_{A-Bj}}{V_A}\right)^2\right) \\ & + n_M Q_{MA}\left(2\left(\frac{\phi_{MA}}{V_A}\right) - \left(\frac{\phi_{MA}}{V_A}\right)^2\right)\Bigg\} \\ & + Q_{AA}\left(2\left(\frac{\phi_{AA}}{V_A}\right) - \left(\frac{\phi_{AA}}{V_A}\right)^2\right) \end{aligned} \tag{2.42}$$

The first three terms are doubled because of the symmetric bonding, and the last term accounts for the A-A bond. An advantage of BOC to atomic valence as stated in Eqs. 2.41 and 2.42 is that it allows one to estimate adsorption enthalpies for rehybridized bonding in olefins. For example, the adsorption enthalpy of di-σ bonded ethene may be approximated using Eq. 2.42, assuming the C–H bond orders do not change and the C–C bond order is 2 in ethene and 1 in di-σ bonded ethene.

$$-\Delta H_{ad}(\text{di-}\sigma C_2H_4) = \tfrac{7}{8} Q_{MC} - \tfrac{5}{16} Q_{CC} \tag{2.43}$$

Equation 2.43 indicates that the stability of di-σ bonded ethene increases with M–C bond energy. In the next section expressions are developed for other types of structures observed when ethene reacts on metal surfaces. In conjunction with the above expression it becomes possible to answer questions about reaction pathways on metal surfaces.

D. Adsorption of Molecular Fragments

Obtaining reliable estimates for adsorption bond energies of molecular fragments is the most useful application of phenomenological approaches as virtually no experimental data are available. The bond additivity approach is conceptually easy to use for obtaining estimates. The adsorption of a molecular fragment is viewed as the chemical reaction of a gas-phase radical and the metal surface with the formation of a chemical bond. The heat of formation of the adsorbed fragment is given by Eq. 2.23. This procedure was first proposed by Benziger and Madix to treat alkyl, alkoxy, and carboxylate species adsorbed on metal surfaces [64]. This simple approximation does a remarkable job at predicting variations in reaction pathways on different transition metals. It does not address structural questions about the surface.

BOC approaches to the estimation of adsorption enthalpies require an approximation for partitioning of bond energies. Shustorovich examined several partitioning schemes, finally recommending the same formalism presented for molecular adsorption [32,33]. An effective A–B bond energy is defined as the difference in energy between AB and fragments A and B in the gas phase, as shown in Eq. 2.35. Shustorovich stressed that this definition is useful only when the decomposition reaction is endothermic [33]. In cases where a radical dissociates into a radical and a stable molecule, the reaction may be exothermic or thermoneutral (e.g., $HCOO \rightarrow H + CO_2$), which suggests negative or zero bond energies that are not meaningful. With this energy partitioning and BOC to unity the adsorption enthalpy should be given by Eq. 2.37; however, Shustorovich found that Eq. 2.37 gave adsorption bond energies that were too low [33]. Shustorovich suggested that radicals would adsorb at the highest coordination sites, and referred to this as "strong adsorption." BOC for "strong adsorption" is written as

$$\chi_{\mathrm{MA}} + \chi_{\mathrm{AB}} = 1 \tag{2.44}$$

This is a variation of Eq. 2.36 where the sum over the M–A bond orders over the metal atoms has been replaced by an effective bond order, x_{MA}, between the surface and atom A. The effective M_n–A bond energy for the Morse potential is then given by the total bond energy of A to the surface, $D(\mathrm{MA})$. Using this set of definitions, the adsorption enthalpy for radical species is given by [33]

$$\Delta H_{\mathrm{ad}}(\bullet\mathrm{AB}) = \frac{-[D(\mathrm{MA})]^2}{D(\mathrm{MA}) + Q_{\mathrm{ad}}} \tag{2.45}$$

Shustorovich applies Eq. 2.45 to "strong adsorption." The criteria used to distinguish "strong adsorption" from "weak adsorption" (described by Eq. 2.37) are based on the electronic structure of AB. If the A atom retains localized unpaired electrons, the adsorption bonding is assumed to be "strong," or "atomic-like," but otherwise the adsorption bonding is assumed to be "weak" [33].

BOC to atomic valence circumvents the need to distinguish between "strong" and "weak" adsorption. Formation of a bond between a radical and a metal surface is assumed to be analogous to radical recombination. The bond to the surface is not done at the expense of weakening other bonds in the molecule, and bond order is not assumed to be conserved during the adsorption process. This approach differs from

Shustorovich's approach by assuming the bond order can change during the adsorption process. The procedure for determining adsorption bond energies for radicals exactly parallels that outlined in Eqs. 2.39 and 2.40. The difference is that in the radical fragment AB_z there is no BOC about the A atom (the unpaired electron represents a decrease in bond order of one). A convenient though not necessary assumption is that the bond orders and bond energies of the other bonds to the A atom do not change much between molecular and radical species (e.g., the C–H bond order and bond energy in a methyl radical are approximately the same as in methane). The unsaturated bond in the radical results in much stronger adsorption bonding for radicals than for molecular adsorption because bonding to the surface occurs without decreasing the A–B bond energies. BOC to atomic valence treats both molecules and radicals by a single formalism.

E. Comparison of Estimated Adsorption Enthalpies for Molecules and Molecular Fragments

In the preceding two sections methodologies were presented to estimate enthalpies of formation for adsorbed species. The three approaches may be denoted as (1) bond additivity, (2) BOC to unity, and (3) BOC to atomic valence. All three were developed to apply to low coverages where adsorbate–adsorbate interactions may be neglected. To our knowledge no reliable thermodynamic data for adsorption of polyatomic species exist except when adsorption is very weak and reversible, such as for alkanes. To compare the three approaches the enthalpies of formation and adsorption bond energies of several small molecules and molecular fragments on nickel have been calculated. Nickel was chosen because reliable data exist for the energies of adsorption of carbon and oxygen. The results are summarized in Table 2.14.

How well do these approaches conform to experimental results? The bond additivity approach is the simplest one to implement. It predicts trends that are quite reasonable for $-CH_x$, giving values in good agreement with the other two methodologies; the values also agree well with calculated values obtained by the *ab initio* "bond preparation" method on nickel clusters [65]. Bond additivity gives low estimates for weakly adsorbed closed-shell systems, such as CH_4, CH_3OH, and H_2CO, where the bonding was assumed to arise only from dispersion forces. The BOC-MP techniques are better suited to estimate the effects of bonding in such systems. The bond additivity approach also fails to distinguish different structural isomers such as the monodentate and bidentate formate species because bond energies are assumed to be linearly proportional to bond order. The approach works well for most cases because adsorbates generally assume adsorption sites where the M–A bond orders are all less than or equal to one.

The advantage of the BOC-MP approaches is that they allow one to compare different structural variations, such as adsorption sites and structural isomers. The BOC to unity approach that Shustorovich advocates [33] cannot be used to compare adsorption of radicals (Eq. 2.45) in different adsorption sites. The approach of conserving bond order to unity works best for closed-shell systems. This is expected because in those systems the adsorption bond to the surface is formed at the expense of the other bonds in the molecule.

Table 2.14. Bond Energies and Enthalpies of Formation of Adsorbed Species on Nickel (kcal/mol)

Adsorbate	Eq. 223		Eq. 2.37 or 2.45		Eq. 2.40				$Q_{M\text{-}A}$[a] *Ab Initio*	
	$Q_{M\text{-}A}$[a]	ΔH_f[b]	$Q_{M\text{-}A}$[b]	ΔH_f[b]	$Q_{M\text{-}A}$[a]	ΔH_f[b]	$Q_{M\text{-}A}$[a] Experiment	Ref.	Calculation	Ref.
CH_4	4	−22	9	−27	8	−26	4	*d*		
$-CH_3$	42	−8	61	−27	42	−8 (on-top)				
					47	−13 (fourfold)			49	*c*
$>CH_2$	84	2	79	6	84	2 (bridge)				
					90	−4 (fourfold)			85	*c*
$>CH$	127	15	114	27	130	12 (fourfold)			112	*c*
CH_3OH	6	−54	16	−64	16	−64	15	*e*		
$-OCH_3$	56	−52	60	−56	48	−44 (on-top)				
					60	−56 (fourfold)				
H_2CO	6	−33	17	−44	17	−44				
···O＼C−H／O···	56	−92	87	−123	56	−92 (on-top)			46	*f*
$-O-C(=O)H$	56	−92	58	−94	48	−84 (on-top)				
$H_2C{=}CH_2$	16	−4	14	−1	15	−3	13	*g*		
H H / HC − CH / ／ ＼	84	−10			84	−10				
H H / − C = C / H	42	27	70	−1	42	27 (on-top)				
					47	22 (fourfold)				
$>C-CH_3$	27	13	110	4	130	10 (fourfold)				

[a] Q_{M-A} is the total bond energy with the nickel surface.
[b] ΔH_f is the enthalpy of formation from the elements in their standard states.
[c] Reference 65.
[d] Reference 67.
[e] Reference 68.
[f] Reference 69.
[g] Reference 70.

For closed-shell systems the BOC to atomic valence MP formalism reduces to the approach advocated by Shustorovich; however, in the case of radical adsorption the BOC to atomic valence method provides an intuitive way to retain the advantages of Shustorovich's BOC methodology and compare adsorption sites and structural isomers. Table 2.14 shows that predicted adsorption energies and enthalpies of formation from the BOC to atomic valence are very reasonable. The bond energies of the $-CH_x$ species are in excellent agreement with the *ab initio* values calculated by Siegbahn and Wahlgren [65]. It was also possible to handle both molecular adsorption and radical adsorption with the same formalism. Variation in adsorption energy with adsorption site is found to be modest, with the variation increasing with increasing total bond order to the surface. The BOC to atomic valence approach also gave a good representation of structural isomers. The bidentate formate was found to be 8 kcal/mol more stable than the monodentate form. (For comparison, the BOC to unity approach gives the value of 12 kcal/mol, as seen from Table 5.1 in Chapter 5.) Scharpf and Benziger found the barrier of conversion of the bidentate to monodentate acetate on Ni(111) to be approximately 12 kcal/mol, which agrees well with the predicted energy differences [66]. Also, the ethylidyne species is found to be thermodynamically more stable than vinyl, which agrees with experimental results on Pt (Pt and Ni are in the same group in the periodic table and show similar catalytic properties for hydrocarbon reactions) [71]. In using BOC to atomic valence it was assumed that each two-center bond could be described by a Morse potential. This worked well for the systems examined here, but further investigations are required to more clearly ascertain its applicability.

V. Thermodynamic Constraints for Surface Reactions

In the previous section procedures were devised to estimate the enthalpies and entropies of formation of surface species. These may be used to evaluate the thermodynamic feasibility of surface reactions. The basic thermodynamic constraint is whether a reaction is associated with a decrease in the free energy of the system. Thermodynamic constraints provide a useful guide to what reactions are feasible and where kinetic barriers exist. If a reaction is thermodynamically prohibited, it is prohibited absolutely. Whereas if a reaction is thermodynamically permissible and it is not observed, one concludes that a barrier exists for that reaction. In this section the thermodynamic constraints of reactions on metal surfaces are examined, focusing on periodic trends.

A. Adsorption of Diatomic Molecules

A key reaction in ammonia synthesis, methanation, and Fischer–Tropsch synthesis is the dissociative adsorption of diatomic molecules such as N_2 and CO. Dissociative adsorption of diatomic molecules such as CO, NO, and N_2 on transition metals has been shown to correlate with position in the periodic table [72]. These correlations were based on UHV adsorption studies where dissociative adsorption meant a molecule dissociated before it desorbed. These correlations may be accounted for from simple thermodynamic considerations [9].

Adsorption of diatomic molecules may be idealized as a competitive adsorption process between molecular and dissociative adsorption. Under the assumption of simple Langmuir adsorption of an AB molecule where adsorption sites may be occupied by either an A molecule, or an A atom, or a B atom, the fractional coverages of molecular (θ_M) and dissociated (θ_D) adsorbed species at pressure P are

$$\theta_M = \frac{K_M P}{1 + 2K_D^{1/2} \mathrm{P}^{1/2} + K_M P} \tag{2.46}$$

$$\theta_D = \frac{K_D^{1/2} \mathrm{P}^{1/2}}{1 + 2K_D^{1/2} P^{1/2} + K_M P} \tag{2.47}$$

The quantities K_M and K_D are equilibrium constants given by

$$K_M = e^{\Delta S_M/k}\, e^{-\Delta H_M/kT} \tag{2.48a}$$

$$K_D = e^{\Delta S_D/k}\, e^{-\Delta H_D/kT} \tag{2.48b}$$

The entropy changes are dominated by the loss of transitional entropy in going from a freely translating gas to an adsorbed state. The entropy changes per mole of adsorbed gas ΔS_M and ΔS_D are both approximately -36 cal/mol-K as given by Eq. 2.3. As dissociative adsorption results in the occupation of two adsorption sites, the entropy change per adsorption site for dissociative adsorption is half that for molecular adsorption.

The enthalpy of molecular adsorption of diatomic molecules may be approximated by Eq. 2.34. The enthalpy of dissociative adsorption is given by the difference in bond energy of two adsorbed atoms less the bond energy of the diatomic molecule:

$$\Delta H_D(\mathrm{AB}) = -\,Q_n(\mathrm{A}) - Q_n(\mathrm{B}) + D(\mathrm{AB}) \tag{2.49}$$

Dissociative adsorption of a diatomic molecule on a transition metal surface will be observed when the coverage of dissociated molecules given by Eq. 2.47 is of order one. This criterion requires $K_D P \gg 1$ and $K_D P^{1/2} \geq K_M P$, which can be translated into inequalities that the enthalpies of adsorption must satisfy.

$$\Delta H_D < T\Delta S + kT \ln P \tag{2.50a}$$

$$2\Delta H_M - \Delta H_D \geq T\Delta S + kT \ln P \tag{2.50b}$$

The first inequality must be satisfied to ensure that the pressure is high enough for appreciable adsorption. The second criterion requires that dissociative adsorption is thermodynamically favored to molecular adsorption. The second inequality can be combined with the first and the thermodynamic criterion for dissociative adsorption is given as $\Delta H_D < \Delta H_M$.

The requirement that dissociative adsorption is thermodynamically permissible only when the dissociative enthalpy of adsorption is less than the molecular enthalpy of adsorption has been previously identified by Benziger [9]. Based on the atomic adsorption energies in Table 2.5 the adsorption enthalpies can be determined using

Eqs. 2.34 and 2.49. The thermodynamic driving force for dissociation of N_2 and CO becomes less favorable as one moves down and to the right in the transition metal series. Among the group VIII metals only the first-row metals show favorable thermodynamics for dissociation.

The adsorption equilibria expressed in Eqs. 2.50a and 2.50b are more restrictive than simply stating $\Delta H_M > \Delta H_D$. At low substrate temperatures where the entropy contributions to the free energy are less significant, molecular adsorption is always favored unless the enthalpy for dissociative adsorption is less than twice the enthalpy for molecular adsorption. Adsorption equilibrium also indicates that molecular adsorption is preferred to dissociative adsorption with increasing pressure. The full thermodynamic criteria should be considered in evaluating the conditions for dissociation as molecular desorption may precede dissociation in a TPD experiment, but at higher pressures and temperatures dissociative adsorption may predominate.

B. Dehydrogenation of Hydrocarbons

The dehydrogenation reactions of simple alkenes, such as ethene, have been used as test reactions in TPR experiments to evaluate C–H bond activation kinetics. However, the desorption of dihydrogen from alkene decomposition may not be indicative of the kinetics of C–H bond activation; in many cases it may be thermodynamically limited. For the decomposition of ethene on a metal surface possible reaction products are ethyne or adsorbed carbon and dihydrogen:

$$C_2H_4 \leftrightarrow C_2H_2 + H_2$$

$$C_2H_4 \leftrightarrow 2C(a) + 2H_2$$

The thermodynamic equilibria in terms of partial pressures of the reacting species for the two reactions are given by

$$K_1 = \frac{P_{H_2} P_{C_2H_2}}{P_{C_2H_4}} = e^{-\Delta G_1^0/kT} \tag{2.51}$$

$$K_2 = \frac{P_{H_2}^2}{P_{C_2H_4}} = e^{-\Delta G_2^0/kT} \tag{2.52}$$

The free energies and enthalpies of formation of all the gas-phase species may be looked up or estimated [6,27]. For adsorbed carbon

$$\Delta G_f^0(C(a)) \approx \Delta H_f^0(C(a)) \approx -Q_n(C) + \Delta H_f^0(C(g)) \tag{2.53}$$

The thermodynamically constrained reaction selectivity for dehydrogenation to ethyne, S_1, and decomposition to adsorbed carbon, S_2, may be written as

$$S_1 = \frac{P_{C_2H_2}}{P_{C_2H_4}} = \frac{K_1}{P_{H_2}} \tag{2.54a}$$

$$S_2 = \frac{P_{H_2}}{P_{C_2H_4}} = \frac{K_2}{P_{H_2}} \tag{2.54b}$$

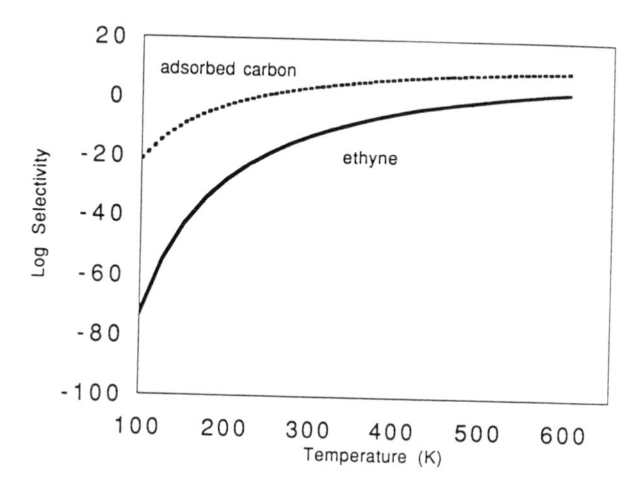

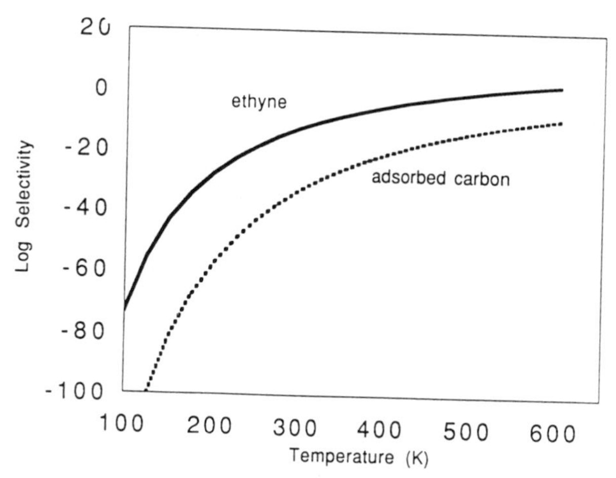

Figure 2.4. Equilibrium reaction selectivity for ethene decomposition to ethyne or adsorbed carbon at a hydrogen pressure of 10^{-} Torr according to Eqs. 2.54a and 2.54b as a function of metal–carbon adsorption bond energy $D(M-C)=150$ kcal/mol **(a)** or 125 kcal/mol **(b)**.

Figure 2.4 shows the reaction selectivity for ethene decomposition equilibria as a function of temperature at a hydrogen pressure of 10^{-9} Torr (typical UHV conditions for a TPR experiment). Comparing Figures 2.4a and 2.4b one sees that the metal–carbon bond energy must be less than 125 kcal/mol for ethyne to be a thermodynamically permissible product. This limits the choice of metals for selective dehydrogenation of ethene to the far right-hand side of the transition series. It is also necessary that S_1 be greater than one ($\log S_1 > 0$) so ethyne is formed before ethene desorbs. This requires the metal–carbon bond to be stronger than 100 kcal/mol. The metals that meet these two requirements are Pd and Pt, which are those used for ethene dehydrogenation.

C. Reaction Pathways for Alcohols and Carboxylic Acids

The thermodynamic constraints for surface reactions of a wide variety of species can be handled in a fashion similar to that outlined above for ethene. An easier way to evaluate periodic trends in surface reactions is simply to evaluate the enthalpy of the reaction. Entropies of surface reactions are generally negligible, so the free energy will be nearly equal to the enthalpy of reaction. Exothermic reactions are thus thermodynamically allowed, whereas endothermic reactions are prohibited at low to moderate temperatures. Comparing enthalpies of surface reactions has previously been proposed by Benziger and Madix to explain trends in the reactivity of oxygenates on metal surfaces [64].

The reactions observed for methanol on metal surfaces in TPR experiments are

$$CH_3OH\,(g) \xrightarrow{1} H(a) + CH_3O(a) \begin{cases} \xrightarrow{3} H_2CO(a) + H(a) \\ \xrightarrow{4} CO(a) + 3H(a) \\ \xrightarrow{5} CH_3(a) + O(a) \end{cases}$$

$$+$$

$$OH(a)$$

$$\uparrow \mathbf{b}$$

$$O(a) + CH_3OH\,(g)$$

On the group VI and VIII metals methanol was found to dissociatively adsorb on the clean metal to form an adsorbed methoxy and adsorbed hydrogen (reaction **1**) [68,73,74]. In contrast, group IB metals do not dissociatively adsorb methanol on a clean surface, but methanol reacts with preadsorbed oxygen to form adsorbed hydroxyl and methoxy (reaction **2**) [75,76]. The adsorbed methoxy reacts differently on the metals as well. On group IB metals, the methoxy yields formaldehyde as the primary product (reaction **3**) [75,76]. On most group VIII metals, CO and hydrogen are the primary products (reaction **4**), and on Fe and group VI metals methane and adsorbed oxygen are observed (reaction **5**) in comparable yields to CO and hydrogen [68,73,74]. The enthalpies for the five reactions observed in TPR of methanol on W, Fe, Ni, and Ag are compared in Table 2.15. It is apparent that only the most exothermic reactions have been observed.

The thermodynamic analysis provides insight into the differences in metal reactivity. The group IB metals form weak M–O and M–H bonds, so the formation of an alkoxy and adsorbed hydrogen is prohibited; however, preadsorbed oxygen permits the coupling of formation of a strong O–H bond in an adsorbed hydroxyl to the breaking of the O–H bond in the alcohol, which results in the formation of the methoxy being thermodynamically favorable. On group IB metals the methoxy decomposition stops at adsorbed formaldehyde because the formation of M–H bonds is not sufficient to offset breaking of the C–H bonds. In contrast, the M–H bonds are stronger for Fe, Ni, and W and provide a driving force sufficient to break the C–H bonds leading to the decomposition of adsorbed methoxy to adsorbed CO and hydrogen. Methane formation on Fe and W is a consequence of the strong M–O and M–C bonds that results in stable oxygen and methyl adsorption.

Table 2.15. Reaction Enthalpies, ΔH (kcal/mol), for Some Reaction Paths of Methanol on Metal Surfaces

Reaction	Metal			
	Fe	Ni	W	Ag
$CH_3OH(g) \rightarrow CH_3O(a) + H(a)$	-23^a	-20^a	-35^a	13
$CH_3OH(g) + O(a) \rightarrow CH_3O(a) + OH(a)$	-7	-7	-7	-7^a
$CH_3O(a) \rightarrow H_2CO(g) + H(a)$	11	11	6	-12^a
$CH_3O(a) \rightarrow CO(a) + 3H(a)$	-19^a	-16^a	-25^a	5
$CH_3O(a) \rightarrow CH_3(a) + O(a)$	-19^a	-12	-50^a	29

[a]The reaction has been experimentally observed by temperature-programmed reaction.

Shown below are the reaction pathways for reaction of formic acid on metal surfaces during TPR.

$$HCOOH(g) \xrightarrow{6} H(a) + HCOO(a) \begin{cases} \xrightarrow{8} H(a) + CO_2(g) \\ \xrightarrow{9} H(a) + CO(g) + O(a) \end{cases}$$

$$+\ OH(a) \xleftarrow{7} HCOOH(g) + O(a)$$

Surface formate is formed from formic acid adsorption on all transition metals other than Ag and Au (reaction **6**). On Ag with preadsorbed oxygen, formic acid adsorbs to form adsorbed formate and adsorbed hydroxyl (reaction **7**) [81]. On the group IB metals and most group VIII metals, adsorbed formate decomposes primarily to CO_2 and adsorbed hydrogen (reaction **8**) [77,79–81]. On the earlier transition metals, formate decomposes to CO and adsorbed oxygen and hydrogen or it is completely decomposed to adsorbed atomic constituents (reaction **9**) [78]. Table 2.16 compares the enthalpies of reaction for formic acid on metal surfaces. Again, the experimentally observed reactions in TPR are the most exothermic.

The decomposition of formic acid on metals is a favorite test reaction for catalytic activity. A classic correlation in catalysis by metals is the "volcano plot" relating the catalytic activity for formic acid decomposition to CO_2 and H_2 with the heat of

Table 2.16. Reaction Enthalpies, ΔH(kcal/mol), for Some Reaction Paths of Formic Acid on Metal Surfaces

Reaction	Metal			
	Fe	Ni	W	Ag
$HCOOH(g) \rightarrow HCOO(a) + H(a)$	-26^a	-26^a	-44^a	12
$HCOOH(g) + O(a) \rightarrow HCOO(a) + OH(a)$	-8	-8	-8	-8^a
$HCOO(a) \rightarrow CO_2(g) + H(a)$	-4^a	-4^a	8	-24^a
$HCOO(a) \rightarrow CO(a) + H(a) + O(a)$	-9^a	-9^a	-24^a	32

[a]The reaction has been experimentally observed by temperature-programmed reaction.

formation of metal formates or metal oxides [82–84]. The maximum rate occurs at an intermediate value of the heat of formation of the metal formate of approximately −80 to −84 kcal/mol [84]. This maximum rate can be explained based on the reaction sequence.

$$HCOOH(g) \rightarrow HCOO(a) + H(a) \rightarrow CO_2(g) + H_2(g)$$

The overall rate will be maximized when both reaction steps are thermodynamically favorable, which requires

$$\Delta H_f(HCOOH(g)) < \Delta H_f(HCOO(a)) + \Delta H_f(H(a)) < \Delta H_f(CO_2(g)) + \Delta H_f(H_2(g)) \quad (2.55a)$$

$$-87 < \Delta H_f(HCOO(a)) + \Delta H_f(H(a)) < -94 \text{ kcal/mol} \quad (2.55b)$$

From Table 2.2, $\Delta H_f(H(a)) \sim -10 \pm 2$ kcal/mol for most transition metals, giving the result

$$-76 < \Delta H_f(HCOO(a)) < -83 \text{ kcal/mol} \quad (2.56)$$

in agreement with the empirical correlation. Furthermore, we can use Eq. 2.42 to estimate $\Delta H_f(HCOO(a))$ to demonstrate that the group VIII metals satisfy this set of inequalities and give the maximum rates. For the group IB metals the left-hand inequality is not satisfied so that adsorption is the rate-limiting step of reaction. On transition metals to the left of the group VIII metals the right-hand inequality is not satisfied, indicating that decomposition is thermodynamically restricted. (For a detailed BOC-MP analysis of CH_3OH and HCOOH decomposition on metal surfaces, the reader is referred to Chapter 5.)

VI. Activation Barriers for Surface Reactions

So far procedures have been outlined for estimating bond energies for surface intermediates and the thermodynamics of surface reactions. The BOC-MP formalism may also be used to estimate activation barriers for surface reactions. A transition state is hypothesized and the BOC method is used to determine the energy of the transition state. In this section activation barriers for surface diffusion and diatomic dissociation are determined within the BOC-MP formalism.

A. Diffusional Barriers

The BOC-MP method gives a very simple straightforward technique to determine activation barriers for surface diffusion. Diffusion requires the movement of an adsorbate from an n-fold coordination site to an adjacent site. The lowest energy path between these sites is to pass a bridge site between the two n-fold sites. For atomic adsorption Shustorovich [29] derived an expression for diffusion at low adsorbate coverages.

$$E_{\text{dif}} = Q_{\text{MA}}\left(\frac{1}{2} - \frac{1}{n}\right) = D(\text{MA})\,\frac{n-2}{4n-2} \tag{2.57}$$

Atomic diffusional barriers based on the BOC analysis vary within 10 to 17% of the adsorption bond energy, where the barrier increases as the coordination number of the adsorption site increases, in good agreement with experimental observations [85].

If BOC to atomic valence is imposed, Eq. 2.57 is valid provided the valence bonds of adatoms in bridge sites do not exceed the bonding capacity of the metal atoms. This can be a problem with nitrogen and carbon atom diffusion on group IB and some group VIII metals. On Cu(100) the excess metal valence is 0.88, so that in the fourfold coordination site the valence on the nitrogen adatom is satisfied, but in a bridge site the nitrogen atom valence cannot be satisfied. This results in a diffusional barrier larger than that given by Eq. 2.57. The barrier can be calculated from the adsorption energy difference using Eqs. 2.13 and 2.20. Equation 2.20 accounts for the weaker adatom adsorption in the bridge site resulting from unsaturated bonding.

Diffusional barriers for molecular species may also be approximated as the difference in bond energy between the n-fold coordination site and a twofold (bridge) site. The bond order to the surface for molecular species is less than that for atomic species and thus the diffusional barrier is less. For example, the calculated barrier for CO diffusion on Ni(100) is 5 kcal/mol, which may be compared to barriers of 20 kcal/mol for oxygen and 25 kcal/mol for carbon on the same surface. The diffusional barriers calculated by this approach represent an upper limit to the diffusional barriers at low coverages. It has previously been mentioned that bonding interactions between donor and acceptor orbitals of the adsorbate and the metal tend to make the differences in adsorption enthalpy between sites smaller than predicted by the BOC-MP formalism [29]. This homogenization of the surface will result in smaller diffusional barriers than predicted. (For a detailed BOC-MP analysis of diffusional barriers, the reader is referred to Chapter 3.)

B. Dissociation of Diatomic Molecules

In Section VA the thermodynamic criterion for dissociation of diatomic molecules was presented. Even if dissociation is permissible thermodynamically, dissociation does not always occur and frequently must be thermally activated. For example, CO adsorption on tungsten surfaces is shown to be molecular at 100 K, and dissociation does not occur until the surface temperature exceeds 250 K [20,21]. The reverse reaction, recombination, is also affected by the barrier for dissociation. This explains why N_2, which is thermodynamically restricted from dissociating on copper surfaces, has a large activation barrier for recombination [86].

Shustorovich treated the dissociation of a diatomic molecule with bond order conserved to unity.

$$\chi_{\text{MA}} + \chi_{\text{MB}} + \chi_{\text{AB}} = 1 \tag{2.58}$$

Within the one-dimensional Lennard–Jones potential energy diagram the transition state corresponds to $\chi_{\text{AB}}=0$, so the dissociation barrier was taken as the intersection of the Morse potential curves for A+B+M and AB+M [31b]. In this one-dimensional representation the barrier for dissociation is given by

$$\Delta E_{\mathrm{dis}}(\mathrm{AB_g}) = D(\mathrm{AB}) - (D(\mathrm{MA}) + D(\mathrm{MB})) + \frac{D(\mathrm{MA})D(\mathrm{MB})}{D(\mathrm{MA}) + D(\mathrm{MB})} \tag{2.59}$$

where the D's refer to total bond energies [80]. For the more realistic multidimensional potential energy diagram, χ_{AB} at the transition state may be nonzero, in which case Eq. 2.59 gives an upper limit to the dissociation barrier. To account for the finite bond order of the AB bond at the transition state, Shustorovich interpolated between the one-dimensional potential energy curve representing an upper limit and a lower limit of a chemisorbed molecule [33], the result being

$$\Delta E_{\mathrm{dis}}(\mathrm{AB_g}) = \frac{1}{2}\left\{ D(\mathrm{AB}) - (D(\mathrm{MA}) + D(\mathrm{MB}) + Q(\mathrm{AB})) + \frac{D(\mathrm{MA})D(\mathrm{MB})}{D(\mathrm{MA}) + D(\mathrm{MB})} \right\} \tag{2.60}$$

Equation 2.60 refers to the direct dissociative adsorption from a gas-phase AB molecule. The dissociation barrier of an adsorbed AB molecule is larger than the barrier of direct dissociation by the adsorption bond energy of $\mathrm{AB_{ad}}$.

$$\Delta E_{\mathrm{dis}}(\mathrm{AB_{ad}}) = Q(\mathrm{AB}) + \Delta E_{\mathrm{dis}}(\mathrm{AB_g}) \tag{2.61}$$

An alternative approach to that outlined above is to postulate adsorbed configurations and use the BOC-MP formalism to calculate the bond energies and thus the barriers for dissociation. Experimentally the most relevant situation is the dissociation of adsorbed molecular species, as this is generally the only case where the AB molecule is in thermal equilibrium with the surface. If the molecule is thermally accommodated to the surface it will always go through a molecularly adsorbed precursor state before dissociation. Within the BOC framework, molecular adsorption of diatomics is expected to be nonactivated. The only barriers that exist are small (<5 kcal/mol) diffusional barriers from one adsorption site to another. The initial state of the system prior to dissociation is taken to be the AB molecule linearly coordinated in an n-fold coordination site. The final state of the system is two adsorbed atoms in separated n-fold coordination sites. To get between these states the molecule will pass through a transition state where the A atom is adsorbed in the n-fold coordination site and the B atom is in a bridge site between two n-fold coordination sites. This represents the highest point on the lowest energy trajectory. The sequence of states for A–B on a (111) surface is shown in Figure 2.5. The postulated sequence is a linearly adsorbed AB molecule in an n-fold site (configuration I), passing through a transition state where the A atom remains coordinated in the n-fold site and the B atom is in a twofold (bridge) site (configuration II). At the transition state the A–B bond order is assumed to be zero. The two atoms then move apart where first they are in adjacent n-fold sites (configuration III) and the final state is two separated n-fold sites (configuration IV).

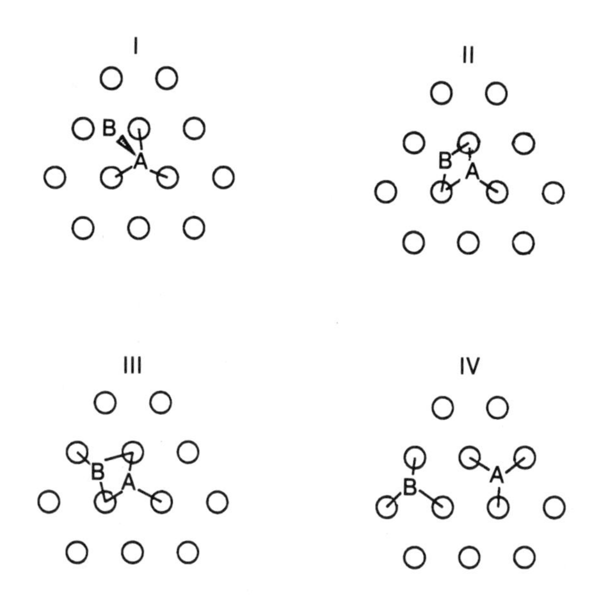

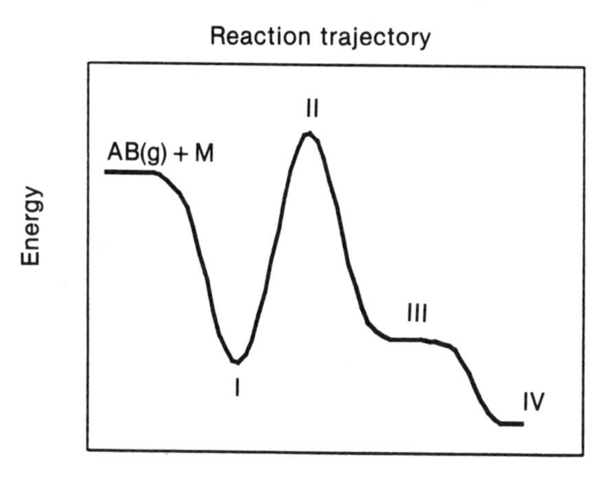

Figure 2.5. Postulated reaction trajectory for diatomic, AB, dissociation on a fcc (111) surface: **(I)** AB linearly adsorbed to a threefold hollow site; **(II)** postulated transition site, A atom in a threefold site and B atom in a twofold bridge site; **(III)** A and B atoms in adjacent threefold hollow sites; **(IV)** A and B atoms in nonadjacent threefold hollow sites.

The BOC approach can be used to determine the energies of the configurations shown in Figure 2.5. To simplify matters we evaluate the barrier in the zero coverage limit. The energy of the AB molecule adsorbed in the n-fold site is given by Eq. 2.26, where the AB bond order is given by

$$\phi_{\mathrm{AB}} = V_{\mathrm{A}} \frac{Q_{\mathrm{AM}} + n(Q_{\mathrm{AB}} - Q_{\mathrm{MA}})}{Q_{\mathrm{MA}} + nQ_{\mathrm{AB}}} \qquad (2.62)$$

At the barrier for dissociation, configuration II, the total energy is that for an A atom adsorbed in an n-fold site and a B atom in a bridge site. It is assumed that at this point the AB bond order is zero. The energy for configuration II is given by Eq. 2.63. The three terms in Eq. 2.63 represent (1) the M–A interaction when the metal interacts only with an A atom, with two-center bond energy Q_{MA}; (2) the M–A interaction when the metal bonds to both the A atom in the n-fold site and the B atom in the twofold site, with two-center bond energy Q'_{MA}; and (3) the M–B interaction when the metal atoms are bound to both the A atom and the B atom, with two-center bond energy Q'_{MB}. The two-center bond energies Q'_{MA} and Q'_{MB} depend on the metal valence, with the two possibilities given by Eq. 2.64 accounting for whether the excess metal valence is sufficient to bond both the A and B atoms (Eq. 2.19).

$$Q_{\mathrm{II}} = (n-2)\, Q_{\mathrm{MA}} \left(\frac{2}{n} - \frac{1}{n^2}\right) + 2\, Q'_{\mathrm{MA}} \left(\frac{2}{n} - \frac{1}{n^2}\right) + \frac{3}{2}\, Q'_{\mathrm{MB}} \qquad (2.63)$$

where

$$Q'_{\mathrm{M}i} = \begin{cases} Q_{\mathrm{M}i} & \text{if } V_{\mathrm{M,ex}} > \dfrac{V_{\mathrm{A}}}{n} + \dfrac{V_{\mathrm{B}}}{2} \\[2ex] Q_{\mathrm{M}i} \left\{ \dfrac{V_{\mathrm{M,ex}}}{V_{\mathrm{A}}/n + V_{\mathrm{B}}/2} \right\} & \text{if } V_{\mathrm{M,ex}} < \dfrac{V_{\mathrm{A}}}{n} + \dfrac{V_{\mathrm{B}}}{2} \end{cases} \qquad (2.64)$$

Lastly, the energies for A and B in adjacent n-coordination sites and separated n-coordination sites are given by Eqs. 2.65 and 2.67, respectively. The two-center bond energies Q''_{MA} and Q''_{MB} must account for the possibility that the excess metal valence is not sufficient to bond A and B atoms in adjacent sites. These two cases are described in Eq. 2.66.

$$Q_{\mathrm{m}} = (n-2)\left(\frac{2}{n} - \frac{1}{n^2}\right)(Q_{\mathrm{MA}} + Q_{\mathrm{MB}}) + 2(Q''_{\mathrm{MA}} + Q''_{\mathrm{MB}})\left(\frac{2}{n} - \frac{1}{n^2}\right) \qquad (2.65)$$

where

$$Q''_{\mathrm{M}i} = \begin{cases} Q_{\mathrm{M}i} & \text{if } V_{\mathrm{M,ex}} > \dfrac{V_{\mathrm{A}} + V_{\mathrm{B}}}{n} \\[2ex] Q_{\mathrm{M}i} \left\{ \dfrac{V_{\mathrm{M,ex}}}{V_{\mathrm{A}}/n + V_{\mathrm{B}}/n} \right\} & \text{if } V_{\mathrm{M,ex}} < \dfrac{V_{\mathrm{A}} + V_{\mathrm{B}}}{n} \end{cases} \qquad (2.66)$$

$$Q_{IV} = (Q_{MA} + Q_{MB})\left(2 - \frac{1}{n}\right) \tag{2.67}$$

The activation barrier for dissociation of AB is given by the transition from adsorbed AB to the transition state, configuration II.

$$\Delta E_{dis}(AB_{ad}) = Q_I - Q_{II} \tag{2.68}$$

The barrier for recombination is given by the transition from two separated adatoms to configuration II.

$$\Delta E_{rec} = Q_{IV} - Q_{II} \tag{2.69}$$

The expressions for the bonding energies derived above may be used with the MA adsorption bond energies to estimate the barriers for dissociation and recombination (the atomic adsorption bond energies in Eqs. 2.62–2.65 refer to the two-center bond energy).

The activation energy for desorption is equal to the barrier for recombination or the bond energy of dissociative adsorption, Q_{IV}, whichever is greater. This alternative approach takes more explicit account of surface structure and provides a physical picture of the transition state for dissociation, which is helpful in visualizing the dissociation process. Of course, critical here is how reasonable are the postulated structures, I–IV.

Some selected results for the barriers for dissociation are compared with experimental values in Table 2.17. The adsorption enthalpies are the same for both BOC-MP approaches. Dissociation is expected when the activation barrier for dissociation is less than $-\Delta H_m$. When barriers are negative they correspond to a zero barrier for dissociation (this means that the assumed transition state has a lower energy than the molecularly adsorbed state). At sufficiently low temperatures experimental results have always found molecular adsorption with no dissociation. The absence of dissociation at low temperature could result from unfavorable thermodynamics rather than a finite barrier for dissociation.

Both approaches give reasonable qualitative predictions, which is all that one should expect as the experimental values in many cases are not very accurate. Some of the experimental values correspond to very low preexponential factors, suggesting that the dissociation reaction occurs at defects. Dissociation and molecular desorption of CO on metal surfaces have generally been assumed to be first-order reaction processes. The preexponential factor for CO dissociation on nickel is 10^{-6} lower than the preexponential factor for molecular desorption [92], whereas on tungsten surfaces the desorption and dissociation preexponential factors are nearly the same [20]. This suggests that the dissociation of CO on nickel may be occurring at defects that have a low concentration. Similar effects may be important for N_2 dissociation on iron surfaces [22].

The approach proposed by Shustorovich [33] (Eq. 2.61) gives very reasonable estimates of the dissociation barriers. The approximation has the advantage of requiring only thermodynamic data. The alternative approach introduced explicitly considers the transition state and gives the same qualitative trends, but suggests

Table 2.17. Energetics of Diatomic Adsorption and Dissociation on Metal Surfaces (kcal/mol)

Adsorption System	$-\Delta H_M$ Eq. 2.26	$-\Delta H_D$ Eq. 2.67	ΔE_{dis}			Ref.
			Eq. 2.61	Eq. 2.68	Experiment	
H_2/Ni(100)	11.6	23.0	10	−2.4	0	17b
H_2/Ni(111)	11.3	17.0	11	0.3	0	17b
H_2/Cu(100)	9.2	9.0	15	8.2	10	87
H_2/Cu(111)	9.0	3.7	15	10.7		
O_2/Ag(100)	13.6	28.0	−1	18.7	8	19
O_2/Ni(100)	30.3	105.0	−7	−41.4		
O_2/Pt(111)	19.1	51.0	6	14.9	10	88
O_2/Fe(100)	37.8	133.0	−15	−77.2		
O_2/W(100)	40.8	143.5	−18	−83.9		
O_2/Ir(111)	22.0	64.3	2	14.0	8	89
N_2/W(100)	30.6	78.5	26	−3.2	10	90
N_2/Fe(100)	26.0	54.0	21	41.0	20	22
N_2/Fe(110)	25.3	40.7	20	102.4	25	22
N_2/Ni(100)	25.4	50.5	22	71.4		
N_2/Re(0001)	33.0	62.0	21.5	87.1	4	91
CO/W(100)	76.4	136.8	15	−41.6		
CO/W(110)	73.3	118.0	15	67.5	12	20,21
CO/Ni(100)	32.8	23.0	37	86.1	20	92
CO/Ni(111)	31.9	9.7	37	148.7		
CO/Fe(100)	32.8	26.5	35	53.0	20	93

much greater variations in the dissociation barriers. The magnitude of the variations is exaggerated by assuming a frozen metal lattice and $\phi_{AB}=0$ at the transition state. Shustorovich's approach uses thermodynamic quantities that factor the surface relaxation and the A–B bonding at the transition state into the approximation so the barriers are not as exaggerated.

Several generalizations can be made from the application of this analysis.

1. The barrier for H_2 dissociation is predicted to be small or nonexistent on the transition metals, except for the group IB metals. This is consistent with experimental findings for hydrogen adsorption [17,87]. The predicted barriers are probably low. The H–H bond distance is much smaller than the M–M distances. This means that the H–H bond distance in configuration II is probably less than those corresponding to the n-fold site and the bridge site. If this were taken into account, the bond energy of configuration II would be reduced, increasing the barrier for dissociation.
2. The barrier for O_2 dissociation is predicted to be negligible except for Ag surfaces and Pt(111). Experimentally, O_2 dissociation on Pt and Ag has been found to be activated [19,88]. This is due to both weak metal oxygen bonding and reduced bonding capacity of surface metal atoms at the far right of the transition metal series.
3. The barrier for N_2 dissociation is predicted to be high on Ni and Fe. Also, the barrier for dissociation is predicted to be greater on Fe(110) than on Fe(100).

Both of these conform to experimental findings [20,51]. On tungsten, N_2 dissociation has a high barrier on (110) surfaces and a much lower barrier on (100) surfaces [90]. These differences in the barrier for dissociation arise as a result of the different bonding capacities of the metal atoms on the (110) and (100) surfaces.

4. CO dissociation has a lower barrier on tungsten than iron [20,93], consistent with the model predictions. The barriers on Ni(100) and Ni(111) are predicted to be significantly greater than the bond energy for molecular adsorption, so dissociation is restricted. The barrier for dissociation of CO on tungsten is predicted to be lower on the (100) than the (110) surface.

The model identifies two important factors for molecular dissociation on metal surfaces, the atomic adsorption bond energies and the metal valence. As the bond energies for the M–A and M–B bonds increase, the barrier for dissociation decreases. This was previously shown by Shustorovich [33], and is evident from Eqs. 2.59–2.61. Hence dissociation barriers on Fe surfaces are greater than those on crystallographically similar tungsten surfaces.

Metal valence is important because of limitations in accommodating adsorbed atoms in adjacent sites on a metal surface. The barrier for N_2 dissociation is much greater on Ni(100) than W(100). In addition to the lower binding energy of N on Ni, the lower excess metal valence at the Ni surface also contributes to the larger barrier for dissociation. On a Ni(100) surface, each metal atom has an excess valence of 4/3 (see Tables 2.7 and 2.8). In the N_2 dissociation transition state, the metal valence is inadequate to fully bond the two nitrogen atoms in the transition state. This results in the transition state having very weak bonding and consequently a large barrier for dissociation. In contrast, on the W(100) surface the metal atoms have an excess valence of 3 (see Tables 2.7 and 2.8) and can accommodate the transition state. The W(110) surface metal atoms have an excess valence of 1.5; hence they do not fully accommodate the nitrogen valence in the transition state, resulting in a larger barrier. The low or negligible barriers for H_2 and O_2 dissociation result from the lower valence of these atoms being more readily accommodated by the excess metal valence of the surface.

For their simplicity, the BOC approaches do well at predicting the trends for dissociation barriers. They account for periodic variations in the dissociation barriers as well as for structural variations. With improved data concerning bond lengths it may be possible to extend this model to give a better structural model of the dissociation process. It would not be necessary to limit the analysis to the bond energy/bond order correlation; in particular, bond lengths could be explicitly accounted for.

The BOC-MP analysis could also be adapted to predict activation barriers for surface reactions of polyatomic molecules. The approach would be similar to that suggested above for dissociation of diatomic molecules; a transition state is hypothesized and the energy differences between different states could be calculated. As the trends for diatomic dissociation barriers are only of limited utility, it would appear that predictions made by a similar approach for molecules will be primarily qualitative. Shustorovich and Bell have extended the BOC at unity approach to catalytic systems [32,33], and the reader is referred to their work and Chapter 5 to judge the value of this approach to more complex systems.

VII. Conclusions

Adsorption and reaction at metal surfaces are truly a complex problem for which a complete theory based on first principles is a long way off. In the interim it is helpful to develop phenomenological approaches to describe adsorption and reaction. Ideally, one desires a simple model that can extrapolate from known physical properties of materials to codify and predict the behavior for unknown systems. The concept of bond additivity for bond energies is a well-known simple concept that has been used to rationalize the formation of many chemical compounds and the energetics of gas-phase reactions. Also, the concept is useful as a first approximation to surface reaction energetics. As a quantitative tool, the approach is limited because it does not account for the structure of the surface.

To account for structural variations it is necessary to use potential functions describing explicitly the interaction between atoms in molecules. Shustorovich pointed out that the simplest potential including both attractive and repulsive forces was the Morse potential [29,30]. The bond order may be equated to an exponential function of bond distance given in Eq. 2.10; however, the analysis in general can be handled only with knowledge of bond order χ and an empirical energy parameter Q (Eq. 2.11).

In treating adsorption and reaction Shustorovich assumed bond order was conserved and normalized to unity. This is analogous to the BOC concept used to treat three-center reactions in the gas phase [44]

$$\text{A}-\text{B} + \text{C} \leftrightarrow \text{A---B---C} \leftrightarrow \text{A} + \text{B}-\text{C}$$

The energy parameter Shustorovich introduced was the atomic adsorption energy at zero coverage [29]. Shustorovich did not explicitly consider any effects of the adsorbate bonding on metal–metal bonding. Such effects are implicitly included through the experimental heats of adsorption. Shustorovich's model does very well for atomic adsorption at low coverages and is a distinct improvement over the bond additivity approach when addressing issues relating to surface structure. Uncertainties with Shustorovich's approach arise when considering diatomic and especially polyatomic adsorbates when one must decide how to partition the total molecular bond energy into two-center fragments. A purely thermodynamic approach was proposed by Shustorovich [33] that handles the majority of problems satisfactorily.

In the treatment of gas-phase reactions the bond energy has been expressed as a power function of the bond order, Eq. 2.16. The activation barrier for the A---B---C transition state requires $p > 1$, theoretical estimates are in the range $p = 1.05$–1.10 [44]. This approach was used to treat adsorption and reaction on Pt surfaces by Weinberg and Merrill [43]. The problem with Eq. 2.16, as pointed out by Shustorovich [29], is that for $p > 1$ it predicts that the bond energy increases greater than linearly with bond order for $\chi > 1$, which contradicts known observations, in particular that $\frac{1}{3}D(\text{C}\equiv\text{C}) < \frac{1}{2}\text{D}(\text{C}=\text{C}) < \text{D}(\text{C}-\text{C})$.

Table 2.18. Carbon–Carbon Bond Energies (kcal/mol)

Bond Type	Q_{expt}[a]	Q^b Eq. 2.18
C–C (ethane)	84	86
C=C (ethene)	145	147
C≡C (ethyne)	194	184

[a]From Ref. 94. Bond energies estimated from total bond energy partitioning assuming C–H bond energies are 98 kcal/mol, regardless of the hybridization state of C.
[b]Assuming the C–C two-center bond energy Q_{CC} = 196 kcal/mol (see Table 2.13).

A combined approach presented here preserves the normal sense of bond order in the Lewis–Pauling sense of shared electron pairs, and uses a Morse-type potential function because of its principal advantage over one-term power functions. This led us to propose a modified potential function based on atomic valence and the bond order, Eqs. 2.17 and 2.18. For a monovalent species such as hydrogen or a halogen this reduces to the Morse potential used by Shustorovich; however, multivalent atoms are handled differently with results that are consistent with the intuitive concept of the electron-pair bond. For example, the strength of carbon–carbon bonds predicted by the modified potential where the two-center bond energy, Q_{CC}, is scaled from diamond may be compared with experimental results, as shown in Table 2.18. The trend in bond energy is remarkably good for such a simple model. The principal advantage of introducing the concept of atomic valence along with BOC is that it provided a convenient way to deal with energy partitioning in polyatomic systems.

A new concept introduced here was the idea of metal valence and excess valence at the metal surface, which is similar to unsaturated bonds at the surface of a covalent crystal. The idea of metal valence has not been used much except for ideas discussed by Pauling [46,47]. It was found to be useful in understanding the capacity of a surface to adsorb species. The barriers for dissociation of diatomics could also be linked to excess metal valence. The concept of metal valence is one that deserves more attention to clarify its application and quantification.

As an instructive tool to understand adsorption and reaction, the phenomenology of Shustorovich [29–33] and that presented here is very useful. It is both simple and intuitive. The natural question to ask is, Where to go from here? The valence BOC approach outlined here is, in a sense, a simple "molecular mechanics"-type model [95–97]. Such approaches allow one to address issues of structure and reactions in large-scale systems that are not amenable to quantum mechanical treatments. The lack of structural data has restricted the use of the Morse potential for bond-length variations. The availability of more experimental data would make the BOC-MP method more quantitative. Potential functions that reflect angular variations could also be introduced to better describe the interactions of molecules with metal surfaces.

Perhaps the greatest limitation to a predictive model for surface reaction energetics is the very limited database of known bond energies and structures. An approximation was suggested relating surface adsorption bonds to bonding in bulk compounds. Although the correlation between the bulk and surface bonding is a reasonable first approximation, it does not fully recognize the differences in coordination that exist in the two cases. Development of improved approaches to predict reaction energetics at metal surfaces will require good experimental measurements of thermodynamic properties of adsorption. No models can be adequately evaluated without such information.

References

1. G. C. Schatz, *Rev. Mod. Phys.* **61** (1989) 669.
2. S. Glasstone, K. J. Laidler, and H. Eyring, *The Theory of Rate Processes*, McGraw-Hill, New York, 1941.
3. H. Eyring, J. Walter, and G. E. Kimball, *Quantum Chemistry*, Wiley, New York, 1944.
4. (a) H. Eyring, *J. Chem. Phys.* **3** (1935) 107. (b) W. F. K. Wynne-Jones, and H. Eyring, *J. Chem. Phys.* **3**(1935) 492.
5. (a) M. G. Evans, and M. Polanyi, (a) *Trans. Faraday Soc.* **33**(1937), 448; (b) *Trans. Faraday Soc.* **33**(1937), 448.
6. S. W. Benson, *Thermochemical Kinetics*, Wiley, New York, 1976.
7. (a) M. Boudart, and G. Djega-Mariadassou, *Kinetics of Heterogeneous Catalytic Reactions*, Princeton Univ. Press, Princeton, N.J., 1984. (b) G. F. Froment, and K. B. Bischoff, *Chemical Reactor Analysis and Design*, Wiley, New York, 1979.
8. T. L. Hill, *An Introduction to Statistical Thermodynamics*, Addison-Wesley, Menlo Park, Calif., 1960.
9. J. B. Benziger, *Appl. Surf. Sci.* **6** (1980), 105.
10. J. C. Tracy, *J. Chem. Phys.* **56** (1972), 2736.
11. (a) K. Christman, O. Schober, and G. Ertl, *J. Chem. Phys.* **60** (1974), 4719. (b) H. Ibach, W. Erley, and H. Wagner, *Surf. Sci.* **92** (1980) 29.
12. (a) T. E. Madey, and D. Menzel, *Jpn. J. Appl. Phys.* (1974), Suppl. 2, 229. (b) J. C. Tracy and P. W. Palmberg, *Surf. Sci.* **14** (1969), 274. (c) H. Pfnür, P. Feulner, H. A. Engelhardt, and D. Menzel, *Chem. Phys. Lett.* **59** (1978), 481.
13. K. Christman, O. Schober, G. Ertl, and M. Newmann, *J. Chem. Phys.* **60** (1974), 4528.
14. P. A. Redhead, *Vacuum* **12** (1962), 203.
15. K. D. Gibson, and L. H. Dubois, *Surf. Sci.* **233** (1990), 59.
16. J. C. Tully, and M. J. Cardillo, *Science* **223** (1984), 445.
17. (a) M. A. Morris, M. Bowker, and D. A. King, in *Chemical Kinetics: Simple Processes at the Gas-Solid Interface* (C. H. Bamford, C. F. H. Tipper, and R. G. Compton, eds.), Elsevier, Amsterdam, 1984. (b) K. Christmann, *Surf. Sci. Rep.* **9** (1988), 1.
18. G. Anger, A. Winkler, and K. D. Rendulic, *Surf. Sci.* **220** (1989), 1.
19. C. T. Campbell, *Surf. Sci.* **157** (1985), 43.
20. E. Umbach, J. C. Fuggle, and D. Menzel, *J. Electron Spectrosc. Relat. Phenom.* **10** (1977) 15.
21. S. Semancik, and P. Estrup, *Surf. Sci.* **104** (1981), 26.
22. (a) F. Bozso, G. Ertl, M. Grunze, and M. Weiss, *J. Catal.* **49** (1977) 18. (b) F. Bozso, G. Ertl, and M. Weiss, *J. Catal.* **50** (1977), 519.
23. G. Ertl, *Adv. Catal.* **21** (1980), 201.
24. L. C. Isett, and J. M. Blakely, *Surf. Sci.* **47** (1975), 645.
25. D. D. Wagman, W. H. Evans, V. B. Parker, R. H. Schumm, I. Halow, S. M. Bailey, K. L. Churney, and R. L. Nutall, *J. Phys. Chem. Ref. Data Suppl. 2* (1982), Vol. II.

26. O. Kubaschewiski, E. L. L. Evans, and C. B. Alcock, *Metallurgical Thermochemistry*, Pergamon, New York, 1967.

27. R. C. Weast (ed.), *Handbook of Chemistry and Physics*, Chemical Rubber Co., Cleveland, Ohio, 1984.

28. C. Kittel, *Introduction to Solid State Physics*, Wiley, New York, 1971.

29. E. Shustorovich, *Surf. Sci. Rep.* **6** (1986), 1.

30. E. Shustorovich, *Acc. Chem. Res.* **21** (1988), 12.

31. E. Shustorovich, (a) *J. Am. Chem. Soc.* **106** (1984), 6479; (b) *Surf. Sci.* **150** (1985), L115; (c) *Surf. Sci.* **163** (1985) L645; (d) *Surf. Sci.* **163** (1985), L730; (e) *Surf. Sci.* **175** (1986), 561; (f) *Surf. Sci.* **176** (1986), L863; (g) *Surf. Sci.* **181** (1987), L205; (h) *Surf. Sci.* **205** (1988), 336; (i) *Catal. Lett.* **7** (1990), 107.

32. (a) E. Shustorovich, and A. T. Bell, (a) *J. Catal.* **113** (1988), 341; (b) *Surf. Sci.* **205** (1988), 492; (c) *Surf. Sci.* **222** (1989), 371; (d) *J. Catal.* **121** (1990) 1; (e) *Surf. Sci.* **235** (1990), 343.

33. E. Shustorovich, *Adv. Catal.* **37** (1990), 101.

34. G. A. Somorjai, *Chemistry in Two Dimensions*, Cornell Univ. Press, Ithaca, N.Y., 1982.

35. J. C. Slater, *Introduction to Chemical Physics*, pp. 452–454, Dover, New York, 1939.

36. (a) P. J. Feibelman, *Phys. Rev. B* **35** (1987), 2626. (b) G. C. Abel, *Phys. Rev. B* **31** (1985), 6184. (c) J. Ferrante, and J. R. Smith, *Phys. Rev. B* **31** (1985), 3427.

37. R. A. VanSanten, *Rec. Trav. Chim. Pay-Bas* **199** (1990), 59.

38. (a) C. Goymour, and D. A. King, *Proc. R. Soc. London Ser. A* **339** (1974), 245. (b) D. A. King, *Surf. Sci.* **47** (1975), 384. (c) D. L. Adams, *Surf. Sci* **42** (1974), 12.

39. (a) T. B. Grimley, *Proc. Phys. Soc.* **90** (1967), 751. (b) T. B. Grimley, *Proc. Phys. Soc.* **92** (1967), 776. (c) T. B. Grimley, and S. M. Walker, *Surf. Sci.* **14** (1969), 395. (d) T. B. Grimley, in *Molecular Processes on Solid Surfaces* (E. Drauglis, R. K. Gretz, and R. I. Jaffe, eds.) McGraw-Hill, New York, 1969.

40. (a) T. L. Einstein, and J. R. Schreiffer, *Phys. Rev. B* **7** (1973), 3269. (b) J. R. Schreiffer, in *The Physical Basis for Heterogeneous Catalysis* (E. Drauglis, and R. I. Jaffe, eds.), Plenum Press, New York, 1975.

41. G. A. Somorjai, and M. A. Van Hove, *Adsorbed Monolayers on Solid Surfaces*, Springer-Verlag, Berlin, 1979.

42. C. M. Chan, M. A. Van Hove, and W. H. Weinberg, *Low Energy Electron Diffraction: Experiment, Theory and Surface Structure Determination*, Springer-Verlag, Berlin, 1986.

43. W. H. Weinberg, and R. P. Merrill, (a) *Surf. Sci.* **33** (1972), 493; (b) *Surf. Sci.* **39** (1973), 206; (c) *Surf. Sci.* **41** (1974), 312; (d) *J. Catal.* **28** (1973), 459; (e) *J. Catal.* **40** (1975), 268.

44. (a) H. S. Johnston, and C. Parr, *J. Am. Chem. Soc.* **85** (1963), 2540. (b) R. A. Marcus, *J. Phys. Chem.* **72** (1968), 891.

45. (a) J. E. Huheey, *Inorganic Chemistry: Principles of Structure and Reactivity*, p. 336, Harper and Row, New York, 1982. (b) A. G. Sharpe, *Inorganic Chemistry*, p. 485, Longwood, New York, 1986.

46. L. Pauling, *The Nature of the Chemical Bond*, Cornell Univ. Press, Ithaca, N.Y., 1960.

47. L. Pauling, (a) *Phys. Rev.* **54** (1938), 899; (b) *Proc. Natl. Acad. Sci. USA* **39** (1953), 551. (c) *J. Solid State Chem.* **54** (1984), 297.

48. H. J. Grabke, and H. Viefhaus, *Surf. Sci.* **112** (1981), L779.

49. M. Kiskinova, and D. W. Goodman, *Surf. Sci.* **108** (1981), 64.

50. J. Benziger, G. Schoofs, and A. Myers, *Langmuir* **4** (1988), 268.

51. Y.-N. Fan, L.-X. Tu, Y.-Z. Sun, R.-S. Li, and K. H. Kuo, *Surf. Sci.* **94** (1980), L203.

52. (a) J. E. Lennard-Jones, *Trans. Faraday Soc.* **28** (1932), 333. (b) J. Bardeen, *Phys. Rev.* **58** (1940), 727. (c) J. O. Hirschfelder, C. F. Curtiss, and R. B. Bird, in *Molecular Theory of Gases and Liquids*, Wiley, New York, 1964.
53. J. B. Benziger, *J. Cataly.*, in press.
54. P. M. Marcus, J. E. Demuth, and D. W. Jepsen, *Surf. Sci.* **53** (1975), 501.
55. S. Andersson, *Surf. Sci.* **79** (1979), 385.
56. (a) J. B. Benziger, and R. E. Preston, *Surf. Sci.* **141** (1984), 576. (b) W. F. Egelhoff, Jr., *Phys. Rev. B* **29** (1984), 3681.
57. F. M. Hoffman, *Surf. Sci. Rep.* **3** (1983), 107.
58. (a) J. B. Benziger, *Surf. Sci.* **17** (1984), 309. (b) W. Erley, H. Wagner, and H. Ibach, *Surf. Sci.* **80** (1979), 612. (c) J. C. Campuzano, and R. G. Greenler, *Surf. Sci.* **83** (1979), 301.
59. (a) K. Horn, and J. Pritchard, *J. Phys.* **38** (1977), 84. (b) B. E. Hayden, and A. M. Bradshaw, *Surf. Sci.* **125** (1983), 787.
60. J. C. Tracy, *J. Chem. Phys.* **56** (1972), 2748.
61. (a) T. W. Haas, J. T. Grant, and G. J. Dooley, in *Proceedings, 2nd International Symposium on Adsorption/Desorption Phenomena* (F. Ricca, ed.), Academic Press, London, 1973. (b) L. E. Davis, N. C. MacDonald, P. W. Palmberg, G. E. Riach, and R. E. Weber, *Handbook of Auger Electron Spectroscopy*, Physical Electronics, Eden Prairie, Minn., 1976. (c) J. B. Benziger, E. I. Ko, and R. J. Madix, *J. Catal.* **54** (1978), 414.
62. J. C. Hamilton, and J. M. Blakely, *Surf. Sci.* **91** (1980), 199.
63. (a) B. A. Sexton, *Surf. Sci.* **88** (1979), 299. (b) B. A. Sexton, *Surf. Sci.* **102** (1981), 271. (c) J. A. Gates, and L. L. Kesmodel, *J. Catal.* **83** (1983), 437. (d) T. E. Felter, W. H. Weinberg, G., Ya. Lastushkina, P. A. Zhdan, G. K. Boreskov, and J. Hrbek, *Appl. Surf. Sci.* **16** (1983), 351.
64. J. B. Benziger, and R. J. Madix, *J. Catal.* **74** (1982), 67.
65. E. M. Siegbahn, and U. Wahlgren, in *Reaction Energetics on Metal Surfaces* (E. Shustorovich, ed.). VCH, New York, 1991
66. E. W. Scharpf, and J. B. Benziger, *J. Phys. Chem.* **91** (1987), 5531.
67. J. D. Beckerle, Q. Y. Yang, A. D. Johnson, and S. T. Ceyer, *J. Chem. Phys.* **86** (1987), 7326.
68. S. W. Johnson, and R. J. Madix, *Surf. Sci.* **108** (1981), 77.
69. T. H. Upton, *J. Chem. Phys.* **83** (1985), 5084.
70. R. A. Zuhr, and J. B. Hudson, *Surf. Sci.* **66** (1977), 405.
71. R. J. Koestner, M. A. Van Hove, and G. A. Somorjai, *J. Phys. Chem.* **87** (1983), 203.
72. (a) G. Broden, T. N. Rhodin, C. Brucker, R. Benbow, and Z. Hurych, *Surf. Sci.* **59** (1976), 593. (b) R. W. Joyner, *Surf. Sci.* **63** (1977), 291.
73. J. B. Benziger, and R. J. Madix, *J. Catal.* **65** (1980), 36.
74. E. I. Ko, J. B. Benziger, and R. J. Madix, *J. Catal.* **62** (1980), 264.
75. I. E. Wachs, and R. J. Madix, *J. Catal.* **53** (1978), 208.
76. I. E. Wachs, and R. J. Madix, *Surf. Sci.* **76** (1978), 531.
77. J. B. Benziger, and R. J. Madix, *J. Catal.* **65** (1980), 49.
78. J. B. Benziger, E. I. Ko, and R. J. Madix, *J. Catal.* **58** (1979), 149.
79. J. B. Benziger, and G. R. Schoofs, *J. Phys. Chem.* **88** (1984), 4439.
80. D. Ying, and R. J. Madix, *J. Catal.* **61** (1980), 48.
81. M. Barteau, M. Bowker, and R. J. Madix, *Surf. Sci.* **94** (1980), 303.

82. B. C. Gates, J. R. Katzer, and G. C. A. Schuit, *Chemistry of Catalytic Processes*, McGraw-Hill, New York, 1979.

83. A. A. Balandin, *Advances in Catalysis and Related Subjects*, Vol. 10, p. 120, Academic Press, New York, 1958.

84. P. Mars, J. J. F. Scholten, and P. Zwietering, in *Proceedings, Symposium on Mechanism of Heterogeneous Catalysis, Amsterdam, Netherlands, 1959* (J. H. deBoer, ed.), p. 66, 1960.

85. (a) R. Gomer, *Vacuum* **33** (1983), 537. (b) C. Dharmadhikari, and R. Gomer, *Surf. Sci.* **143** (1984). (c) R. Butz, and H. Wagner, *Surf. Sci.* **63** (1977), 448. (d) A. Polak, and G. Erlich, *J. Vac. Sci. Technol.* **14** (1977), 407.

86. D. Heskett, A. Baddorf, and E. W. Plummer, *Surf. Sci.* **195** (1988), 94.

87. (a) M. Balooch, M. J. Cardillo, D. R. Miller, and R. E. Stickney, *Surf. Sci.* **46** (1974), 358. (b) M. J. Cardillo, M. Balooch, and R. E. Stickney, *Surf. Sci.* **50** (1975), 263.

88. (a) J. L. Gland, *Surf. Sci.* **93** (1980), 487. (b) C. T. Campbell, G. Ertl, H. Kuipers, and J. Segner, *Surf. Sci.* **107** (1981), 220. (c) A. Winkler, X. Guo, H. R. Siddiqui, P. L. Hagans, and J. T. Yates, Jr., *Surf. Sci.* **201** (1988), 419.

89. J. L. Taylor, D. E. Ibbotson, and W. H. Weinberg, *Surf. Sci.* **79** (1979), 349.

90. (a) D. A. King, and M. G. Well, *Surf. Sci.* **29** (1972), 454. (b) S. M. Ko, and L. D. Schmidt, *Surf. Sci.* **47** (1975), 557.

91. G. Hasse, and M. Asscher, *Surf. Sci.* **191** (1987), 75.

92. L. J. Whitman, L. J. Richter, B. A. Gurney, J. S. Villarrubia, and W. Ho, *J. Chem. Phys.* **90** (1989), 2050.

93. (a) D. W. Goodman, R. D. Kelley, T. E. Madey, and J. M. White, *J. Catal.* **64** (1980), 479. (b) C. Astaldi, A. Santoni, F. Della Valle, and R. Rosei, *Surf. Sci.* **220** (1989), 322.

94. A. L. Ternay, Jr., *Contemporary Organic Chemistry*, Saunders, Philadelphia, 1979.

95. N. L. Allinger, *Adv. Phys. Org. Chem.* **13** (1976), 1.

96. (a) F. H. Westheimer, and J. E. Mayer, *J. Chem. Phys.* **14** (1946), 733. (b) F. H. Westheimer, *J. Chem. Phys.* **15** (1947), 252.

97. U. Burkert, and N. L. Allinger, *Molecular Mechanics*, Am. Chem. Soc., Washington, D.C., 1982.

CHAPTER 3

Surface Diffusion of Atomic and Molecular Adsorbates

Roger C. Baetzold

I. Introduction

A. General Statements

The motion of atoms or molecules determines the possibility of diverse phenomena such as the nucleation and growth of crystals and films, heterogeneous chemical reactions, and redox processes at crystal surfaces. Diffusion is a step that is intrinsic for these processes. A number of technologies including silver halide photography, semiconductor fabrication, and the wide range of commercialized heterogeneous catalytic processes use aspects of surface diffusion. Despite this important role for surface diffusion, the body of knowledge concerning this phenomenon is in its early stage of development [1–7]. Experimentally, surface diffusion is rather difficult to measure and it is only in recent years that a number of reliable techniques have been developed that include field emission spectroscopy [1], thermal He scattering [8], and laser-induced thermal desorption [9].

This chapter deals with conceptual understanding of the energetics of diffusion of chemisorbed atoms and molecules on metal surfaces. We concentrate our attention on well-ordered low-Miller-index surfaces where the temperature is sufficiently high to permit a hopping motion of the atom or molecule from site to site. Clearly several matters must be considered in approaching this area. How does the surface couple to the adsorbed species to achieve thermal equilibrium? Does the adsorbed species occupy special sites on the surface that might be used to define a trajectory for diffusion? How do interactions between adsorbed species influence the diffusion rates? To shed light on each of these questions we refer to a range of theoretical approaches.

B. Adsorption Sites

We first consider low-Miller-index surfaces and the available sites for adsorbates. The (110), (100), and (111) face-centered cubic (fcc) surfaces are sketched in Figure 3.1. At low coverage, where we assume the adsorbate species do not interact significantly with one another, we can consider the adsorbate to occupy the symmetric

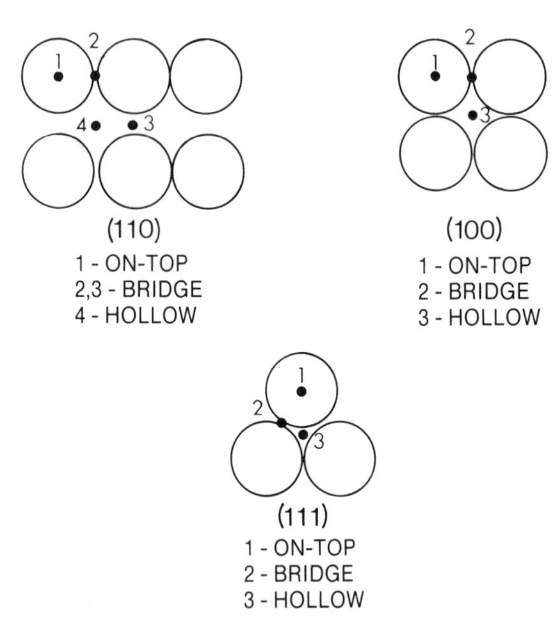

Figure 3.1. Sketches of fcc **(110)**, **(100)**, and **(111)** surfaces showing the variety of possible symmetric adsorption sites.

surface sites. These sites consist of hollow, bridging, or on-top where the adsorbate species possesses up to four, two, or one metal atom nearest neighbors, respectively. We need to be able to compute the energy as the adsorbate species moves from site to site on the surface. The method of analysis should anticipate that the metal-to-adsorbate distance will vary with surface site. This variation is well documented experimentally in metal clusters [10].

Most of our discussion will not take into account the geometric effects introduced by the second layer of metal atoms. On fcc (111) and hcp (001) surfaces two kinds of threefold hollow sites exist. At one there is a metal atom directly beneath and at the other there is none. These sites alternate across the surface. These two inequivalent sites can have different chemisorption energies especially for a small adsorbate atom like H, which in turn could influence the diffusion characteristics. There have not been experimental reports of such effects at this time, but one simulation [11] has shown that the diffusion coefficient should be coverage independent in the absence of adsorbate–adsorbate interactions.

Experimentally, on flat low-index surfaces the site of adatom adsorption is almost always the hollow site. Experiments such as low-energy electron diffraction (LEED), high-resolution electron energy loss (HREELS), and surface-extended x-ray absorption fine structure (EXAFS) have demonstrated this point in a variety of systems [12,13] including H, O, N, C, S, and halogens. Sometimes the bridge coordination site involving two metal atoms is favored, but the on-top coordination never occurs. Hydrogen is a particularly difficult species to study, but a number of examples have been reported with H occupying the threefold hollow site on Ni(111) [14], Ru(001) [15,16], Rh(111) [17], Pd(111) [18,19], and Pt(111) [20]. These

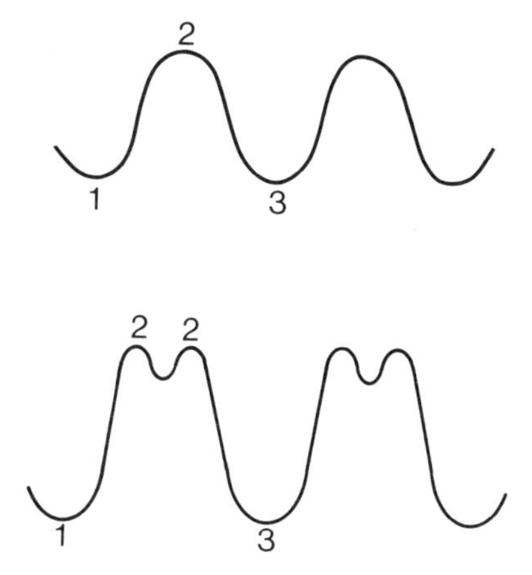

Figure 3.2. Potential energy curves for particular trajectories of a diffusing particle: **(1)** ground state, **(2)** saddle point, **(3)** final state. The top curve is regular and the bottom curve has local energy minima between two saddle points.

results have been generally deduced from vibrational and electron diffraction studies. Also, for H on Pd(100) ion-channeling studies [21] point out the fourfold hollow site to be occupied [22,23].

By contrast, molecular chemisorption shows a wide range of preferred site occupancies. The most extensively studied molecule is CO. This molecule is found with its axis perpendicular to the surface in most cases [22,23]. Exceptions are noted for Fe(100), where the axis is tilted 55° to the normal [24]; Cr(110), where the axis parallel to the surface has been reported [25,26] for some adsorbate coverages; and Mo(100) [27], where the axis is tilted. Adsorption sites for CO bound perpendicular to the surface include a number of different possibilities. On Pt(111), the on-top site is occupied at low coverage with the appearance of twofold coordinated bridge sites at higher coverage [28,29]. Recent studies [30] have shown that the binding energy difference of these two sites is very small and of the order of 1 kcal/mol, although it varies with coverage and temperature. We note that on Ni(111) [31] the pattern is the reverse of that on Pt(111) where the site preference is hollow $>$ bridge $>$ on-top, but the binding energy differences are less than 1 kcal/mol. The most important conclusion seems to be that for CO adsorption the binding energy differences among the various possible sites is small so that as a function of coverage and temperature some or all of these site occupancies might be observed [32].

C. Langmuir Layers

We begin by considering diffusion in a Langmuir layer where the adsorbed particles do not interact with one another, and only one particle can occupy a given binding site. Mobility involves hopping from one site to another adjacent site and can be formulated in terms of a product of a frequency factor (Γ) and an energy barrier (ΔE). Figure 3.2 sketches two possibilities for the potential energy diagram. In the first, the saddle point represents binding at a particular site on the surface, so that ΔE is just the difference in chemisorption energy of the sites. In the second possibility, if a local energy minimum is found at the binding site, the difference in energy of the two binding sites is not the diffusion activation energy. We explore in what circumstances these two possibilities are likely to be found.

Consider the case [3] of no local minima on the diffusion profile. The frequency of jumps is

$$\Gamma = \nu e^{\Delta S/k} \cdot e^{-\Delta E/kT} \tag{3.1}$$

where ν is approximated as a vibration frequency parallel to the surface, ΔS is the entropy change between ground state and the saddle point, T is the temperature, k is Boltzmann's constant, and ΔE is the energy barrier associated with a successful hop from site to site. If one assumes that there is no interaction between adsorbates and only one species may occupy a site, the diffusion coefficient can be shown to be [33]

$$D = \Gamma\lambda^2 = \nu\lambda^2 e^{\Delta S/k} \cdot e^{-\Delta E/kT} = D_0 e^{-\Delta E/kT} \tag{3.2}$$

where λ is the distance of a hop. This formula will be used in interpreting behavior in a wide range of different diffusing systems. The diffusion coefficient is defined in terms of Fick's first law [34]

$$J = -D\nabla C \tag{3.3}$$

where J is the flux, ∇C is the concentration gradient, and D is the diffusion coefficient.

When local minima are present on the energy profile, as sketched in Figure 3.2, the Arrhenius expression in Eq. 3.2 holds, but the frequency factor and activation energy may have different meaning. For example, the effective frequency could be reduced or increased, respectively, depending on whether the trajectory follows the local minima or can avoid these energy states and permit hops of greater length than one nearest neighbor. The activation energy must be at least enough to overcome the maximum in the diffusion profile, but it is possible that some coupling between the jumps on different parts of the trajectory may take place, leading to a more complicated form for the microscopic activation energy. In general, deviations from the simple frequency factor expression in Eq. 3.2 may be expected.

D. Transition State Theory

A useful model for expressing the diffusion coefficient is transition state theory. We consider the energy surface connecting the ground state of an adspecies to the saddle point as motion to an adjacent surface site takes place. Standard transition state theory assumes a local equilibrium between the transition state species and the ground state adspecies. Second, there is the assumption that the system passes through the transition state only once before it is stabilized or thermalized as a reactant or product [35]. With these assumptions, the frequency factor (Γ) in Eq. 3.2 is

$$\Gamma = 4\sigma\left(\frac{kT}{h}\right)\frac{Q^{\neq}}{Q}\, e^{-E_0/kT} \tag{3.4}$$

where σ is the symmetry factor, h is Planck's constant, $Q^{\neq}$ and Q are the partition functions for transition state and reactants, respectively, and E_0 is the potential energy difference between transition state and reactants [36]. Expressions of this general sort may be used along with ensemble average techniques [37,38] to evaluate the diffusion coefficient.

The transition state expression for diffusion is typically evaluated in terms of partition functions for each degree of freedom (electronic, rotational, vibrational, and translational) along with an adsorbate–surface potential function. In one case [39] of H diffusion on Ru(001), the vibrational partition functions were computed to be the dominant terms. The diffusion coefficient was expressed as

$$D = \frac{1}{4}\sigma\lambda^2\left(\frac{kT}{h}\right)\frac{q_y^{\neq}q_z^{\neq}}{q_xq_yq_z}\,e^{-E_0'/kT} \tag{3.5}$$

where $q^{\neq}$ and q represent the vibrational partition functions and E_0' includes the zero-point vibrational energy corrections. In this case the preexponential term of the diffusion coefficient was computed to be 2.2×10^{-3} cm²/s, in good agreement with experiment. With the aid of well-parameterized Morse pair functions the energy barrier computed was 3.84 kcal/mol, in good agreement with the experimental value of 4.0 ± 0.5 kcal/mol.

The preexponential part of the diffusion coefficient expected on the basis of transition state theory is in the range 10^{-2} to 10^{-4} [40,41]. This estimation assumes single-lattice-distance hops and that the diffusing species has negligible interaction with surrounding adspecies. The experimental data in Tables 3.1 and 3.2 show that many systems fit this pattern, but there are exceptions where additional factors are needed to explain the behavior.

We note that the vibrational partition functions vary with the site of adsorption. Consider CO diffusion from the on-top to bridge site on a typical fcc (111) surface. For CO/Pt(111) the Pt–C and C–O vibration frequencies measured [29] are 470 and 2100 cm^{-1} for on-top and 380 and 1850 cm^{-1} for bridge, respectively. Using the vibrational partition function we compute a term 0.71 as the ratio of partition functions appearing in an expression such as Eq. 3.5. The effect is not negligible for this vibrational mode normal to the surface.

II. Experimental Diffusion Data

A. Methods

Recent experiments of surface diffusion have provided a great deal of information concerning the frequency factor and energy barrier of chemisorbed species on metal surfaces. The experimental techniques have been primarily laser-induced thermal desorption (LITD) [11,42–47] and field emission microscopy (FEM) [48–50], with application of the beam-scattering [8], time-resolved spectroscopy [51] and second harmonic generation [52]. These techniques have been applied to a number of metal surfaces such as Pt(111), Ru(001), Ni(100), and W(110), where molecules such as H_2, O_2, and CO are the adsorbates. Patterns of the mechanism of diffusion as well as thermal parameters are beginning to emerge from these studies.

Table 3.1. Experimental Atomic Surface Diffusion Data

System	Coverage, θ	ΔE (kcal/mol)	D_0 (cm^2/s)	$\Delta E/Q$	Ref
D_2/Pt(111)	0.001	12	3×10^4	0.18	42
	0.33	7	0.5	0.11	42
H_2/Pt(111)	0.24	7	1.0	0.11	42
D_2/Rh(111)	0.02	3.7	8×10^{-4}	0.06	43
	0.33	4.3	8×10^{-4}	0.06	43
H_2/Rh(111)	0.33	4.3	1×10^{-3}	0.06	43
H_2/Ru(001)	0.15	3.7	7.9×10^{-4}	0.06	45
	0.95	3.7	7.9×10^{-4}	0.06	45
H_2/W(110)	0.1	4.09	1.7×10^{-7}	0.06	50
	0.9	5.13	2.9×10^{-5}	0.08	50
D_2/W(110)	0.1	3.95	3.5×10^{-5}	0.06	50
	0.9	5.35	5.9×10^{-2}	0.08	50
H_2/W(110)	0.1	4.80	3.3×10^{-4}	0.06	50
	0.9	5.84	6.2×10^{-4}	0.07	50
O_2/W(110)	0.3	14	4.5×10^{-4}	0.12	84
	0.8	22	–	0.19	84
H_2/W(111)	Low	7.1	9×10^{-6}	0.10	49
	High	2.4	4×10^{-8}	0.03	49
D_2/W(111)	Low	7.2	1×10^{-5}	0.10	49
	High	2.9	6.8×10^{-8}	0.04	49
D_2/Ru(001)	Low	4.1	4.6×10^{-4}	0.01	45
H_2/Ni(100)	–	3.5	2.5×10^{-3}	0.06	48
D_2/Ni(100)	–	4.4	8.5×10^{-3}	0.07	48

A key question for each experimental method is how to properly interpret the experimental results in microscopic physical terms. There are also experimental problems associated with each technique, and we illustrate this point for the LITD technique.

LITD is a recent technique that has been used extensively. A clean and oriented single crystal is exposed to an adsorbing gas in ultrahigh vacuum. Generally, expo-

Table 3.2. Experimental Molecular Surface Diffusion Data

System	Coverage	ΔE (kcal/mol)	D_0 (cm^2/s)	Ref.
CO/Pt(111)	0.025	7.0	1×10^{-4}	43
CO/Rh(111)	0.01	7.0	2×10^{-3}	43
	0.40	7.0	3×10^{-2}	43
CO/Rh(100)	0.27	11.0	3.8×10^{-1}	46
	0.58	6.2	6.0×10^{-2}	46
CO/Ni(100)	0.25	6.4	2.5×10^{-1}	47
	0.66	4.6	3.0×10^{-2}	47
CO/Ni(111)	–	6.9	1.2×10^{-5}	52
CO/Ni(111)	0.0	6.8	1.0×10^{-3}	48
	0.6	6.8	1.0×10^{-3}	48
CO/Pt(111)	0.006	4.4	–	51
CO/Pt(111)	–	7.0	–	8

sure is carried out to achieve some predetermined surface concentration. At this point a laser pulse is focused onto a small area of the sample so as to heat it and desorb species within this area. After a time delay a second laser pulse is focused onto the same small areas so as to desorb all species that have diffused into that area from the surrounding region on the surface. These desorbed species are detected by a mass spectrometer and their concentration is determined at various time delays of the second laser pulse. This information is then analyzed to determine the surface diffusion coefficient. The diffusion coefficient is analyzed in an Arrhenius form as in Eq. 3.2.

Although the LITD method enjoys considerable popularity, some experimental concerns have been raised that may limit its application. One such point is whether the laser desorption creates defects on the crystal surface. One study [53] has shown this to be possible as more adsorbate was found in the laser-exposed areas following repopulation than in the unexposed areas. Of course, the presence of defects or impurities can strongly influence the binding and thus the value of the diffusion coefficient of adsorbates. It is almost impossible to prepare a defect-free surface, so at low concentration this effect should be particularly manifest. Another point is the presence of impurities on the surface. As demonstrated in the case of sulfur coverage on Ru(001) [54], a site blocking for hydrogen diffusion was observed. Clearly, any inadvertent impurities present on a surface could influence the diffusion coefficient at low concentration. These experimental issues must be dealt with, and we return to issues of data interpretation in Section IVG.

B. Atomic Diffusion

An excellent summary of the diffusion parameters for a variety of systems is provided in the work of Seebauer *et al.* [43]. Parts of their summary and recent data are collected in Table 3.1 where we show the activation energy, preexponential factor, and ratio of diffusion activation energy to chemisorption energy. A variety of patterns appear in the coverage (θ)-dependent data. Activation energies increase markedly with coverage for O/W (110), whereas the reverse is true for D/Pt (111) and many systems show only a mild dependence. Likewise, the preexponential factor shows a variety of coverage dependencies that often deviate markedly from the standard value of 1×10^{-3} cm^2/s expected on the basis of transition state theory. The ratio of diffusion activation energy to heat of adsorption ($\Delta E/Q$) varies up to about 0.2, but often its value is about 0.1.

We focus attention on thermally activated hopping motion, but note that a tunneling mechanism is often found for H and D adsorbates. In a typical Arrhenius plot [49] for H/W(111) there is a thermally activated regime and temperature-independent regime. Table 3.1 shows that H and D diffusion on W(111) has a much higher activation energy at low coverage than at high coverage. The D/Pt(111) system is another example where the diffusion activation energy is much greater at low than high coverage [42].

The comparison of diffusion parameters for different isotopes in Table 3.1 shows larger activation energies for the heavier isotope. This behavior can be understood from the zero-point vibrational energies in Eq. 3.5 (see Section ID). The preexponential factor is related to the vibrational frequency of a mode parallel to the surface. This factor should lead to an approximate $m^{-1/2}$ dependence of the preexponential factor on mass, m. Experimental preexponential factors usually mirror this dependence, although inverse relationships have been observed and discussed [49,55].

C. Molecular Diffusion

Experimental studies of molecular diffusion have been most extensively performed for CO within the range of chemisorption. The majority of the measurements have been reported at a single coverage where the $\Delta E/Q$ value was found to be near 0.20. The coverage-dependent data show an activation energy independent of coverage or decreasing with coverage. In the latter cases the preexponential factor decreases with coverage.

A number of different experimental techniques [43,46,47,51,52] have placed the diffusion energy of CO/Pt(111) in excess of 4.4 kcal. Rather comprehensive recent infrared measurements [30] have shown that the difference in binding energy of on-top and bridge sites is 60 meV, or about 1.4 kcal/mol. Thus a diffusion trajectory linking these two sites must involve a barrier located between the two sites, as in Figure 3.2b. In Section IIID we present some theoretical results suggesting that this behavior is quite probable for acceptor-type molecules.

We note the variety of patterns for the coverage dependence of the CO diffusion energy in Table 3.2. On Rh(111), the activation energy is constant up to $\theta = 0.40$, whereas it decreases with coverage on Ni(100) and Ru(001) up to $\theta = 0.66$. The measurement [48] of CO diffusion on Ni(111) and stepped Ni(115) surfaces gives an activation energy of 6.8 kcal/mol which is constant with coverage up to $\theta = 0.6$ and then decreases are observed on the stepped surface. A particularly large effect is found [46] on Ru(001), but on this surface the activation energy is constant up to $\theta = 0.33$. For $\theta = 0.33$, the CO molecules bind to on-top metal sites. Above this coverage repulsive CO–CO interactions become important and force the occupation of some nearest-neighbor sites by CO. This also leads to the forcing of some CO molecules slightly off the on-top site and these factors seem to be responsible for the lower activation energy. Later in Section IVD, we show how repulsive adsorbate–adsorbate interactions may lead to a lower activation barrier for diffusion.

III. Models Useful at Low Coverage

A. General Statements

The energy barriers to diffusion may be determined from total energy values specified along the appropriate surface trajectory. We approach this problem from a static framework to illustrate the conceptual features of diffusion. Thus, we consider energy barriers in the average field of the neighboring adspecies and substrate. This approach may be contrasted with dynamic approaches in which the time evolution in the instantaneous field of the adspecies and substrate is considered. In the latter case, substrate atom displacements and energy exchange between the substrate and adspecies are allowed. We defer consideration of these effects until Section IVF. In principle, any type of total energy calculation capable of locating equilibrium distances can be used in the static problem. These may include methods of varying sophistication as applied appropriately to finite (cluster) or infinite (band structure) model systems. In the next paragraphs we consider some of these approaches before turning to a specific conceptual model within the framework of bond order conservation models and pair potentials.

While computing diffusion barriers for adsorbed species, the first thing to consider is whether the diffusion is operating on the thermally activated or tunneling regime. Several different computations [56–59] have treated H atoms on a metal surface which are most prone to tunneling. Here the importance of motion of the metal surface atoms in coupling phonons to the adsorbate has been shown [58]. In the case of H/Cu(100) it was concluded [57] that below 160 K the diffusion is phonon assisted and that it is essential for quantitative accuracy to include movable metal atoms.

The interactions of adsorbate species with metal surfaces can be treated by many different methods. Our emphasis is on energy surfaces; we are not concerned with other properties of the adsorbed species. Classical pair functions such as we discuss below can be fit to experimental data to give information on the potential surface. These often employ two-center Morse functions. The functions can be used within the London–Eyring–Polanyi–Sato (LEPS) method to construct the polycenter potentials [60–65]. This formalism goes beyond pairwise additivity. A rather extensive set of computations for H diffusion barriers on various Ni surfaces have been presented [56] with this method. On a (111) surface the classical barrier for diffusion from a threefold to bridge site is computed to be 3.4 kcal/mol which is lowered to 2.7 kcal/mol by zero-point energy corrections. On the (100) surface a barrier of 14.2 kcal/mol is computed which is considerably at variance with the experimental value of 4 kcal/mol. It was concluded that the LEPS potential may underestimate the binding at a twofold site.

Electronic structure methods such as the local density functional method [66–68] have been used to determine potential surfaces for H diffusion on Ru(001). The low-energy pathway involves hollow-to-bridge diffusion, giving an effective barrier of 4.8 kcal/mol as compared with the experimental value of 3.6 ± 0.5 kcal/mol over a comparable temperature range. This application seems to be promising. A recent example of *ab initio* treatment of surface diffusion for Al adatom on Al(001) [69] is essentially in accord with the above examples, but shows the importance of concerted bond formation and breaking.

B. Pairwise Energy Calculations

Probably the simplest approach to computing the energy of adsorbed atoms to a flat surface is through the use of a sum of pairwise potential energy functions [70–75]:

$$E = \sum_{i>j} E_{ij} \tag{3.6}$$

In the case of weak adsorption such an approximation may be very accurate because the angle-dependent terms associated with covalent bonding are weak. In particular, Lennard–Jones potential energy functions

$$E_{ij} = 4\varepsilon_{ij}\left(\left(\frac{\sigma_{ij}}{r_{ij}}\right)^{12} - \left(\frac{\sigma_{ij}}{r_{ij}}\right)^{6}\right) \tag{3.7}$$

are often used where ε_{ij} and σ_{ij} are constants and r_{ij} is the distance between atoms i and j. Experimental heats of adsorption and vibration frequencies may be used to scale these constants. Another choice is a Morse function, where the energy (E_{ij}) of the bond between atoms i and j is described as

$$E_{ij} = D_0\,(\chi_i^2 - 2\chi_i) \tag{3.8}$$

where

$$a^{-1} = 2\pi\nu\sqrt{\mu/2E_0}$$

$$\chi_i = \exp(-(r_i - r_{i0})/a)$$

ν is the vibration frequency, μ is the reduced mass, D_0 is the equilibrium bond energy, r_{i0} is the equilibrium bond length, r_i is the actual bond length, and a is a scaling parameter. Figure 3.3 shows a simulation of surface trajectories [74] where parameters are chosen to represent O chemisorption to the (111) surface of Pt. We note that the most stable site of adsorption is the hollow site on both (100) and (111) surfaces. In addition, there are no local minima at positions other than the highest symmetry sites. The trajectory hollow to bridge to hollow is the preferred direction for diffusion. The plot also shows two other general features. The adsorption energy is greater at the (100) hollow site versus the (111) hollow site and the diffusion barrier is larger on the (100) surface. Another case where pairwise models were used to treat potential surfaces involved in surface diffusion is the work of Mak and George [39] for hydrogen on Ru(001). Here the Morse potential function was parameterized by using experimental adsorption energies and vibrational frequencies of the adsorbate. Potential energy surfaces were constructed and the preferred diffusion trajectory was found to be hollow to bridge to hollow. The vibration frequency parallel to the surface was shown to sensitively influence the diffusion barrier, whereas the barrier was insensitive to the vibration frequency normal to the surface.

C. Bond Order Conservation: Analytic Model

The bond order conservation analytical model [32,76–79] is a valuable approach to treating surface diffusion. In this formalism the bond order χ is defined as

$$\chi = exp(-(r - r_0)/a) \tag{3.9}$$

The similarity to χ_i in Eq. (3.8) makes the Morse function particularly simple and useful for expressing the energy of a bond.

An analytical model of surface diffusion has been developed by Shustorovich [76–79] based on the bond order conservation principle, and we follow those considerations in this section. The model treats M (metal)–A (adsorbate) interactions

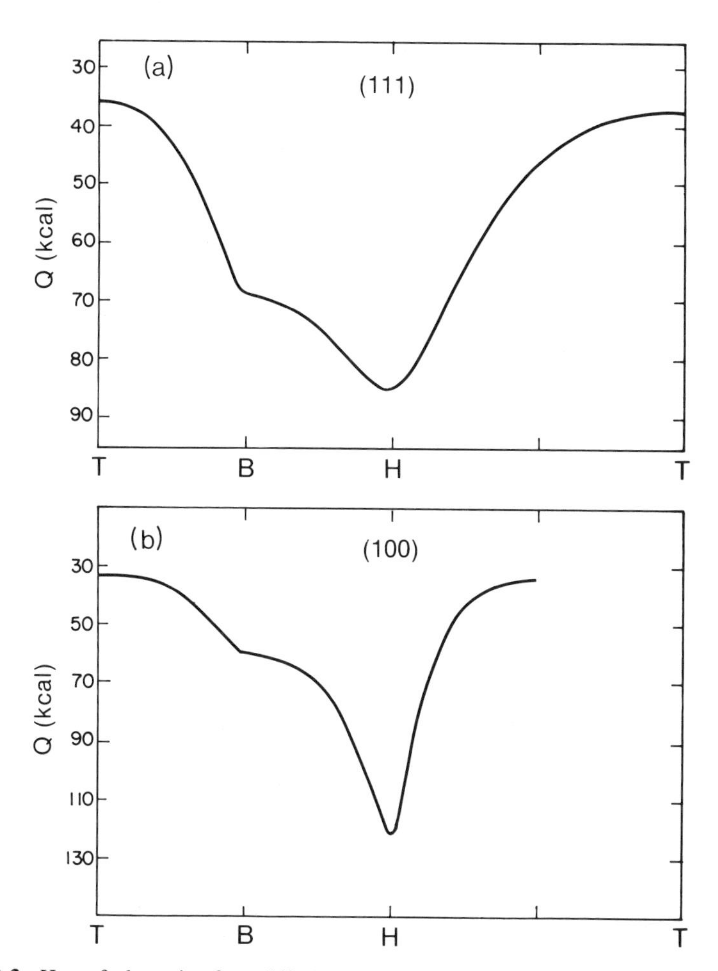

Figure 3.3. Heat of adsorption for a diffusion profile connecting high-symmetry sites on **(a)** fcc (111) and **(b)** fcc (100). The energy-minimum computation is used. Parameters are $a = 0.391$ Å^{-1}, $D_0 = 26$ kcal/mol, and $r_0 = 2.10$ Å. For the (111) surface we compute a heat of adsorption of 85 kcal/mol versus the experimental heat of adsorption for O on Pt(111) of 85 kcal/mol [G. T. Campbell, G. Ertl, H. Kuipers, and J. Segner, *Surf. Sci.* **107** (1981), 220]. The computed equilibrium bond length of 2.09 Å at the hollow site on Pt(111) compares with a value of 2.05 Å for the sum of atomic radii of O and Pt. Reprinted with permission from Ref. 74.

under the following assumptions: (1) Each two-center M–A interaction is described by the Morse-type potential, Eq. (3.8). (2) The M_n–A interactions are limited to n nearest-neighbor metal atoms. [For instance, for the fcc (100) surface (cf. Figure 3.1), $n = 4$, 2, and 1 for the hollow (C_{4v}), bridge (C_{2v}), and on-top ($C_{\infty v}$) sites.] (3) For a given M_n–A, n two-center M–A interactions are additive. (4) The total bond order for M_n–A interactions is normalized to unity and conserved along the diffusion path regardless of the values of n, namely,

$$\sum_{i=1}^{n} \chi_i = 1 \tag{3.10}$$

Table 3.3. Analytical Bond Order Model Relationships[a]

n	Q_n/Q_0	$\Delta E/Q_n$	Surface
1	1.00		
2	1.50		
3	1.67	0.10	hcp (001)
			fcc (111)
			bcc (110)
4	1.75	0.14	fcc (100)
5	1.8	0.17	bcc (100)
6–9	1.83–1.89	0.18–0.21	Stepped, kink
12	1.92	0.22	

[a]See text for notation.

1. Atomic Adsorbates

Consider an atom A chemisorbed on a surface with a C_{nv} unit mesh M_n, for example, C_{3v} ($n = 3$) for hcp (001) or fcc (111) and C_{4v} ($n = 4$) for fcc (100). The bond order χ_{Ai} ($i = 1, 2, \ldots, n$) associated with the position of A at an M_n site relates to the M_n–A bond energy $Q_n(i)$ as

$$Q_n(i) = Q_0 \sum_{i=1}^{n} \left(2\chi_{Ai} - \chi_{Ai}^2\right) = Q_0 \left(2 - \sum_{i=1}^{n} \chi_{Ai}^2\right) \tag{3.11}$$

It follows *rigorously* from Eqs. 3.10 and 3.11 that, for a given n, the maximum M_n–A bonding energy Q_n is reached for equivalent two-center M–A interactions having equal bond order $1/n$. Thus, Q_n monotonically increases with n, namely,

$$Q_n = Q_0\,(2 - 1/n) \tag{3.12}$$

where Q_0 is the maximum M–A bond energy in the on-top site. Table 3.3 gives the ratios of Q_n/Q_0 for some typical ($n = 1$–5) and possible ($n > 5$) values of n. We see that Q_n increases as the surfaces become more open and rough, so that the stepped and kink surfaces where an adatom has the increased coordination number will have the largest values of Q_n. Quantitatively, however, the difference ΔQ between the smoothest ($n = 3$) and roughest ($n > 5$) surfaces is small; more specifically, $\Delta Q/Q_n$ does not exceed 0.1. There can be only one (global) total energy minimum corresponding to the highest coordination number (hollow) site because Q_n increases monotonically with n. Thus, the observed heat of chemisorption Q can be identified with Q_n. In other words, for the surfaces with C_{nv} unit meshes such as fcc (111) or fcc (100), the bridge (C_{2v}) sites are not stationary points and the on-top site is always the total energy maximum. For such surfaces, atomic migration will traverse a twofold axis path, namely, hollow (C_{nv}) → bridge (C_{2v}) → hollow (C_{nv}). The total energy along this path first monotonically increases (up to the bridge site) and then monotonically decreases. In such a case, the migration barrier ΔE (Eq. 3.13) can be identified with

$$\Delta E = Q_n - Q_2 \tag{3.13}$$

the difference between the M_n-A (Q_n) and M_2-A (Q_2) bonding energies. For a given C_{nv} unit mesh, the interrelation of ΔE with Q_n can be easily found:

$$\Delta E = \frac{n-2}{4n-2} Q_n \tag{3.14}$$

The ratio $\Delta E/Q_n$ (Eq. 3.14) monotonically increases with n when the surfaces become more open and rough. Some estimates are given in Table 3.3. For the smoothest surfaces such as hcp (001) and fcc (111), C_{3v}, we have $\Delta E = 0.10Q_3$; for fcc (100), C_{4v}, $\Delta E = 0.14Q_4$; for bcc (100), C_{4v} (assuming effective M_5-A coordination), $\Delta E = 0.17Q_5$. For $n = 6$–9, on some rough (stepped, kink) surfaces, the ratio $\Delta E/Q$ can reach 0.2.

For highly symmetric surfaces (unit meshes M_n-A), the ratios Q_n/Q_0 (Eq. 3.12) and $\Delta E/Q_n$ (Eq. 3.14) are structural constants dependent only on n. For less symmetric surfaces, the preferred site and, therefore, the migration path and the migration barrier will depend on the nature of M and A as well.

The preference of the hollow site is based on the validity of Equation 3.10 for all sites. The situation will change, however, if there are steric constraints. When the metal–metal lattice constant does not allow the bond order to be conserved at sites of high coordination, an adatom will prefer to occupy the site of lower coordination. It is reasonable to assume that the bridge sites can always satisfy Eqs. 3.9 and 3.10. If we consider the hypothetical migration strictly along the M_2 bridge (from the minimum at the C_{2v} site to the maximum at the on-top site), we can estimate the upper limit of the migration barrier, namely, $\Delta E/Q = 0.33$ (cf. Table 3.3). In summary, we find the migration barrier ΔE to be directly proportional to the heat of chemisorption (Eq. 3.15).

$$\Delta E = kQ, \qquad k = 0.1\text{–}0.3 \tag{3.15}$$

The coefficient k may be either a constant or some parameter with values typically in the range 0.10–0.25. Because this model does not take into account the adsorbate–adsorbate interactions, the above values of k and $\Delta E/Q$ should be assigned to the extreme case of zero coverage (of single isolated adatoms).

2. *Molecular Adsorbates*

Consider the surface migration of a diatomic molecule AB such as H_2 or CO. Again, we want to know what metal site M_n will yield the global energy minimum and how the activation barrier ΔE will vary along the migration path. As the model assumptions and conclusions for admolecules should be reduced to those for adatoms if we neglect the internal structure of an admolecule, the fundamental assumptions such as pairwise additivity (limited to nearest neighbors) and conservation of bond order (normalized to unity) are preserved in the molecular case. The new assumptions reflecting the specificity of molecular migration include the general case in which the A–B distance and thus bond order change across a trajectory [76,77]. The signs of the two additive contributions to the M–AB bond energy, M–A (Q_A) and M–B (Q_B) may be the same ($Q_A > 0$, $Q_B > 0$) or opposite ($Q_A > 0$, $Q_B < 0$), depending on the nodal structure of the relevant LCAO MO of AB, namely, in-phase or out-of-phase, respectively [80]. The in-phase structure corresponds to the donor M ← AB bonding, the out-of-phase structure to the acceptor M → AB bonding.

$$Q_A > 0, \quad \begin{array}{l} Q_B > 0 \text{ for } donor \text{ AB} \\ \\ Q_B < 0 \text{ for } acceptor \text{ AB} \end{array} \tag{3.16}$$

The derivations of the analytic model have been presented [76–79] and are not repeated here. We note that when an admolecule diffuses on a surface between sites of different coordination number n, the bond order components M–A, M–B, and A–B change differentially because the relevant bond lengths change differentially. The preferred chemisorption site is no longer exclusively determined by the coordination number, but rather by the complex interrelationship of the bond order model parameters.

The chemisorption energy pattern of molecules may be visualized as resulting from a sum of two terms. The first is atomic-like and predominately involves Q_A as discussed above. The second term is more complex and may be either attractive or repulsive. Donor molecules typically have an attractive term promoting the atomic-like pattern of chemisorption. Acceptor molecules typically have repulsive terms leading to a preference for the sites of lower coordination such as on-top.

Because of such an opposing behavior of the two terms for an acceptor molecule the energy along the migration path can change nonmonotonically. Thus, for molecular chemisorption (contrary to atomic chemisorption), the M_n–AB energy profile can have more than one minimum. In this case, the migration barrier ΔE is larger than ΔQ between the minima, which may correspond to both symmetric and asymmetric sites. We conclude that adatoms and acceptor admolecules will typically prefer different chemisorption sites (sites of different coordination).

D. Bond Order Conservation: Computational Model

Both analytical [76–79] and computational models [74] are based on two assumptions in common: (1) the Morse potential (Eq. 3.8) for each two-center M–A interaction and (2) pairwise additivity of these interactions in the multicenter M_n–A site. The computational assumptions different from the analytical ones are as follows: a five-layer metal film is considered; the coordination number n is no longer limited to the nearest neighbors but includes all the neighbors within several bond lengths of the adsorbate (for example, for a fcc (111) film, 127 metal atoms per layer have been included); an A/M_n energy profile is determined by conserving the total M_n–A bond order as in Eq. 3.10 with the summation running over all metal atoms, except that the total bond order need not be 1.0; the parameters employed are defined by overall energy computations Q_0 and r_0 to give realistic values of the computed heat of chemisorption Q_n and bond length r_n (a was chosen from the realistic ranges discussed earlier in connection with the analytical formalism).

We stress that the computed values of Q_n (and r_n) include additive contributions from many neighbors, and each contribution is determined by the whole parameter

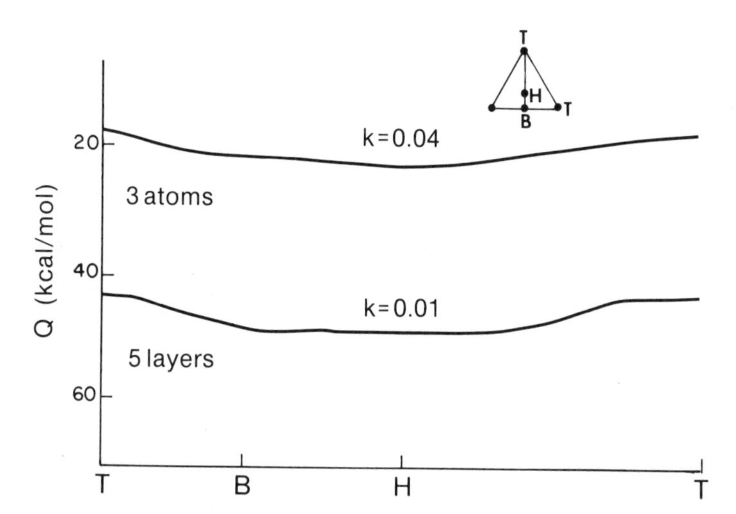

Figure 3.4. Computed energy profiles for 25 uniformly spaced points describing atomic migration simulation H/fcc (111) chemisorption [the parameter set used (r_0 = 2.3 Å, a = 0.87 Å^{-1}, D_0 = 8 kcal/mol) gives Q = 55 kcal/mol]. Compared are the results obtained within the nearest-neighbor ("3 atoms") versus all-neighbor ("5 layers") approximations and by using the bond order conservation criteria. Note the monotonic increase in the M_n–A bond energy as the effective coordination number n increases along the series T < B < H. The values of $k = \Delta E/Q$ are also shown.

set (Q_n, r_n, a), which makes it difficult to specify for a given system what might be the optimal value of each parameter. For this reason, our computational approach is more suitable to treat trends rather than to calculate specific systems. Here we discuss only typical findings that are most relevant to a conceptual framework for surface migration. It should be noted that the total bond order is conserved at a value, unequal to one, that gives a good representation of the heat of adsorption and bond lengths in the lowest-energy chemisorbed state. This procedure differs from the bond order conservation analytic model where the bond order is conserved at one, but only nearest-neighbor M–A interactions are considered. Because as experience has shown, a rescaling of parameters in the computational procedure could bring the total conserved bond order close to one, the analytical and computational approaches are complementary.

1. Atomic Adsorbates

Figure 3.4 compares energy profiles computed at 25 uniformly spaced points for atomic migration on a fcc (111) surface, which is simulated by a three-atom C_{3v} unit mesh (nearest-neighbor approximation) and a five-layer film model. The parameter set (Q_0, r_0, a) was chosen to give Q close to the experimental value for H chemisorption (Q = 55 kcal/mol) for a five-layer film. We see that the profiles are qualitatively similar, persistently showing a monotonic increase in the A–M_n bond energy in the order on-top < bridge < hollow. Obviously, the five-layer curves

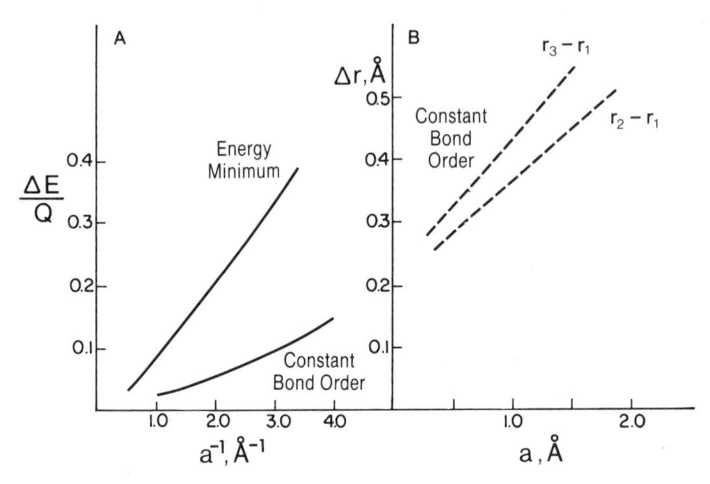

Figure 3.5. (A) Plot of computed $\Delta E/Q$ for five-layer films with an atomic adsorbate versus a^{-1}. **(B)** Plot of increase in equilibrium bond length for three-center (r_3) and two-center (r_2) versus one-center (r_1) bonding of an atomic adsorbate to fcc(111) surfaces versus a. Reprinted with permission from Ref. 74.

correspond to larger values of Q (and to smaller values of $k = \Delta E/Q$) than the three-atom curves, but the differences in Q are only twofold, i.e., three nearest neighbors contribute to Q as much as all other neighbors. These findings lend support to the nearest-neighbor approximation used in the analytical model and support the possibility of rescaling the three-atom case to represent an infinite surface model. Also, note that the five-layer potential curve is more shallow, which leads to a smaller anisotropy ΔQ among the on-top (T), bridge (B), and hollow (H) sites, as well as to the smaller migration barrier k. Neither curve in Figure 3.4 is parameterized for a specific system, so that only the trends are emphasized. Qualitatively, the migration patterns shown in Figure 3.4 have been found persistently for all the adatoms studied, regardless of the parameters employed. We conclude that a monotonic increase in Q_n, with n leading to the preferred site of the highest coordination along with the identity $\Delta E \equiv \Delta Q$ (between the hollow and bridge sites), is the fundamental parameter-independent pattern of atomic migration on metal surfaces.

The general pattern for energy barriers in atomic chemisorption resulting from the five-layer film calculations on a fcc (111) surface is shown in Figure 3.5. Here we observe a dependence of $\Delta E/Q$ on the a parameter which controls the range of the interaction. As the range increases, the activation barrier becomes smaller. Likewise, the metal–adsorbate bond length depends on this parameter. We show the increase in threefold versus onefold (r_3–r_1) or twofold versus onefold (r_2–r_1) distance in the figure.

2. *Molecular Adsorbates*

In the case of diatomic (AB) molecular adsorption the heat of adsorption can be written with three terms just as in the case of the analytic model:

$$Q = \Gamma_A + \Gamma_B + \Gamma_{AB} \tag{3.17}$$

The three terms are expressed as Morse functions (Eq. 3.8) and we conserve the total bond order

$$\sum_M (\chi_{AM} + \chi_{BM}) + \chi_{AB} = \text{constant} \tag{3.18}$$

where we are allowing for changes in the A–B length. The total bond order must be conserved in this procedure where both the metal–A and A–B distances are varied simultaneously to achieve a particular value of the heat of adsorption consistent with the other parameters chosen on the basis of the Morse parameters (Eq. 3.8). In our procedure we pick the lowest-energy configuration that satisfies Eq. 3.18.

We investigate molecular surface migration recalling that the analytical model expects atomic-like patterns for donor admolecules but foresees quite different migration behavior for acceptor heteronuclear admolecules such as CO. To better comprehend the reasons behind these effects, let us begin with a homonuclear donor molecule A_2. Figure 3.6 shows the energy profiles for a five-layer fcc (111) film calculated within the constant bond order (BO_{con}) procedure. We see the complete atomlike pattern; namely, Q_n monotonically increases with n, the preferred site is hollow, and $\Delta E = \Delta Q$.

Now let us turn to a heteronuclear admolecule AB on a five-layer fcc (111) film. If A is not much different from B, the AB/M_n energy profiles are still atomlike. On the other hand, if A is rather different from B, the AB/M_n energy profile shows nonmonotonic changes in Q_n, resulting in the inequality $\Delta E > \Delta Q$, as seen in

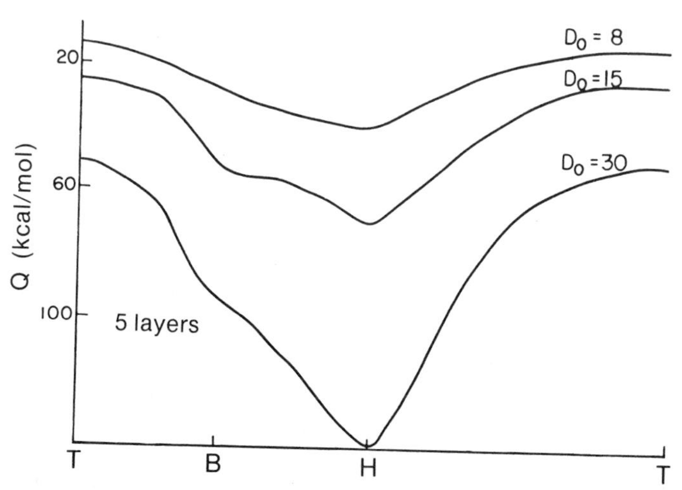

Figure 3.6. Computed energy profiles for 25 uniformly spaced points describing a homonuclear donor molecule A_2 on a five-layer fcc(111) film. The upright geometry is kept throughout the migration path. The results were obtained for fixed values of r_0 (1.9 Å) and a (0.43 Å^{-1}) with E_0 varied. Note the completely atomlike patterns (cf. Figure 3.4). Reprinted with permission from Ref. 74.

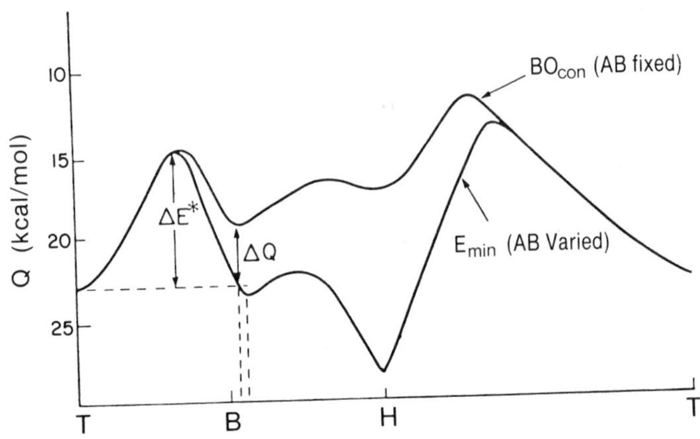

Figure 3.7. Computed energy profiles for 25 uniformly spaced points describing a heteronuclear acceptor molecule AB (the upright configuration) on a five-layer fcc (111) film where A and B are rather different. Both E_{min} and BO_{con} procedures show nonmonotonic changes of the M_n–AB bond energy along the path T → B → H, the preferred site being either H (for E_{min}) or T (for BO_{con}), and nonsymmetric local minima can exist. Reprinted with permission from Ref. 74.

Figure 3.7. (Here the parameters were chosen so as to simulate the energetics of CO/M chemisorption: $Q_{AB} \simeq 1$ eV, Q_A, $Q_B \simeq 4$–6 eV [26–32].) Moreover, the BO_{con} curve reveals the Q_n order reversed as compared with that of adatoms, namely, on-top > bridge > hollow. Finally, we find the local minima in both symmetric and nonsymmetric sites to be energetically favorable in some regions. The correspondence between the analytical and computational results is excellent.

IV. Models Treating Interactions between Adsorbate Species

A. Adsorbate–Adsorbate Interactions

We now can consider the interactions that take place between adsorbed species on a metal surface and their consequences for diffusion. First we can consider the evidence that interactions do take place. The best evidence for this comes from LEED patterns which show that different ordered phases exist in the submonolayer region as the coverage is changed. Good examples are noted in the paper by Ertl [81] where stability of disordered and ordered phases is described by a phase diagram of temperature versus coverage. Likewise, the heat of adsorption shows changes with coverage that are related to the interactions between adsorbate species. For example, in a typical system such as CO/Pd(111) the heat of adsorption is roughly constant up to one-third monolayer [81] coverage and then drops rather sharply with increased coverage. This effect is interpreted in terms of a net repulsive interaction energy between the adsorbed CO species. As the CO coverage increases the adsorption site may change. Mixed on-top and bridge adsorption of CO on Pt(111) is observed at one-third monolayer coverage [28]. The on-top site is occupied at low coverage. This example is a case where adsorbate–adsorbate interactions are operating.

The origin of the adsorbate interactions with one another has often been discussed in terms of direct or indirect interactions [82]. The direct interactions between adsorbate species may be due to van der Waals or dipole–dipole-type interactions. The indirect interactions are through the metal and should be more important in chemisorption systems. It is not our purpose to investigate further the origin of these forces, but instead to look at their influence on diffusion.

A convenient means of expressing the interactions between adsorbate species is in terms of pairwise neighboring interactions. For nearest-neighbor interactions (ω_1) and next-nearest-neighbor interactions (ω_2), the heat of adsorption (Q) can be considered to be

$$-Q = E = E_0 + \tfrac{1}{2}\omega_1 \sum_i n_i + \tfrac{1}{2}\omega_2 \sum_j n_j \tag{3.19}$$

where E_0 is the energy of the isolated adsorbate and n_i and n_j are the occupation numbers for the appropriate nearest and next-nearest neighbors, which have the value 1 if their site is occupied and 0 if not occupied. We call attention to the fact that repulsive interactions have $\omega_i > 0$ and attractive interactions have $\omega_i < 0$. The summations in Eq. 3.19 run only over the appropriate nearest or next-nearest-neighbor sites, respectively. This type of formalism has been used to fit experimental phase diagram data [81], chemisorptive energy data [83] and diffusion data [84], to name a few.

Although Eq. 3.19 is a convenient mathematical expression of considerable practical and pedagogical value, we need to note some reservations. The expression is of an ad hoc nature. It considers the metal–adsorbate interaction as constant with coverage and treats only direct adsorbate–adsorbate interactions, omitting interaction of the indirect type. These are clearly assumptions that would not be expected to hold over the broad coverage ranges included in most experiments. Thus, the parameters in Eq. 3.19 have no clear physical meaning or transferability from one situation to another. In fact, an analysis [85] of the formalism in Eq. 3.19 concluded that the ω_i parameters are not unique in magnitude or even in sign.

The reservations noted above are widely recognized, yet Eq. 3.19 remains in popular application because of its convenience. The bond order conservation formalism represents a well-defined physical model capable of treating the interactions of adsorbate species with one another and the surface. Below we show how the bond order formalism can be cast into a form of Eq. 3.19 and, thus, lend some physical insight into the parameters.

B. Interaction Parameters

Consider an atomic adsorbate A on the fcc (111) or hcp (001) surfaces with N_1 nearest neighbors A and N_2 next-nearest neighbors A. For these A–A interactions, the corresponding bond orders are χ_1 and χ_2, respectively. For each atom A chemisorbed in the threefold (C_{3v}) hollow site, we can write the normalized total bond order

$$1 = 3\chi_{MA} + N_1\chi_1 + N_2\chi_2 \tag{3.20}$$

where χ_{MA} is the bond order of the adsorbate to a metal atom.

Now we employ the Morse potentials defined in Eq. 3.8, where D_{AA} refers to the adsorbate–adsorbate bond and D_{MA} refers to the metal–adsorbate bond. Here the value of $D_{MA} = \frac{3}{5} Q_A$ as follows from Eq. 3.12 and D_{AA} is the effective energy of the A–A bond which we shall treat as a parameter. The energy of the chemisorbed species A is the sum of metal–adsorbate and adsorbate–adsorbate components:

$$E = 3D_{MA}\chi_{MA}(-2 + \chi_{MA}) + D_{AA}\,[N_1(-2\chi_1 + \chi_1^2) + N_2(-2\chi_2 + \chi_2^2)] \tag{3.21}$$

This equation is used along with Eq. 3.20 to determine the change in energy ΔE (where $\Delta E = E + 5D_{MA}/3$) resulting only from mutual interactions of the adsorbate species. The ω_i interactions are the partial derivatives of the energy ΔE:

$$\omega_1 = \left(\frac{\partial \Delta E}{\partial N_1}\right)_{N_2} = \frac{4}{3}D_{MA}\chi_1 + D_{AA}\chi_1\,(-2+\chi_1) + \frac{2}{3}D_{MA}\chi_1(N_2\chi_2 + N_1\chi_1)$$

$$\omega_2 = \left(\frac{\partial \Delta E}{\partial N_2}\right)_{N_1} = \frac{4}{3}D_{MA}\chi_2 + D_{AA}\chi_2\,(-2+\chi_2) + \frac{2}{3}D_{MA}\chi_2(N_1\chi_1 + N_2\chi_2) \tag{3.22}$$

The full expression may be employed but the nonlinear χ_i terms are rather small and this equation can often be reduced to

$$\omega_i \cong \chi_i\left(\frac{4}{3}D_{MA} - 2D_{AA}\right) \tag{3.23}$$

for most interactions. Remember that the heat of adsorption for isolated adsorbates is $5D_{MA}/3$ and D_{AA} represents the effective Morse interaction term between the appropriate adsorbate species. This simple derivation has not accounted for bond order changes in the metal lattice but these will presumably not be large. Thus, we now have a means of specifying the adsorbate–adsorbate interaction parameters within a pairwise model framework.

The adsorbate–adsorbate interaction parameters can be derived for molecular species using considerations similar to these. Here we must also consider effects resulting from changes in bond order of internal molecular bonds on chemisorption, in addition to the terms important in atomic adsorption. For the case of a diatomic molecule (AB) adsorbed on-top with atom A next to the metal surface and having the bond order χ_{MA} we can write the bond order conservation equation in normalized form as

$$1 = \chi_{AB} + \chi_{AM} + N_1\chi_1 + N_2\chi_2 \tag{3.24}$$

where we consider N_1 and N_2 nearest- and next-nearest admolecule neighbors, respectively, with corresponding bond orders χ_1 and χ_2. The total energy can then be expressed in terms employing Morse functions. Comparison of the energy at zero and arbitrary coverage gives an expression for the change in energy with coverage where we retain only terms linear in the bond order.

$$\Delta E = 2(D_{AM} - D_{AA})(N_1\chi_1 + N_2\chi_2) \tag{3.25}$$

The interaction parameters ω_i become

$$\omega_i = 2(D_{AM} - D_{AA})\chi_i \tag{3.26}$$

where we note that D_{AA} represents an effective interaction parameter between the admolecules. The importance of these derivations is that now the bond order conservation formalism may be cast in a pairwise form easily tractable for computation. Finally, we stress that this is a mathematical device and not a statement on the relative importance of direct versus indirect interactions.

C. Computation of Activation Barriers

We have performed some calculations within the bond order framework using Eqs. 3.23 and 3.26 to estimate the pairwise parameters that will determine activation barriers for diffusion in the presence of adsorbate–adsorbate interactions. A fcc (111) surface is modeled by 46 metal atoms arranged in two layers about the diffusion trajectory. Here we treat nearest- and next-nearest-neighbor adsorbate–adsorbate interactions and all of the metal–adsorbate interactions. We compute an adatom energy profile at several points along the trajectory hollow to bridge to hollow site so that the adsorbate–adsorbate interactions can influence the energy of both the ground state and saddle point.

We have conserved the total bond order to 1 in this calculation as in Eq. 3.20. We accomplish this by performing the calculation at various distances of the adsorbate above the surface and then select the distance giving bond order conservation. Several variables are involved in any single energy profile we compute. These include the local concentration of adsorbate species near the adatom undergoing diffusion. Depending on whether the net interaction potentials determined by Eq. 3.23 are repulsive or attractive the potential energy for diffusion decreased or increased, respectively. Figure 3.8 shows typical energy profiles for different local concentrations and interaction parameters. The adsorbate particles attract one another and we observe that they increase the barrier relative to the zero-coverage case (curve 1) compared with the high-coverage case (curve 2 or 3) in Figure 3.8. The value of the barrier for the zero-coverage case depends strongly on the *a* value which influences the range of the bond order effects. When the *a* value is larger, the important interactions are short range and behavior similar to Eq. 3.14 is found. As the *a* value is decreased, the interaction becomes longer range and the barrier becomes smaller.

The case of repulsive nearest- and next-nearest-neighbor interactions is treated in Figure 3.9. Here we see that the diffusion activation energy decreases with coverage. This behavior is found for various examples of different local concentration. We note that even though the energy of the saddle point configuration is determined by local adsorbate–adsorbate interactions, the overall schematic behavior shown in Figure 3.10 seems to be observed for both attractive and repulsive interactions [2]. Finally we note that because of the adsorbate–adsorbate interactions, the saddle point for atomic diffusion may sometimes be shifted from the bridge position. This is particularly apparent in Figure 3.8.

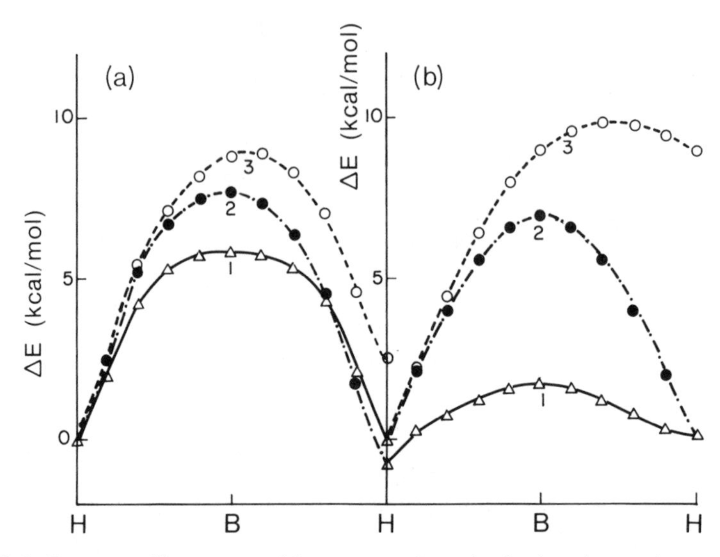

Figure 3.8. Energy profiles computed for two sets of metal–adsorbate interaction parameters: **(a)** $a = 0.2$ Å^{-1} and **(b)** $a = 0.5$ Å^{-1}. The local coverages are **(1)** no neighboring adsorbates, **(2)** all adsorbate sites full, **(3)** only adsorbate sites full on the half-plane nearest the starting adsorbate site. The diffusion trajectory is hollow to bridge to hollow on the fcc (111) surface and the saddle point is at the bridge site or close to it for each set of initial conditions. The interaction parameters are $\omega_1 = -5.3$ kcal/mol and $\omega_2 = -0.15$ kcal/mol where we employ $D_{AA} = 60$ kcal/mol and $D_{MA} = 37$ kcal/mol and computed $\chi_1 = 0.075$ and $\chi_2 = 0.0022$.

This computation has shown that with the bond order conservation parameterization of a pairwise scheme the dependence of the energy at each point on a diffusion trajectory may be specified. Thus the ground state and transition state energies are each modified by adsorbate–adsorbate interactions. The pattern generally expected [2] for repulsive-versus-attractive interactions is observed. A new feature of adatom diffusion appears in that now the saddle point position is modified by adsorbate–adsorbate interactions. The high-symmetry site [bridge, C_{2v}, on these fcc(111) models] is not necessarily the saddle point.

D. Chemical versus Tracer Diffusion Coefficients

A rather subtle distinction needs to be drawn between chemical and tracer diffusion coefficients [1–3,33,84]. The chemical diffusion coefficient is measured when there is a gradient in the chemical potential of the diffusing species as opposed to the tracer diffusion coefficient measured at equilibrium. The tracer diffusion constant can be expressed in two dimensions as

$$D = \lim_{t \to \infty} \frac{\langle r(t)^2 \rangle}{4t} \tag{3.27}$$

where r is the displaced position at time t, and $\langle \; \rangle$ represents an average value over many trajectories. No gradient in adsorbate concentration exists. The chemical

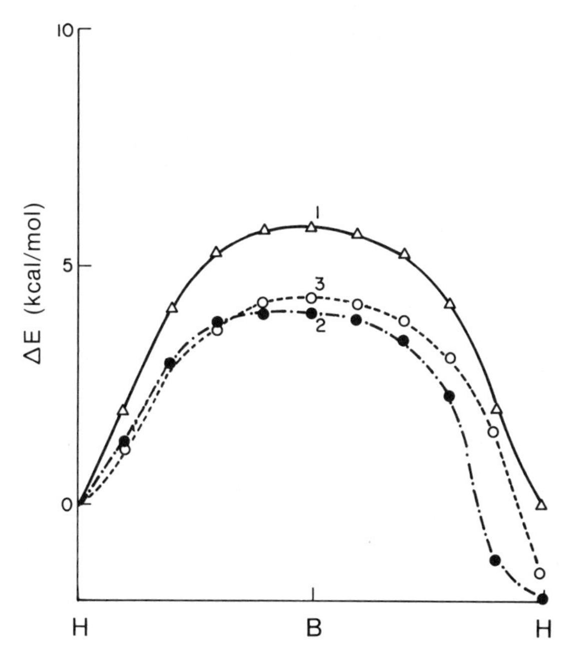

Figure 3.9. Energy profiles computed for the metal–adsorbate parameter $a = 0.2$ Å^{-1}. The local coverages are **(1)** no neighboring adsorbates, **(2)** all adsorbate sites full, **(3)** only adsorbate sites full on the half-plane nearest the starting adsorbate site. The diffusion trajectory is hollow to bridge to hollow on the fcc (111) surface. The interaction parameters are $\omega_1 = 1.44$ kcal/mol and $\omega_2 = 0.04$ kcal/mol where we employ $D_{AA} = 15$ kcal/mol and $D_{MA} = 37$ kcal/mol and compute $\chi_1 = 0.075$ and $\chi_2 = 0.0022$.

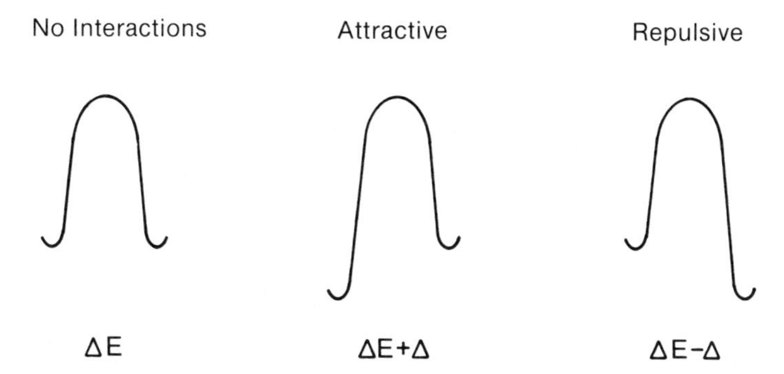

Figure 3.10. Sketch of potential energy curves for diffusion for no adsorbate–absorbate interactions, attractive interactions, and/or repulsive interactions.

diffusion coefficient represents application of Fick's diffusion laws to the case of a gradient in the chemical concentration. In these systems material transport takes place because of a gradient in the chemical potential (μ) of the adsorbed atoms [33]. As a result the chemical and tracer diffusion coefficients need not be the same or show the same dependence on coverage. Explicitly, the preexponential and activation energy factors for chemical diffusion are related to the corresponding factors for tracer diffusion by [33]

$$\Delta E = \Delta E^{T} - \left(\frac{\partial \ln \theta}{\partial \mu / kT}\right)_{T} \left(\frac{\partial \bar{E}}{\partial \ln \theta}\right)_{T}$$

$$D_0 = D_0^{T}\left(\frac{\partial(\mu / kT)}{\partial \ln \theta}\right)_{T} \exp\left[-\left(\frac{\partial \ln \theta}{\partial \mu}\right)_{T} \cdot \left(\frac{\partial \bar{E}}{\partial \ln \theta}\right)_{T}\right] \tag{3.28}$$

where $\bar{E}$ is the partial molecular internal energy of the adsorbed material, θ is the coverage, and k and T have their usual meaning. These equations derived by Reed and Ehrlich [33] are for the case of a jump diffusion coefficient involving the quantities D_0 and ΔE as related to the tracer diffusion coefficient involving D_0^T and ΔE^T.

The importance of the chemical-versus-tracer diffusion coefficient has been emphasized in Monte-Carlo studies [84] of oxygen diffusion on W(110). At sufficiently low coverage the diffusion coefficients become similar when only the dynamics of the interaction of a single adsorbed species with the surface are involved. At higher coverage the coefficients as well as their Arrhenius parameters may differ. Tringides and Gomer [84] showed that the slope of the activation energy dependence on coverage may even be different for tracer and chemical diffusion coefficients over certain coverage ranges for the O/W(110) system. Similar behavior is also found for the preexponential factor.

Further simulations [86] of the LITD experiment have shown significant differences in the coverage-dependent behavior of the chemical and tracer diffusion coefficients. For an interaction model including only site exclusion among the adsorbed species the diffusion coefficient determined from the LITD simulation is constant with coverage, suggesting it is the chemical diffusion coefficient. On the other hand, the tracer diffusion coefficient decreases with coverage because fewer sites are available to jump to as coverage increases. The chemical diffusion coefficient is constant because the decreased jumping rate is compensated by a decrease in overlayer compressibility as the coverage increases [86].

E. Quasichemical Models

A simplified model generally termed the quasichemical model [33] has proven very valuable for understanding the behavior of diffusion parameters as a function of adsorbate concentration. This model considers only pairwise nearest-neighbor adsorbate interactions (ω_i) and permits only one adsorbate per surface site. In addition, the adsorbate species are treated as a lattice gas on a rigid substrate where the energy of the saddle point during diffusion is independent of adsorbate concentration.

This lattice gas model can be employed to treat either the chemical or the tracer diffusion coefficient which we described earlier. The behavior predicted by this

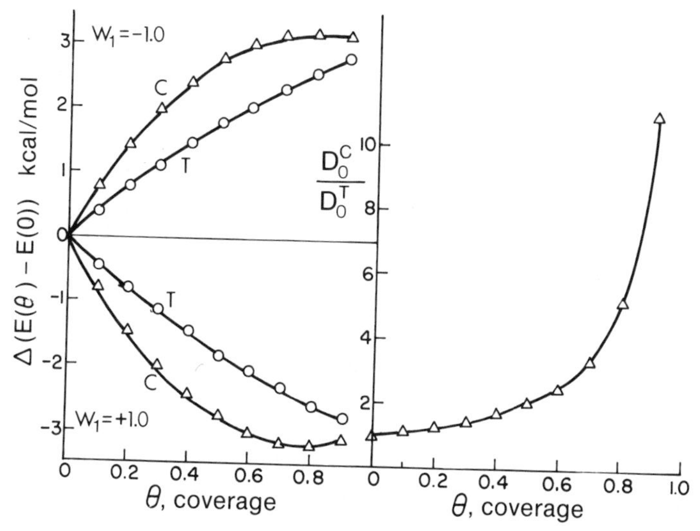

Figure 3.11. Change in activation energy **(a)** or preexponential factor **(b)** for the quasichemical model with attractive (1 kcal/mol) nearest-neighbor adsorbate interactions and repulsive (1 kcal/mol) interactions is shown versus adsorbate coverage (θ). The chemical (C) and tracer (T) parameters are indicated.

model is shown in Figure 3.11. Attractive nearest-neighbor interactions cause an increase in activation energy for chemical and tracer diffusion coefficients, whereas the reverse is true for repulsive interactions. The ratio of chemical to tracer diffusion coefficient varies considerably with coverage. The value of this approach is its analytic form not requiring a computer simulation to understand qualitative effects. The difficulty is that a much more complicated interaction model is typically required between adsorbed species to fully describe their behavior.

F. Molecular Dynamics

Molecular dynamics models are often used to simulate the diffusion of adsorbate species on surfaces [6,38,87–92]. Several important model features included in this type of treatment are direct adsorbate–adsorbate interactions, energy exchange between the adsorbate and the surface, and rearrangement of the surface atoms. Clearly, these model features represent advantages over the static models that were discussed earlier.

In a typical molecular dynamics simulation the classical equations of motion are solved. This allows simultaneous treatment of the movement of adsorbed particle as perturbed by the neighboring surface and adsorbed particles. One of the requirements for realistic calculations is that a large enough ensemble of surface atoms and adsorbed particles be treated in the simulation. Furthermore the simulation must be run for a sufficient number of time steps so that realistic averages are obtained. This clearly demands that computer-intensive resources be brought to bear on the problem. In the case of thermally activated diffusion the time scale between adsorbate hops is quite large, implying that the length of molecular dynamics trajectories required to study surface diffusion is, in some cases, impractically large [6].

An important point to consider in molecular dynamics is the interatomic potential. Progress has been made in specifying interatomic potentials in ionic [93] and metallic [94] systems, but realistic potentials are not always available for adsorbate–surface systems. Typically, Lennard–Jones or Morse potentials are employed with parameters determined by experimental information or cluster calculations. In any case the results strongly depend on the interatomic potentials used, and this is an area where further work is needed.

An example of the molecular dynamics method used to treat diffusion was the treatment of O on Pt(111) [87]. Here Lennard–Jones potentials were parameterized on the basis of Pt bulk properties, vibrational properties, and adsorption energy properties of chemisorbed oxygen. The surface was represented as 12 layers each containing 32 atoms and one adatom species with periodic boundary conditions. Relaxation of the top layers was allowed. The Arrhenius-type behavior was found for the diffusion constant and good agreement with other computations was reported.

Surface diffusion of H and its isotopes on metal surfaces has been a topic of interest on Cu(100) surfaces [95,96]. The importance of metal motions has been found to be strongly related to isotope effects. This is particularly true at low temperature where for these light atoms tunneling behavior becomes apparent in the Arrhenius plots. Increases in the H/D kinetic isotope effect were found compared with rigid lattice calculations. It is not clear how to extrapolate these results to heavier adsorbate systems. Intuitively, those systems where a surface reconstruction induced by adsorbate is known to occur [97] would seem most prone to show the largest effect in the diffusion coefficient.

G. Monte-Carlo Simulations

1. Methodology

Monte-Carlo simulations are a popular way of representing surface diffusion. This is because they can properly account for the local concentration fluctuations not apparent in an overall concentration specification. The contribution to diffusion for these local concentrations is weighted by an appropriate function of their total energy. This leads to the possibility of disordered or ordered regions of adsorbate particles as a function of total concentration or temperature.

The Monte-Carlo simulation typically assumes a rigid metal lattice on which an adsorbed gas can diffuse by a series of site-to-site jump processes. Only one particle can occupy each lattice site. The particles are allowed to interact with one another in a pairwise fashion and a random walk procedure is begun. This overall procedure has been employed in a number of surface diffusion studies [84,98–104].

A typical Monte-Carlo simulation expresses the Hamiltonian of the lattice gas in the form of Eq. 3.19. More complete Hamiltonians containing three-body interactions can be considered [84,100].

The procedure for simulation now randomly places a number of particles on the substrate model to achieve a particular concentration level. Typical calculations may involve grids containing 100×100 sites or more. In a prediffusion phase, the particles are allowed to rearrange to achieve their low-energy configuration consistent with the particular temperature, concentration, and interaction parameters. This is achieved by a Metropolis walk [105] in which a particle and a jump direction are picked randomly. If the site to be jumped to is occupied, no jump occurs. If the site is unoccupied the probability of a jump (P) is computed from the change in energy (ΔE) if the jump occurs:

$$P = e^{-\Delta E/kT} \tag{3.29}$$

If P is greater than a random number picked between 0 and 1 the jump occurs; otherwise it is rejected. This procedure continues until the number of attempts equals the number of particles and then a time unit consisting of one Monte-Carlo time step has elapsed. This unit is used to relate the simulated time to real-time limits. The equilibration phase continues until equilibrium is reached. This can usually be determined by computing the energy of the system and noting when it becomes constant. Now the diffusion simulation can proceed. The probability of a jump now involves the difference in energy between the saddle point and the initial state. Many simulations assume that the energy of the saddle point is constant and only account for the energy changes of the initial state by way of adsorbate interactions. Here the energy change from the ground state to saddle point is

$$\Delta E = \varepsilon_0 + \Delta\varepsilon \tag{3.30}$$

where ε_0 is the diffusion barrier in the absence of adsorbate–adsorbate interactions and $\Delta\varepsilon$ is the change in initial state energy.

An interesting detailed analysis of the dynamics of Monte-Carlo simulations has been provided by Kang and Weinberg [106]. They have emphasized the need to treat the energy barrier in the dynamics of thermally excited processes. The important point they make is that using the energy barrier makes ΔE a function of the local adsorbate environment. As an extension of this important concept they have shown that variations in the distribution of local configurations can be responsible for experimentally observed compensation effects in the reaction rate, activation energy, and preexponential factor.

A tracer diffusion coefficient may be computed from Eq. 3.27. We note that one may want to compare this value with a chemical diffusion coefficient. Here the diffusion coefficient is taken from Reed and Ehrlich [33]:

$$D(\theta) = \Gamma(\theta)\lambda^2 \left(\frac{\partial(\mu / kT)}{\partial \ln \theta}\right) \tag{3.31}$$

The thermodynamic factor is expressed as

$$\left(\frac{\partial(\mu / kT)}{\partial \ln \theta}\right)_T = \frac{\langle n \rangle}{\langle \Delta n^2 \rangle} \frac{N_s'}{N_s + N_s'} \tag{3.32}$$

where n is the number of adatoms contained within a small region A which fluctuates about the mean value $\langle n \rangle$. The mean square of these fluctuations is

$\langle \Delta n^2 \rangle = (n - \langle n \rangle)^2$. Here N_s and N_s' are the number of sites inside and outside the area A, respectively.

2. *Results*

The usefulness of the Monte-Carlo procedure can be illustrated by an example concerning oxygen diffusion on W(110). Tringides and Gomer [84,100] employed pairwise and three-body oxygen–oxygen interaction parameters determined from LEED studies of phase diagrams. The experimental increase in activation energy of approximately 8 kcal/mol in the coverage range 0.2 to 0.6 was reproduced almost quantitatively. The preexponential factor also increases with coverage, but somewhat less than that observed experimentally. This work provides a good justification of the Monte-Carlo simulation procedure and a good estimate of the interaction parameters.

Monte-Carlo simulations have provided a good means of understanding the effect of impurity traps on the diffusion coefficient [99]. Interestingly, in the case of a surface in which a fraction of the sites is blocked and thus inaccessible to the diffusing species, the diffusion coefficient of the mobile species becomes independent of its concentration. As the block concentration increases, the diffusion coefficient decreases. These results are especially important when considering the case of surface diffusion on impurity-covered surfaces. Specific simulations of hydrogen diffusion on sulfur-covered Ru (001) have been reported [99]. Excellent agreement of the simulation with experiment concerning reduction of the hydrogen diffusion coefficient with sulfur coverage was found when each sulfur atom was allowed to block ten hydrogen threefold adsorption sites.

An important consideration in diffusion experiments is whether phase changes might be occurring over the temperature range of the study. This point has been illustrated nicely in a Monte-Carlo study of interacting particles [104,106]. In the case of attractive interactions the diffusion coefficient drops with increasing temperature and the linear Arrhenius plot is found. At low temperature the diffusion coefficient becomes small because the particles are tied up in dense phase domains. In the case of repulsive interactions the diffusion coefficient increases with a decrease in temperature.

3. *Laser-Induced Thermal Desorption Simulations*

A LITD experiment has been simulated explicitly by Monte-Carlo methods [107]. Here a fcc (111) hexagonal site arrangement is considered within the framework of the pairwise interaction model. The diffusion coefficient is measured by computing the concentration of adsorbate species within the repopulating region and then fitting this profile to the solution of the diffusion equation for circular boundary conditions [108]. A fit to the simulation is shown in Figure 3.12 from which a diffusion coefficient can be extracted. These repopulation curves are determined at successive time intervals as the concentration within the vacated area increases. It has been found [107] that the diffusion coefficient changes during repopulation. This behavior is caused by the concentration changes taking place within the repopulating

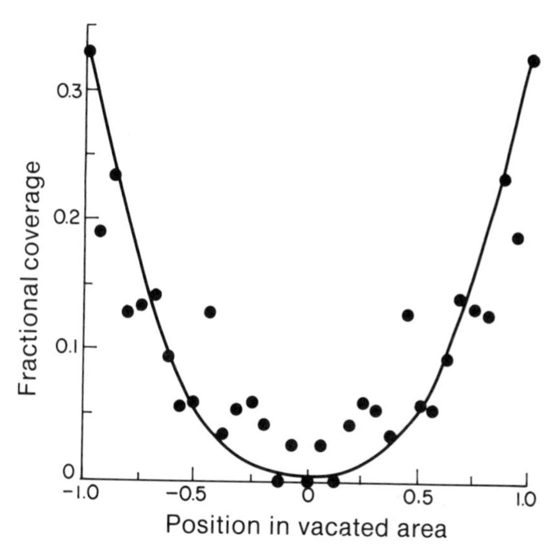

Figure 3.12. Plot of the refilling of LITD vacated area as simulated by Monte-Carlo methods. The simulation for H/Rh(111) at coverage equal to 0.33 is at 300 K and at 100 time steps.

region and is found for both attractive and repulsive interactions. This effect clearly indicates the difficulty with the analysis of LITD data where customarily one assumes a constant diffusion coefficient.

Arrhenius-type behavior is found from the diffusion coefficients extracted from the LITD simulation [107]. Figure 3.13 shows this behavior and a comparison with experimental LITD data. It is interesting that good agreement is observed in the simulation and experimental results when the larger preexponential factor is used. The normal preexponential factor of 1×10^{-3} cm^2/s, derived from transition state theory, corresponds to jump distances of one nearest neighbor. Thus, the larger preexponential factor required in Figure 3.13 (see also Table 3.1) may indicate jump distances much greater than one nearest-neighbor length.

Diffusion coefficients determined by analyzing the mean square displacement and LITD simulation may be compared [107]. The diffusion coefficients are generally twice as large when determined by mean square displacement as the LITD simulation. This difference may result from the fact that LITD measures a chemical diffusion coefficient, as has been pointed out in a theoretical Monte-Carlo analysis of this experiment [86].

H. Relation to Other Surface Experiments

Surface diffusion plays a key role in a number of phenomena currently under active study in surface science such as adsorbate-induced reconstruction of surfaces, surface reaction kinetics, and thermal desorption, to name a few. A strong case has been made that surface diffusion may frequently be the rate-limiting step in surface

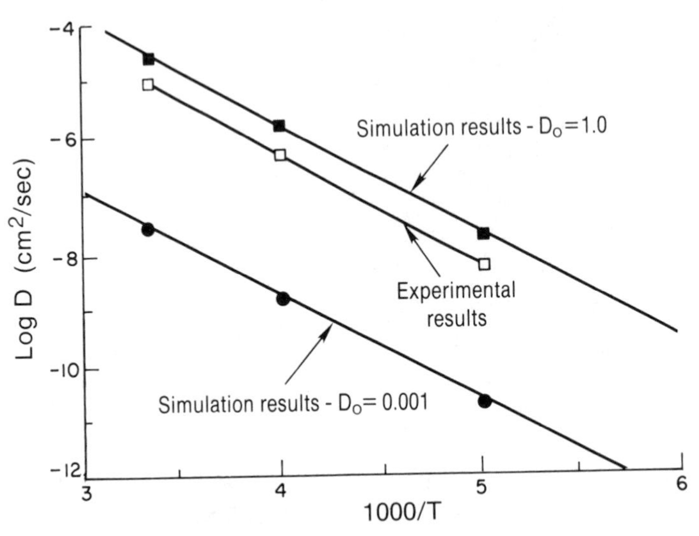

Figure 3.13. Arrhenius plot of diffusion coefficient versus reciprocal temperature for 0.24 coverage H/Pt(111). The diffusion coefficient is determined by the simulation using Monte-Carlo methods.

reactions [35]. In particular, it may be important in periodic transformations of surface structure during catalytic reactions. A good example of this is CO oxidation on Pt surfaces [81].

The hex (100) surface of Pt is well known to be stable as a hexagonal arrangement of atoms in contrast to the square arrangement expected from the bulk structure. On CO chemisorption the surface rearranges to the square structure because of the larger heat of adsorption on the square than hexagonal surface. On removal of CO the surface rearranges back to the hexagonal structure with an activation energy of 27 kcal/mol, and this phase is 10 kcal/mol more stable than the square phase [81].

The oxidation of CO involves a surface reaction between adsorbed CO and O atoms leading to CO_2 which rapidly desorbs. Under some conditions the steady-state rate of this reaction varies periodically with time when the Pt(100) surface is used as a catalyst. The hexagonal and square surfaces differ in their adsorption characteristics for CO, as discussed above, and for O_2. The adsorbed CO leaves only a few sites for O_2 adsorption, but once O_2 is adsorbed it dissociates and reacts readily with a molecule of CO, leaving two empty sites on the square Pt surface. As the CO is consumed by further reaction with O, the concentration of CO becomes very low and the surface transforms back to the hexagonal phase. This phase allows essentially only CO adsorption so a nearly complete CO monolayer is formed and the cycle begins again.

Clearly, in all of these processes the diffusion of metal atoms in the surface restructuring and the diffusion of chemisorbed atoms in the surface oxidation reaction play a key role. The nonperiodic oxidation of CO on various metal surfaces proceeds by the reaction with an adsorbed O atom. The adsorbed CO and O are clustered into islands, with reaction taking place at the phase boundaries. This phenomenon has been modeled well by various Monte-Carlo simulations [109–111]. The preexponential part of the rate constant is much smaller than a typical bimolecular preexponential and this fact is believed to be consistent with the reaction at the edges of islands. The activation energy of this reaction at high coverage on a variety of metal surfaces is close to the value 7 kcal/mol which is similar to the diffusion activation energy for CO (see Table 3.1). Thus there is the conjecture that many surface reactions are limited in rate by a diffusion step.

V. Summary

We have attempted to focus on conceptual aspects of surface diffusion involving adsorbed species on metal surfaces. The field is quite active and a number of experimental and theoretical techniques are proving to be useful. Adsorbate–adsorbate interactions cause the diffusion coefficient to vary with surface concentration. Thus it is convenient to separate our analysis along a concentration line below which we assume that these interactions can be neglected. We find that at low enough coverage atomic diffusion persistently follows a trajectory of hollow adsorption site to bridge saddle point. This result is found for adatoms treated by several methods including pairwise energy methods and bond order conservation methods. The energy barrier computed in the bond order conservation analytical model [76–79] directly relates to the heat of adsorption. This relationship is supported by diverse experimental and various pairwise energy and bond order conservation calculations.

Molecular chemisorption trajectories are more complex than atomic patterns. Donor molecules seem to follow atomic patterns, but acceptor molecules possess local energy minima along the diffusion trajectory. This means that the energy barrier to diffusion is not the energy difference between high-symmetry sites along the trajectory as in atomic diffusion. This result is projected by both the bond order conservation analytic model and computations and is quite consistent with experiment for CO chemisorption on Pt(111) surfaces.

In summary, the bond order conservation method offers a convenient analytic formulation that consistently predicts diffusion behavior. It provides a clear distinction between the monotonic energy trajectories of atoms and donor molecules as opposed to nonmonotonic energy trajectories of acceptor molecules. The method is flexible enough to treat adsorbate–adsorbate interactions and in particular leads to convenient formulas for the interaction terms. Application of these formulas to straightforward trajectory computations shows that the saddle point energy is dependent on local configuration of adsorbed species and may be shifted from the symmetric twofold sites expected at low concentration.

Much of the modeling, computation, and conceptual thinking treating surface diffusion involves a rigid substrate lattice. This assumption is clearly inadequate in some cases such as those involving chemisorption-induced transformations of the surface structure. It appears that the chemisorption energy may be a cause of a surface restructuring which would have a drastic effect on diffusion. An example of a periodic surface reaction involving CO oxidation on Pt(100) is understood in these terms.

The experimental manifestations of adsorbate–adsorbate interactions are varied in surface science. These interactions can be derived from the bond order conservation model and reduced to a pairwise formulation. The possible attractive or repulsive interactions are expressed in terms of the interaction energies for various two-center bonds. These terms can be used in bond order calculations where the explicit energy barrier between ground state and saddle point is computed. In this computation the energy of both the ground state and saddle point can change as a result of adsorbate–adsorbate interactions. It is found that the barrier increases for attractive interactions and decreases for repulsive interactions. This behavior is qualitatively similar to that produced by the quasichemical model.

Molecular dynamics has proven useful in understanding surface behavior. The instantaneous adsorbate–surface and adsorbate–adsorbate interactions are considered as opposed to interactions in a static field. Even rather crude adsorbate–surface potentials have shown the ability to uncover qualitatively important details of the interaction. These include the importance of energy exchange between the adsorbate and surface and the need for nonrigid surface atoms.

Monte-Carlo simulations have played an important role in relating adsorbate–adsorbate interactions to diffusion coefficients measured experimentally. Here the chemical diffusion coefficient measured in response to a gradient in the chemical potential and measured in most experiments is contrasted with the tracer diffusion coefficient which is measured at equilibrium. These coefficients show different dependence on adsorbate concentration.

References

1. R. Gomer, *Vacuum* **33** (1983), 537.
2. D. A. King, *J. Vac. Sci. Technol.* **17** (1980), 241.
3. G. Ehrlich, and K. Stolt, *Annu. Rev. Phys. Chem.* **31** (1980), 603.
4. G. Ehrlich, *CRC Crit. Rev. Solid State Mater. Sci.* **10** (1982), 391.
5. A. G. Naumovets, and Yu. S. Vedula, *Surf. Sci. Rep.* **4** (1985), 365.
6. J. D. Doll, and A. F. Voter, *Annu. Rev. Phys. Chem.* **38** (1987), 413.
7. S. J. Lombardo, and A. T. Bell, *Surf. Sci. Rep.* (1991).
8. B. Poelsema, L. K. Verheig, and G. Comsa, *Phys. Rev. Lett.* **49** (1982), 1731.
9. R. Viswanothan, D. R. Burgess, P. C. Stair, and E. Weitz, *J. Vac. Sci. Technol.* **20** (1982), 605.
10. E. L. Muetterties, T. N. Rhodin, E. Band, C. F. Brucker, and W. R. Pretzer, *Chem. Rev.* **79** (1979), 91.
11. C. H. Mak, J. L. Brand, B. B. Koehler, and S. M. George, *Surf. Sci.* **191** (1987), 108.
12. G. A. Somorjai, *Chemistry in Two Dimensions, Surfaces*, Cornell Univ. Press, Ithaca, N.Y., 1981.
13. G. A. Somorjai, and M. A. Van Hove, *Struct. Bonding (Berlin)* **38** (1979), 1.
14. W. Ho, N. J. DiNardo, and E. E. Plummer, *J. Vac. Sci. Technol.* **17** (1980), 134.
15. H. Conrad, R. Scala, W. Stenzel, and R. Unwin, *J. Chem. Phys.* **81** (1984), 6371.
16. M. A. Barteau, J. Q. Broughton, and D. A. Menzel, *Surf. Sci.* **133** (1983), 443.
17. C. M. Mate, and G. A. Somorjai, *Phys. Rev. B* **34** (1986), 7417.

18. H. Conrad, M. E. Kardesch, R. Scala, and W. Stenzel, *J. Electron Spectrosc. Relat. Phenom.* **38** (1986), 289.
19. T. E. Felter, E. C. Sowa, and M. A. Van Hove, *Phys. Rev. B* **40** (1989), 891.
20. A. M. Baro, H. Ibach, and H. D. Bruchman, *Surf. Sci.* **88** (1979), 384.
21. F. Besenbacher, I. Stensgaard, and K. Mortensen, *Surf. Sci.* **191** (1987), 288.
22. F. M. Hoffman, *Surf. Sci. Rep.* **3** (1983), 107.
23. J. P. Biberian, and M. A. Van Hove, *Surf. Sci.* **118** (1982), 443.
24. R. S. Saiki, G. S. Herman, M. Yamada, J. Osterwalder, and C. C. Fadley, *Phys. Rev. Lett.* **63** (1989), 283.
25. N. D. Shinn, and T. E. Madey, *Phys. Rev. Lett.* **53** (1984), 2481.
26. N. D. Shinn, *Langmuir* **4** (1988), 289.
27. F. Zaera, E. Collin, and J. L. Gland, unpublished.
28. G. S. Blackman, M. L. Xu, D. F. Ogletree, M. A. Van Hove, and G. A. Somorjai, *Phys. Rev. Lett.* **61** (1988), 2352.
29. I. J. Malik, and M. Trenary, *Surf. Sci.* **214** (1989), L237.
30. (a) E. Schweizer, B. N. J. Persson, T. Tushaus, D. Hoge, and A. M. Bradshaw, *Surf. Sci.* **213** (1989), 49. (b) W. D. Micher, L. J. Whitman, and W. J. Ho, *Chem. Phys.* **91** (1989), 3228.
31. S. L. Tang, M. B. Lee, Q. Y. Yang, J. D. Beckerle, and S. T. Ceyer, *J. Chem. Phys.* **84** (1986), 1876.
32. E. Shustorovich, *Acc. Chem. Res.* **21** (1988), 183.
33. D. A. Reed, and G. Ehrlich, *Surf. Sci.* **102** (1981), 588.
34. P. G. Shewmon, *Diffusion in Solids*, McGraw-Hill, New York, 1963.
35. D. G. Truhlar, and B. C. Garrett, *Annu. Rev. Phys. Chem.* **35** (1984), 159.
36. J. G. Lauderdale, and D. G. Truhlar, *Surf. Sci.* **164** (1985), 558.
37. A. F. Voter, and J. D. Doll, *J. Chem. Phys.* **80** (1984), 5814.
38. A. F. Voter, and J. D. Doll, *J. Chem. Phys.* **80** (1984), 5832.
39. C. H. Mak, and S. M. George, *Chem. Phys. Lett.* **135** (1987), 381.
40. R. C. Baetzold, and G. A. Somorjai, *J. Catal.* **45** (1976), 94.
41. V. P. Zhdanov, J. Pavlicek, and Z. Knor, *Catal. Rev.-Sci. Eng.* **30** (1988), 501.
42. E. G. Seebauer, and L. D. Schmidt, *Chem. Phys. Lett.* **123** (1986), 129.
43. E. G. Seebauer, A. C. F. Kong, and L. D. Schmidt, *J. Chem. Phys.* **88** (1988), 6597.
44. S. M. George, A. M. DeSantolo, and R. B. Hall, *Surf. Sci.* **159** (1985), L425.
45. C. H. Mak, J. L. Brand, B. G. Koehler, and S. M. George, *Surf. Sci.* **188** (1987), 312.
46. A. A. Deckert, J. L. Brand, M. V. Arena, and S. M. Georges, *Surf. Sci.* **208** (1989), 441.
47. (a) B. Roop, S. A. Costello, D. R. Mullins, and J. M. White, *J. Chem. Phys.* **86** (1987), 3003. (b) D. R. Mullins, B. Roop, S. A. Costello, and J. M. White, *Surf. Sci.* **186** (1987), 67.
48. T.-S. Lin, H.-J. Lu, and R. Gomer, *Surf. Sci.* **234** (1990), 251.
49. C. Dharmadhikari, and R. Gomer, *Surf. Sci.* **143** (1984), 223.
50. S. C. Wang, and R. J. Gomer, *Chem. Phys.* **83** (1985), 4193.
51. J. E. Reutt-Robey, D. J. Doren, Y. J. Chabal, and S. B. Christman, *Phys. Rev. Lett.* (1988), 2778.
52. X. D. Zhu, Th. Rasing, and Y. R. Shen, *Phys. Rev. Lett.* **61** (1988), 2883.
53. J. N. Russell, Jr., and R. B. Hall, *Surf. Sci.* **203** (1988), L642.
54. J. L. Brand, A. A. Deckert, and S. M. George, *Surf. Sci.* **194** (1988), 457.

55. A. Auerbach, K. F. Freed, and R. Gomer, *J. Chem. Phys.* **86** (1987), 2356.
56. T. Truong, G. Hancock, and D. G. Truhlar, *Surf. Sci.* **214** (1989), 523.
57. T. Truong, and D. G. Truhlar, *J. Phys. Chem.* **91** (1987), 6229.
58. J. G. Lauderdale, and D. G. Truhlar, *J. Chem. Phys.* **84** (1986), 1843.
59. K. B. Whaley, A. Nitzan, and R. B. Gerber, *J. Chem. Phys.* **84** (1986), 5181.
60. J. H. Lin, and B. J. Garrison, *J. Chem. Phys.* **80** (1984), 2904.
61. A. Gelb, and M. J. Cardillo, *Surf. Sci.* **75** (1978), 199.
62. V. Avdeev, T. H. Upton, W. H. Weinberg, and W. A. Goddard, *Surf. Sci.* **95** (1980), 391.
63. J. C. Tully, *J. Chem. Phys.* **73** (1980), 6333.
64. A. R. Gregory, A. Gelb, and R. Silbey, *Surf. Sci.* **74** (1978), 497.
65. A. Kara, and A. E. Depristo, *Surf. Sci.* **193** (1988), 437.
66. M. Y. Chou, and J. R. Chelikowsky, *Phys. Rev. Lett.* **59** (1987), 1737.
67. M. Y. Chou, and J. R. Chelikowsky, *Phys. Rev. B* **39** (1989), 5623.
68. T. N. Truong, D. G. Truhlar, J. R. Chelikowsky, and M. Y. Chou, *J. Phys. Chem.* **94** (1990), 1973.
69. P. J. Feibelman, *Phys. Rev. Lett.* **65** (1990), 729.
70. J. R. Banavar, M. H. Cohen, and R. Gomer, *Surf. Sci.* **107** (1980), 113.
71. G. Wahnstrom, *Phys. Rev. B* **33** (1986), 1020; *J. Chem. Phys.* **84** (1986), 5931.
72. K. Haug, G. Wahnstrom, and H. Metiu, *J. Chem. Phys.* **92** (1990), 2083.
73. J. B. Moffat, and D. Boerner, *Surf. Sci.* **114** (1982), 109.
74. R. C. Baetzold, *Surf. Sci.* **150** (1985), 193.
75. D. Boerner, and J. B. Moffat, *Surf. Sci.* **122** (1982), L608.
76. E. Shustorovich, *J. Am. Chem. Soc.* **106** (1984), 6479.
77. E. Shustorovich, *Surf. Sci.* **150** (1985), L115.
78. E. Shustorovich, *Surf. Sci.* **175** (1986), 561.
79. E. Shustorovich, *Surf. Sci. Rep.* **6** (1986), 1.
80. E. M. Shustorovich, R. C. Baetzold, and E. L. Muetterties, *J. Phys. Chem.* **87** (1983), 1100.
81. G. Ertl, *Langmuir* **3** (1987), 4.
82. T. L. Einsten, and J. R. Schrieffer, *Phys. Rev. B* **1** (1973), 3629.
83. R. C. Baetzold, *Langmuir* **2** (1986), 64.
84. M. Tringides, and R. Gomer, *Surf. Sci.* **145** (1984), 121.
85. E. D. Williams, S. L. Cunningham, and W. H. Weinberg, *J. Chem. Phys.* **68** (1978), 4688.
86. M. Tringides, *Surf. Sci.* **204** (1988), 345.
87. S. M. Levine, and S. H. Garofalini, *Surf. Sci.* **167** (1986), 198.
88. V. P. Zhdanov, *Surf. Sci.* **214** (1989), 289.
89. J. C. Tully, G. H. Gilmer, and M. Shugard, *J. Chem. Phys.* **71** (1979), 1630.
90. J. D. Doll, and D. L. Freeman, *Surf. Sci.* **134** (1983), 769.
91. A. F. Voter, and J. D. Doll, *J. Chem. Phys.* **82** (1985), 80.
92. S. H. Garofalini, and Halicioglu, *Surf. Sci.* **104** (1981), 199.
93. A. M. Stoneham, and J. H. Harding, *Annu. Rev. Phys. Chem.* **37** (1986), 53.
94. R. Biswas, and D. R. Hamman, *Phys. Rev. Lett.* **55** (1985), 2001.
95. T. N. Truong, and D. G. Truhlar, *J. Chem. Phys.* **88** (1988), 6611.

96. S. M. Valone, A. F. Voter, and J. D. Doll, *Surf. Sci.* **155** (1985), 687.
97. T. Gritsch, D. Coulman, R. J. Behm, and G. Ertl, *Phys. Rev. Lett.* **63** (1989), 1086.
98. M. Bowker, and D. A. King, *Surf. Sci.* **71** (1978), 583.
99. C. H. Mak, H. C. Andersen, and S. M. George, *J. Chem. Phys.* **88** (1988), 4052.
100. M. Tringides, and R. Gomer, *Surf. Sci.* **166** (1986), 419.
101. A. Sadiq, and K. Binder, *Surf. Sci.* **128** (1983), 350.
102. D. A. Reed, and G. Ehrlich, *Surf. Sci.* **105** (1981), 603.
103. M. Stiles, and H. Metiu, *Chem. Phys. Lett.* **128** (1986), 337.
104. X.-P. Jiang, and H. Metiu, *J. Chem. Phys.* **88** (1988), 1891.
105. M. Metropolis, A. W. Rosenbluth, A. N. Teller, and E. Teller, *J. Chem. Phys.* **21** (1953), 1087.
106. H. C. Kang, and W. H. Weinberg, *J. Chem. Phys.* **90** (1989), 2824. H. C. Kang, T. A. Jachimowski, and W. H. Weinberg, *J. Chem. Phys.* **93** (1990), 1418.
107. L. A. Ray, and R. C. Baetzold, *J. Chem Phys.* **93** (1990), 2871.
108. J. Crank, *The Mathematics of Diffusion*, Oxford Univ. Press, Oxford, 1975.
109. M. Silverberg, and A. Ben-Shaul, *J. Chem. Phys.* **87** (1987), 3178.
110. M. Silverberg, A. Ben-Shaul, and F. J. Rebentrost, *Chem. Phys.* **83** (1985), 6501.
111. M. Silverberg, and A. Ben-Shaul, *Surf. Sci.* **214** (1989), 17.

CHAPTER 4

Dissociative Chemisorption of Diatomic Molecules

David Halstead and Stephen Holloway

I. Introduction

A. General Background

The theoretical modeling of molecular scattering from a solid surface is a difficult problem. Because of the many degrees of freedom involved, there is an awesome array of channels into which the collision may redistribute energy and scattering probability. For a diatomic, with six internal degrees of freedom, the possibilities are more limited. The diatomic can reflect from the surface with a redistribution of internal energy, or remain trapped on the surface either as a molecule or as separate atoms. The surface itself, consisting of $\sim 10^{23}$ atoms, presents a subsystem that can be treated either by statistical methods or by division into a small localized zone, at the collision site, coupled to the remainder of the crystal in some way.

With the advent of ultrahigh vacuum experimentation, high-quality beam sources, single-crystal surfaces, and laser diagnostic instrumentation, gas–surface scattering is becoming a state-to-state spectroscopy. Many excellent reviews exist to which the interested reader is referred [1–7]. Coupled with these experimental developments, advances in theory and computational power have enabled both classical and quantum simulations to increase the number of degrees of freedom explicitly considered [2,8–12].

In the calculation of quantities that may be observed experimentally, care must be taken that the degree of parameterization in the interaction potential is not so great that *any* observed characteristic may be incorporated by "fine tuning" a meaningless variable in parameter space. *The use of simulations should be to conceptualize experimental findings and provoke further investigation rather than merely to corroborate experimental data using hindsight as the major input into the fitting process.* The key to modeling any interaction is to understand and reproduce the important energetics encountered during the interaction. Given the "correct" potential energy surface (PES) and an accurate simulation methodology, time-dependent dynamics may be performed and analyzed to gain insight into the microscopic reality.

In this review we restrict our attention to certain classes of reactive phenomena. One of the most interesting (and most studied!) adsorbates is the hydrogen molecule, probably because of its fundamental position in the periodic table and its importance in many industrially important reactions. Coupled to this, H_2 has an additional, intriguing quality in that for many purposes it must be treated as a quantum particle. From a theoretical viewpoint this entails a rethinking of methodology, as much of the existing simulation machinery involves integration of Hamiltonian equations of motion, which for this particular case could be inappropriate. In this chapter we concentrate on the interaction of H_2 with a variety of surfaces and explore the dependence of experimental observables on potential energy topologies. It is shown that considerable progress is being made in dealing with time-dependent quantum dynamic systems where the degrees of freedom are strongly coupled.

Although we do not *explicitly* address the dissociative adsorption of heavier molecules (CO, NO, O_2, etc.), conclusions drawn on direct reaction mechanisms for the light species do, to some extent, carry over. The main difference involves the degree to which heavier molecules transfer energy to surface excitations (electron-hole pairs and phonons) [13,14]. This, in turn, makes the problem richer in phenomena, as exemplified by the occurrence of precursor states in low-energy collisions [15]. Here, sufficient energy is lost to the substrate that molecular adsorption can occur; the molecule then diffuses over the surface until it finds a suitable site to dissociate. This problem is extremely difficult to simulate at a microscopic level as it involves long residence times over which the surface temperature must be accurately maintained. At higher energies a direct mechanism takes over which may be well understood from the findings related to H_2 dissociation. Coupled to these *dynamic* problems, detailed potential energy surfaces for such systems are far more complicated than for H_2, as is discussed in Chapter 1.

Since 1980 a significant body of experimental information has been collated on the interaction of H_2 with the transition metals. Results display a wide range of experimentally interesting scattering phenomena such as physisorption [16,17], activated dissociative chemisorption [18,19], diffraction [20], reactivity that depends on the population of internal degrees of freedom [21], and isotope and crystal face sensitive results [22]. Additionally there are a wide range of calculations for the interaction of hydrogen atoms and molecules with metal surfaces in general [23] and, in particular, cluster [24] and effective medium [25] calculations for both H_2/Cu and H_2/Ni systems. By using these calculations as a starting point, we examine the dynamic effects that should arise from differing potential energy surface topologies and demonstrate how it might be possible to link an initial reaction geometry with an observed scattering distribution.

To begin, we present a brief discussion of time-dependent quantum methodology and, in particular, its application to surface dynamic simulations.

B. Multidimensional Quantum Dynamics on a Grid

The full solution of the time-dependent Schrödinger equation may be evaluated in many different ways, employing a wide variety of classical, semiclassical, and quantum techniques. As this work is concerned with light molecules interacting with activated dynamic systems, the most direct approach is to solve the time-dependent Schrödinger (TDS) equation for a given initial set of conditions. This is most easily done by meshing out the region of space in which the interaction occurs, and then employing some iterative technique to propagate the system through time. The viability of such a methodology depends on how efficiently the wavefunction and associated operators are represented. This sampling of one point in phase space to be representative of a cluster of other, slightly displaced trajectories, is critical in classical trajectory calculations. The onset of chaos is predicted from two infinitesimally different initial trajectories resulting in uncorrelated final states. The meshed quantum representation requires at least one sampled point per unit of phase space volume h^D (Planck's constant, where D is the dimensionality of the problem). The closer to one sampling point per volume element (this is before the method breaks down), the more efficient the iteration scheme. In all grid methods, the maximum resolvable momentum depends on how finely spaced the sampled points in real space are; the most negative momentum resolvable is $-2\pi N/L$, whereas the most positive is $2\pi(N-1)/L$, where N is the number of mesh points and L is the length of real space being meshed.

C. The Split Operator Method

The split operator method was initially formulated for solving the propagation of laser pulses through the atmosphere [26], but was later generalized for the time-dependent Schrödinger equation [27]. The simplest way of stating the time development of the TDS equation is integrating for a time interval Δt, that is,

$$\int_{\psi_0}^{\psi\Delta_t} \frac{d\psi}{\psi} = -i \int_0^{\Delta t} (\mathfrak{T}+\mathfrak{V})dt, \tag{4.1}$$

leading to

$$\psi(\Delta t)=\exp\left(-i\Delta t\,(\mathfrak{T}+\mathfrak{V})\right)\psi_0 \tag{4.2}$$

To use Eq. 4.2, the operators must be applied individually and by symmetrically splitting the kinetic energy operator, giving

$$\psi(\Delta t) = \exp\left(-\frac{i\Delta t\mathfrak{T}}{2}\right)\exp\left(-i\Delta t\mathfrak{V}\right)\exp\left(-\frac{i\Delta t\,\mathfrak{T}}{2}\right)\psi_0 \tag{4.3}$$

which leads to integration correct to Δt^3. The key advantage this provides is that the total number of time steps needed for accurate propagation may now be reduced to as few as 250, giving a factor of at least 4 over earlier methods [28,29]. It is also interesting to note that for a constant potential, Eq. 4.3 becomes exact; thus errors accumulate only when the packet is interacting with the spatially varying region of the potential.

Examination of Eq. 4.3 shows there to be a half-time-step free space propagation, followed by a full-time-step potential-induced phase change, then a further half-step free space propagation. In practice, if multiple cycles of these three operations are used then sequential half-free-space steps are run, together resulting in the scheme shown in Figure 4.1. From this it can be seen that each time iteration again requires

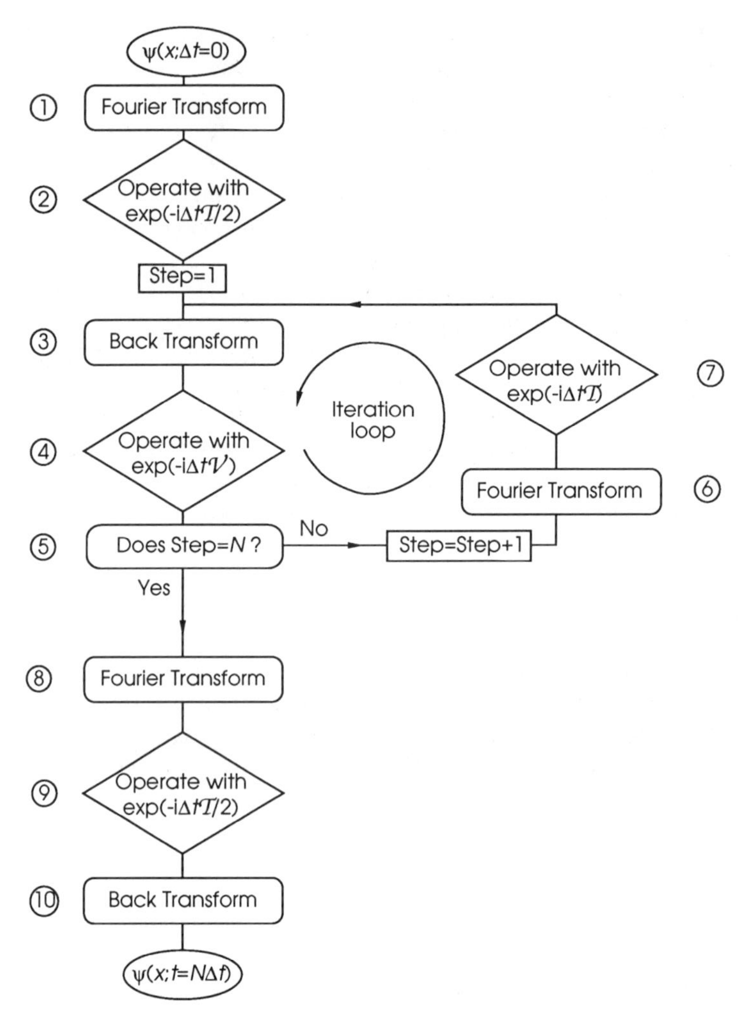

Figure 4.1. Flow diagram of the split operator propagation scheme showing the way in which the half-time steps are concatenated in the iteration loop to give an efficient cycle requiring two Fourier transforms and two multiplications.

only two N-dimensional Fourier transforms. Application of the potential operator is simple, requiring the multiplication of each configuration space mesh point by $\exp(-i\Delta t\,\mathcal{V}(z))$, where $\mathcal{V}(z)$ is the value of the potential at position z.

The accuracy with which the free space propagation operator maps onto the momentum space wavefunction is the key to the increase in efficiency. Examination of the operator in momentum space shows that it too is only a phase change, multiplying each mesh point by a constant, imaginary exponential, dependent only on the momentum value at each sampled point. This may be expected to be the case because the momentum space probability function $(\psi_p{}^*\psi_p)$ is time independent in the absence of a varying potential. For a particle of mass m, the momentum space propagator may be expressed as

$$\psi_n(t+\Delta t) = \psi_n(t) \exp\left(-\frac{i\Delta t p^2}{2m}\right) \tag{4.4}$$

which in grid coordinates corresponds to

$$\psi_n(t+\Delta t) = \psi_n(t) \exp\left(-\frac{i\Delta t \pi n^2}{mL_x}\right), \tag{4.5}$$

where the index n counts from $-N/2$ to $N/(2\text{-}1)$. The extension of this method to higher dimensions is trivial, with the kinetic energy operator becoming

$$\psi_{n_x n_y}(t+\Delta t) = \psi_{n_x n_y}(t) \exp\left(-i\pi\,\Delta t\left(\frac{n_x^2}{m_x L_x} + \frac{n_y^2}{m_y L_y}\right)\right), \tag{4.6}$$

for a two-dimensional grid spanning the interval L_x and L_y.

This method was first applied to surface dynamics by Jackson and Metiu [30], who studied the dissociative adsorption of H_2 on a Ni surface modeled by a London–Eyring–Polanyi–Sato (LEPS) [12] potential. This study demonstrated the feasibility of the method for surface dynamic problems and led the way for a number of similar studies on a wide range of other potential surfaces [31–33]. More recently, the method has been used to study systems of extended dimensionality and it is here that we begin. In the next section we present some recent results for a four-dimensional simulation of H_2 scattering from a Cu surface. In the ensuing sections we dissect the principal results from this work and question the validity of limited dimensionality studies. It should be borne in mind that at this point in time it is not computationally feasible to perform a six-dimensional time-dependent quantum study of the dissociation of a diatomic molecule on even a *static* surface. Consequently, it is important to be able to assess the extent of coupling involved between the various degrees of freedom, particularly near the transition state, to judge whether limited dimensionality studies are at all useful.

II. Dissociative Adsorption of Hydrogen: Rotation, Vibration, and Diffraction

A. Potential Energy Surface and Simulation Details

To treat the collision of the hydrogen molecule with a static surface, six degrees of freedom must be considered. Of these six dimensions, three are internal degrees of freedom, two rotations and a vibration, and the other three are center-of-mass translations. As stated in the Introduction, because of computational limitations only four of these six dimensions may be treated explicitly using exact quantum dynamics. To this end, both hydrogen atoms are constrained to lie in a plane perpendicular to the metal surface connecting the bridge and center sites of the (100) face of copper (Figure 4.2). Although this does not imply that azimuthal angle (ϕ) effects will be negligible or even uninteresting, there will be stronger dynamic coupling to the dissociation and rotational excitation probability for variations in the polar angle θ, which is included in the calculation [34]. This may be expected as dissociation cannot occur for a molecule constrained to have its axis normal to the surface

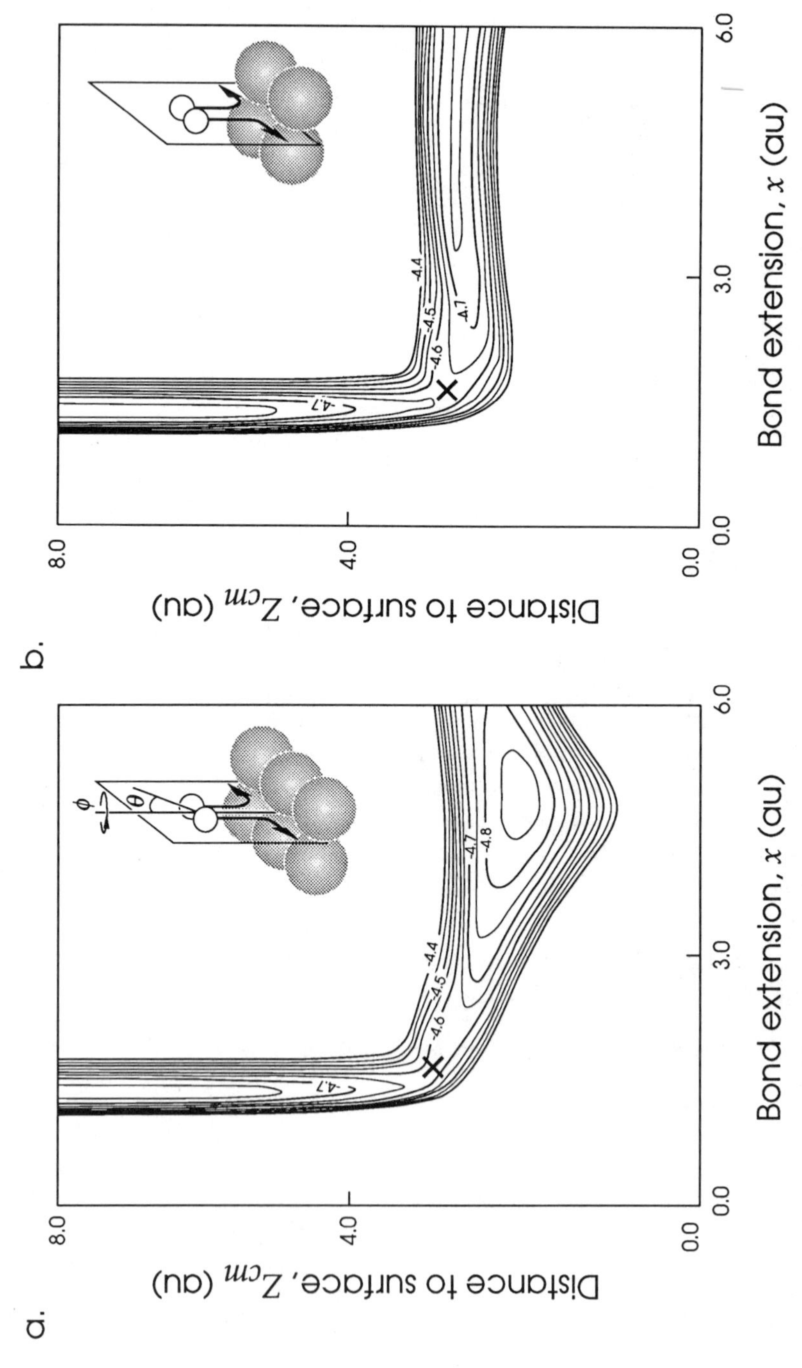

Figure 4.2. Two cuts of the PES used in Section II. Both are for the restricted geometry that keeps the molecular axis parallel to the surface. Contour values are in electron volts, and zero energy is taken to be two dissociated hydrogen atoms far from the surface. The gas-phase well depth for H_2 is taken as -4.79 eV. **(a)** Bridge-to-center slice. The location of the activation barrier is shown by a cross and its value is $E_{act} \sim 175$ meV. **(b)** Center-to-bridge dissociation with an activation barrier of 155 meV.

($\theta = 0$) and impulsive collisions will predominantly excite cartwheel rotations with J parallel to the surface [35]. The two protons are located at $X_1\ Z_1$ and $X_2\ Z_2$, where X is in the surface plane and Z is normal to the surface. For propagation, a second choice of coordinates is employed that refers to the center of mass $X_{cm} = (X_1 + X_2)/2$ and $Z_{cm} = (Z_1 + Z_2)/2$ and the relative atomic separations $x = X_1 - X_2$ and $z = Z_1 - Z_2$, which results in a simple expression for the Hamiltonian of the system.

The potential energy surface chosen for this quantum simulation has been obtained from an effective medium calculation [23]. Even though the value of the activation barrier to adsorption is at variance with current experimental observations, it has the virtue that all degrees of freedom are treated on an equal footing, thus enabling a multidimensional analysis to be attempted.

The usual method for depicting the reaction potential is to constrain the molecular axis in the surface plane, then map out the potential energy as a function of distance above the surface and bond extension. This gives the ideal configuration for dissociation to occur and thus shows the minimum activation barriers for the PES. In this case there is a 20-meV difference between the value of the dissociation barrier at the bridge (155 meV) and center (175 meV) sites. These two slices of the PES are shown in Figure 4.2. The finding that H_2 adsorption on Cu(100) is activated is in good agreement with experiment but, as mentioned, the calculated barrier appears too low (experimental findings vary between 0.5 and 1 eV [19,20]). This PES is, however, representative of a particular class of systems having a barrier that is low compared with the $0 \rightarrow 1$ vibrational excitation in the free molecule (this point is discussed at length in Section IV). This PES is used in investigating the effects of the rotational degree of freedom on reactivity and also on the rotational population of the scattered flux, to which ends it is ideally suited.

Using the four coordinates X_{cm}, Z_{cm}, x, and z, the mapping of the split operator propagation scheme becomes trivial, with the mass in the two relative coordinates being the reduced mass of H_2 (918 au) and the molecular mass (3672 au) in the remaining two. The number of mesh points required to resolve each dimension and the various box lengths are presented in Table 4.1. An absorbing boundary [29] was placed in the outermost 16 points along each edge of the x coordinate to deal with flux lost through surface-mediated dissociation. No boundary is needed in the z coordinate, because for this dimension the potential rises steeply toward the box edges as amplitude in this region would physically correspond to desorption of a single atom, a process that is forbidden for the energies used in this work.

Table 4.1. Grid Parameters Used in the Four-Dimensional Calculation of H_2/Cu(100)[a]

Coordinate (au)	Grid Size (points)
$-6.0 < x < +6.0$	128
$-3.0 < z < +3.0$	32
$-2.33 < X_{cm} < +2.33$	8
$-0.5 < Z_{cm} < +17.5$	128

[a]See Section II.

The initial state is chosen to be that of a beam of molecules incident normally to the surface in the ground rotational and vibrational states,

$$\psi_n^i \sim \phi_0\left(\sqrt{x^2 + z^2}\right) g\left(Z_{cm} - Z_{cm}^i, k_Z^i\right) \tag{4.7}$$

where $\phi_0\left(\sqrt{x^2+z^2}\right)$ is the ground eigenstate of the gas-phase H_2 molecule and $g(Z_{cm} - Z_{cm}^i, k_Z^i)$ is a Gaussian wave packet in Z with mean position Z_{cm}^i and mean momentum k_Z^i. The value of Z_{cm}^i is chosen so that initially there is no overlap between the molecular wavefunction and the surface potential.

The accessible scattering channels are dissociation into the atomic well on the surface, reflection from the surface with rotational excitation, and Bragg diffraction as a result of the PES variation within the unit cell. No vibrational excitation can occur for the scattered flux as the translational energies used in this work lie well below the 512-meV threshold for the $0 \rightarrow 1$ transition in gas-phase H_2. Flux is lost through dissociation and interaction with the absorbing boundary at large x, but because the propagator is unitary, the complement of the integrated intensity reflected from the surface is equal to the sticking fraction, S_0.

B. Results

a. Diffraction and Dissociation in the Four-Dimensional Simulation

The underlying periodicity of the surface unit cell imposes the Bragg scattering condition on the reflected wavefunction which restricts the change in parallel momenta to be $\pm 2\pi/L_x$. By summing over rotational satellites, the diffraction intensities as a function of translational energy for a normally incident beam are shown in Fig. 4.3.

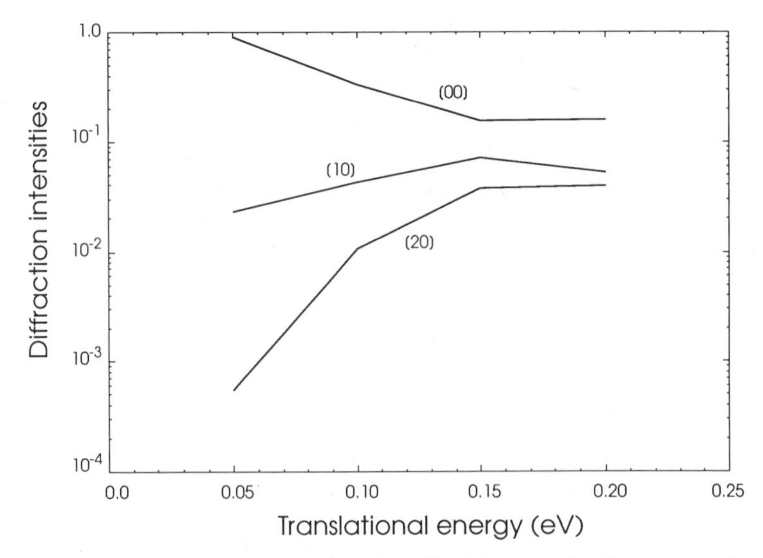

Figure 4.3. Diffracted intensities as a function of initial translational energy obtained from the four-dimensional calculation described in Section II.

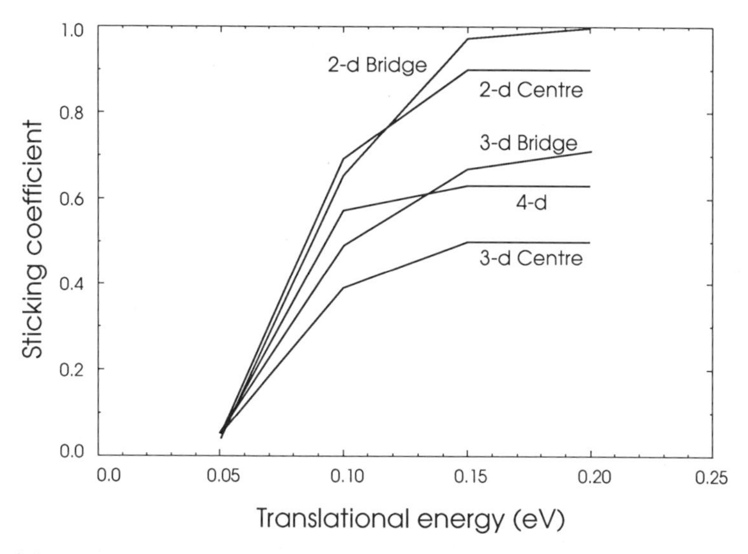

Figure 4.4. Sticking coefficient as a function of translational energy of the molecule. Shown in this figure are the two-dimensional results for a dissociation geometry that restricts the molecular axis to be parallel to the surface and the center of mass to be above either a bridge site or a center site (see Fig. 4.2). Additionally, three-dimensional results are shown whereby rotational motion is included. Finally, results for a four-dimensional calculation are given where, in addition to the three-dimensional simulations, we allow the center of mass to translate within the plane of rotation.

As might be expected at low energies, before the barrier energy is reached, the surface appears smooth and little diffraction occurs. As the translational energy is raised and exceeds the activation barrier, the behavior of the curves differs little from that which may be obtained from a nonreactive surface; that is, the specular falls and the higher-order beams gain intensity. The only difference is that here the total intensity is modified by the fraction that is dissociated by the surface which does not reflect coherently. The surprising feature is that the diffraction intensities are not sensitive enough to reflect the fact that a major feature in the PES is changing and that a new decay channel is opening. Figure 4.4 shows S_0 as a function of the initial translational energy E_i. The saturation value for the four-dimensional sticking probability lies well below unity at ~ 0.63. This nonunity probability may derive from several sources such as variation in the potential in the coordinate across the surface unit cell and orientational considerations. To check this, a series of lower-dimensionality test simulations were run.

b. Coupling to Translational Motion: Two-Dimensional Simulations

When the molecular axis is constrained to lie in the plane of the surface and the center of mass is fixed over a high-symmetry site, the scattering problem reduces to two dimensions, that is, Z, the center-of-mass distance above the surface and the bond length x. These two-dimensional results are shown in Figure 4.4 for constrained geometries at both the bridge and center sites. It is clear that in both cases the curves rise to a much higher value before saturating, implying that the lower value of S_0 found in the four-dimensional simulations does not arise from the coupling of the center-of-mass motion to the reaction coordinate.

The onset of dissociation is far below the value of the activation barrier (~ 165 meV) for either site. This effect has been seen on several previous occasions [30,33] and has its origins in two effects: (1) vibrational-to-translational energy transfer, V → T, as the molecule traverses the "corner" of the reaction zone (Figure 4.2), and (2) tunneling through the activation barrier. Curves for both geometries rise rather steeply, indicating a wide barrier, and saturate near E_i ~ 170 meV, with the bridge result leveling near unity sticking probability and the center simulation at 0.9. What this implies is that there is very little difference between the *dynamics* at these two sites in the surface. Looking at the immediate contrast between the two plots shown in Figure 4.2 this may seem surprising; however, while the barrier at the center site is certainly less than that at the bridge site, to dissociate at the center, the molecule must make a very sharp turn, which results in unfavorable dissociation dynamics. This is exactly the opposite of what occurs at the bridge site, where the dissociating molecule simply runs down in the direction of the chemisorption minimum, located above the center site. At the highest values of E_i, the sharp turn required at the center site cannot be made, as the turnaround time in Z_{cm} is too quick for dissociation and hence the saturation value never reaches unity. As the translational degree of freedom does not give rise to the suppressed sticking coefficient in the four-dimensional case, the effect must be due to the coupling of rotational motion with the reaction coordinate.

c. Coupling to Rotational Motion: Three-Dimensional Simulations

By constraining the center of mass to lie over a high-symmetry site, but allowing it to rotate and vibrate in a plane, the problem becomes three-dimensional in nature. The results for two different surface sites are given in Figure 4.4 and clearly show that the inclusion of rotation normal to the plane of the surface results in the depressed saturation value for the sticking coefficient. For the bridge site a value of ~0.7 is found and for the center site, a value of 0.5. This implies that it is the interplay between the rotational degree of freedom and the reaction coordinate that accounts for the suppression of the sticking coefficient seen in four-dimensional calculations. To probe deeper into the dynamics of the interaction, Figure 4.5 shows the asymptotic rotational state distributions for the scattered fraction of the flux as a function of translational energy. As the H_2 molecule is a homonuclear diatomic, the excitation of odd j states is forbidden by symmetry. At very low energies all of the flux emerges in the $j = 0$ state purely from conservation of energy, as the thresholds for rotational excitation for the $0 \rightarrow 2$ and $0 \rightarrow 4$ transitions lie at 30 and 120 meV, respectively. At these low translational energies it has been shown that "mechanical" collision with a corrugated wall is not capable of exciting such a degree of rotational excitation in the H_2 molecule [36]. The only possibility left is that in the course of the reaction a selectivity occurs that, in turn, accounts for the perturbation in the state distributions.

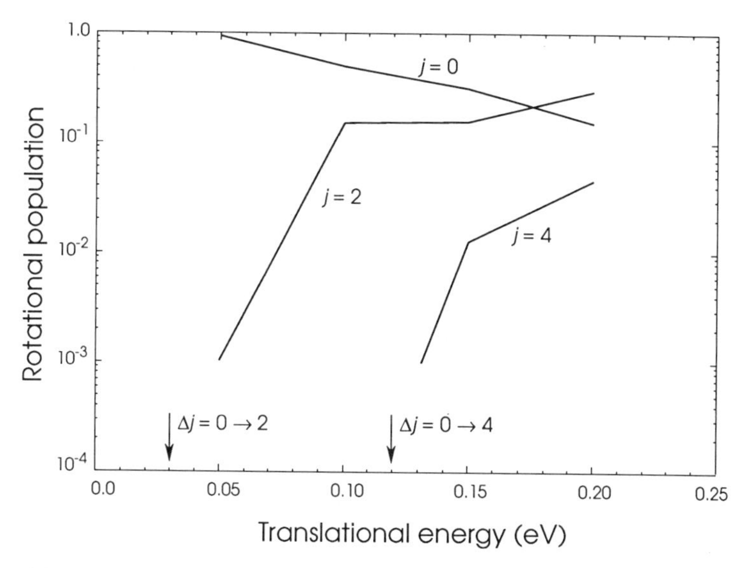

Figure 4.5. Rotational excitation probabilities for the three-dimensional calculation described in Section II. This occurs as a consequence of the stereospecificity in the dissociation process. Those molecules with their molecular axis parallel to the surface normal do not dissociate, and, as a consequence, the scattered flux has lost a particular angular fraction (see Eq. 4.8). Translated into a final state distribution, this implies rotational excitation. Alternatively one can think of this as arising from a strongly anisotropic gas–surface potential. The thresholds for the first two rotational excitations are indicated by arrows.

To investigate this effect further, consider a model [37] where the molecular center of mass is restricted to lie over one particular surface site and, furthermore, assume that the vibrational coordinate responds adiabatically as the transition state is crossed (this approximation will be discussed in detail in Section IV). If the molecule rotates only in a plane with J parallel to the surface then, because the potential energy for the homonuclear diatomic is periodic in the interval $0 \leq \theta \leq \pi$, scattered molecules will obey the selection rule $\Delta j = \pm 2$. Suppose that the initial rotational state distribution is isotropic with $j = 0$ labeling the two-dimensional quantum number. For the scattering of H_2 from surfaces, it has been frequently observed that almost no rotational excitation occurs [18]. Let us assume that only those molecules having their molecular bond axis within a specific range of angles $\pm\alpha$ are able to dissociatively chemisorb. This is in accord both with simple intuition and with total energy calculations [38]. In the extreme limits, molecules with their axis normal to the surface will fragment into an adsorbed atom and an atom in the gas phase, and those dissociating parallel to the surface will chemisorb two H atoms, the former process being energetically less favorable by $\sim$2.7 eV. Defining the surface normal to be $\theta = 0$, preferential dissociation will occur for molecules having $\theta = \pi/2 \pm \alpha$. Clearly, α will depend markedly on the substrate chosen and also the initial translational energy. For a system with an activation barrier to dissociation, it might be expected that at the (classical) threshold α would be very small, increasing as the barrier energy is surpassed. The scattered flux of molecules will therefore be depleted of those molecules lying flat and this selection, by the highly anisotropic surface potential, will result in a rotational distribution with an unusually high occupation of j states. This distribution will, of course, be modified by any attractive steering forces and the impulsive collision, but for the H_2 potential these are expected to be small.

To model this behavior, consider as an ansatz a scattered wavefunction whose angular distribution, originally isotropic, has two Gaussian-shaped portions removed around $\theta=\pi/2$ and $3\pi/2$,

$$\psi_f(\theta) = \frac{1}{\sqrt{2\pi}}\left[1-A\left(\exp\left\{-\left(\frac{\theta-\pi/2}{\alpha}\right)^2\right\}+\exp\left\{-\left(\frac{\theta-3\pi/2}{\alpha}\right)^2\right\}\right)\right] \quad (4.8)$$

where A is the amplitude of the Gaussian function. Normalization is chosen such that the integral of $\psi_f{}^*\psi_f$ over the interval $0 - 2\pi$ is $(1 - S_0)$. For $\alpha < 90°$, S_0 is approximately[1]

$$S \approx \frac{A\alpha}{\sqrt{2\pi}}\,[2\sqrt{2}-A] \quad (4.9)$$

Assuming that the angular width is small enough so that the removed fractions do not overlap, the resulting probability distributions of angular momenta in the scattered beam are found by projecting Eq. 4.8 onto the rotational wavefunctions $e^{ij\theta}/\sqrt{2\pi}$:

$$|P_0|^2 = \left[1 - \frac{A\alpha}{\sqrt{\pi}}\right]^2 \quad (4.10)$$

and

$$|P_j|^2 = \frac{(A\alpha)^2}{2\pi}\exp\{-j^2\alpha^2/2[1 + \cos(j\pi)]\} \qquad (j\neq 0)\,. \quad (4.11)$$

Figure 4.6 shows plots for the rotational distributions, $|P_j|^2$, as a function of α for the cases $A = 0.5$ and $A = 1.0$. These results show quite clearly that as the dissociative adsorption channel opens and α increases, the scattered fraction exhibits significant rotational excitation, a conclusion borne out in more detailed calculations where the actual energetics have been included [34,39]. The $|P_j|^2$ tend to bend over at the higher values of α; this is simply a consequence of the $\psi_f{}^*\psi_f$ reverting to an isotropic distribution as more particles stick.

It is clear from the figure that as the acceptance angle increases, the amount of excitation of the $j = 2$ state in particular is radically altered. The reason for this is that because the PES has θ periodic in π, the resulting wavefunction will adopt this functionality and consist predominantly of $j = 0$ and $j = 2$ states. In addition the greater the dissociation probability for a given orientation, altered using A, the greater is the excitation of all the higher states, as seen when comparing Figures

---

[1]This obtains for $0° < \alpha < 120°$ for $A=0.5$ and $0° < \alpha < 78.5°$ for $A=1.0$.

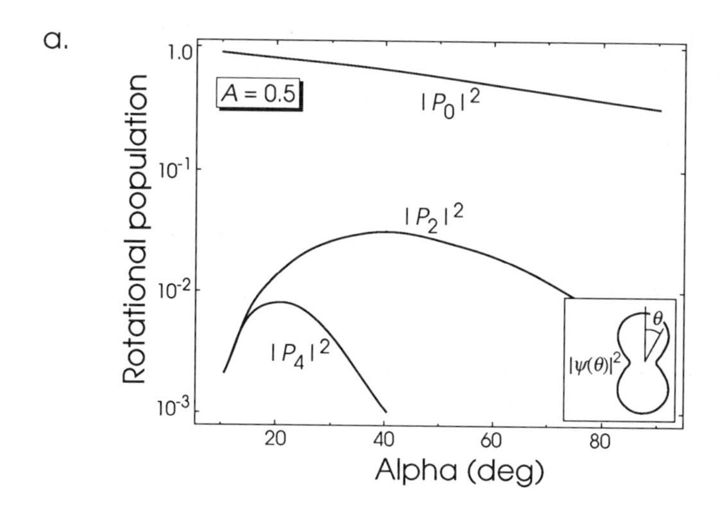

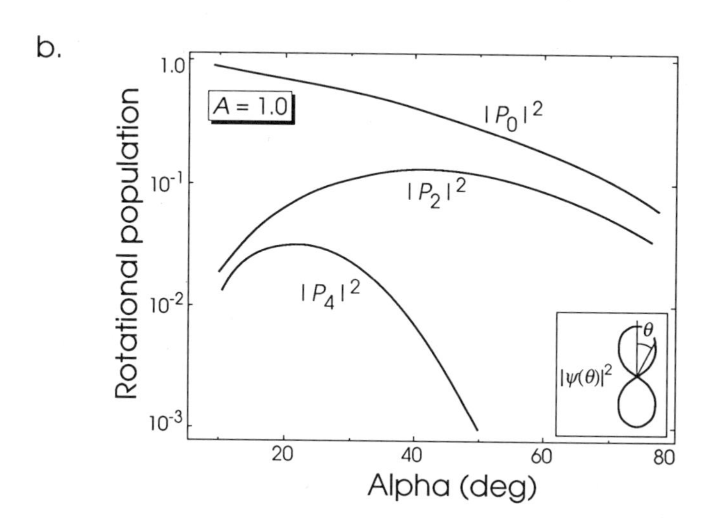

Figure 4.6. Variation of the rotational state populations with the width of the Gaussian function used in the ansatz Eq. 4.8. Two values for the amplitude have been chosen for illustration: **(a)** $A = 0.5$ and **(b)** $A = 1.0$. Within the insets are shown the probability distributions for the two cases with $\alpha = 45°$.

4.6a and 4.6b. With reference to the data from the four-dimensional calculation presented in Figure 4.5, it is possible to obtain an approximate value for A by equating it with the value of S_0 found in the two-dimensional calculations shown in Figure 4.4. For $A = 0.5$ ($E_i = 0.09$ eV) and $A = 0.9$ ($E_i = 0.15$ eV), the values of α that give the correct sticking fraction in the four-dimensional simulations (Eq. 4.10) are $\sim 60°$ and $70°$, respectively. For these parameter values the rotational populations from Eqs. 4.10 and 4.11 are $|P_0|^2 \sim 50\%$ and $|P_2|^2 \sim 2\%$ at 0.09 eV and $|P_0|^2 \sim 7\%$ and $|P_2|^2 \sim 4\%$ at 0.15 eV, which compare favorably with values shown in Figure 4.5.

It is important to note that *no* energetics have been included in this calculation and the abscissa in Figure 4.6 is the width of the acceptance angle for dissociation. If the total energy of the diatomic molecule lies below the energy needed to populate the $j = 2$ state, then no rotational excitation will be seen in the scattered flux even though dissociation may still occur but with an isotropic loss of all orientations.

III. Diffraction of H_2 from Reactive Metal Surfaces

A. Potential Energy Surface H_2/Cu[110]

In the four-dimensional simulation presented in Section II, the diffractive scattering arising from the surface corrugation- and site-dependent reaction probability showed remarkably little sensitivity to the opening of the channel to dissociation (Figure 4.5). This observation will now be examined in more detail by employing a model PES that allows for control over the key features of the interaction potential while capturing the interesting features of the full effective medium PES [31].

The basic principle fundamental to the design of this PES was simulation of the corrugation functionality known to exist from atom scattering data [40], introduced by Harris and Liebsch [41] in their study of the He/Cu(110) system. In addition, the product channel of adsorption, either into a molecular precursor or directly to surface-activated dissociation, must be accounted for. The physisorption region of the potential is based on the traditional van der Waals z^{-3} attraction,

$$V_{vW} = -\frac{C_{vW}}{(z-z_{vW})^3} f(k_c[z-z_{vW}]) \tag{4.12}$$

where C_{vW} and z_{vW} are the van der Waals constant and origin, respectively, and $f(q) = 1\text{-}[2q(1+q)+1]e^{-2q}$, which simulates the saturation of the interaction as z tends to z_{vW}. As the localized states on the molecule begin to overlap with the outermost (Fermi level) electrons, an exponential repulsion is encountered, which shows increasing site dependence as the molecule moves deeper into the metal electron states

$$V_{\text{rep}}(x,y,z) = V_0 e^{-\alpha z}\,[1 - \alpha h_{\text{rep}}(x,y,z)] \tag{4.13}$$

where

$$h_{\text{rep}}(x,y,z) = \frac{1}{2} e^{-\beta(z-z_0)} \left\{ h_x\left(1\text{-}\cos\left[\frac{2\pi x}{a_x}\right]\right) + h_y\left(1\text{-}\cos\left[\frac{2\pi y}{a_y}\right]\right)\right\}. \tag{4.14}$$

The parameters used are as follows: α and β are related to the electron decay length outside the surface and determine the steepness of the potential, V_0 and z_0 may be estimated from bound-state physisorption data [42], and a_x and a_y are the dimensions of the surface unit cell. The remaining parameters, h_x and h_y, are the corrugation amplitudes and depend on the detailed electronic structure of the crystal face under investigation.

Having defined the H_2/surface diabatic potential there only remains generation of a functional form for the interaction of atomic hydrogen with the surface. As was assumed in the previous section, the vibrational coordinate is again treated by a one-dimensional effective potential. As a consequence, the vibrational coordinate evolves adiabatically, allowing the wave packet to follow the minimum energy pathway to dissociation. The attractive potential experienced by an H atom follows approximately the corrugated effective electronic potential, while the repulsion will be proportional to the local electron density [43]. Thus, the entire potential will display a marked site dependence and asymptote as $z \rightarrow \infty$ to the gas phase dissociation energy. A detailed description of its behavior close to the surface is, however, not required as it is assumed that, following dissociation, any subsequent desorption will be incoherent. This flux will thus not be propagated for more than a few atomic units beyond the barrier (i.e., a distance greater than the wavelength of the probe molecule, typically 1 au). The explicit form used for the corrugated exponential attraction is

$$V_{att}(x,y,z) = V_{diss} - V_1 \exp[-\gamma\,(z - g_{att}\,(x,y,z))] \tag{4.15}$$

where

$$g_{att}\,(x,y,z) = \frac{1}{2}\left\{ g_x\left(1\text{-}\cos\left[\frac{2\pi x}{a_x}\right]\right) + g_y\left(1\text{-}\cos\left[\frac{2\pi y}{a_y}\right]\right)\right\} \tag{4.16}$$

V_{diss} is the dissociation energy for gas-phase hydrogen (including zero-point energy), and V_1, γ, g_x, and g_y are varied to position the minimum activation barrier, of the required height, at different points within the unit cell. Zero potential is taken to be that of ground state, gas-phase H_2.

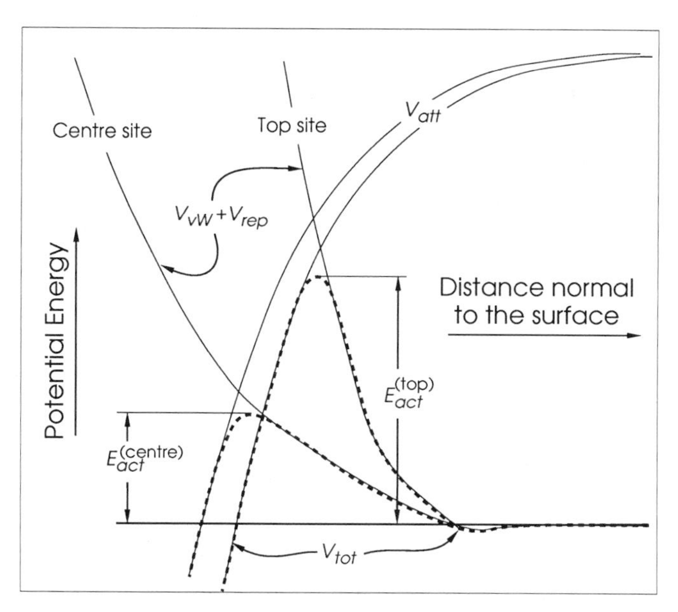

Figure 4.7. Two slices through the activated PES at the center and top sites in the diffraction study presented in Section III. Both of the diabatic potentials are shown for each site (thin solid line) and their adiabatic combination is shown by the thick dotted line.

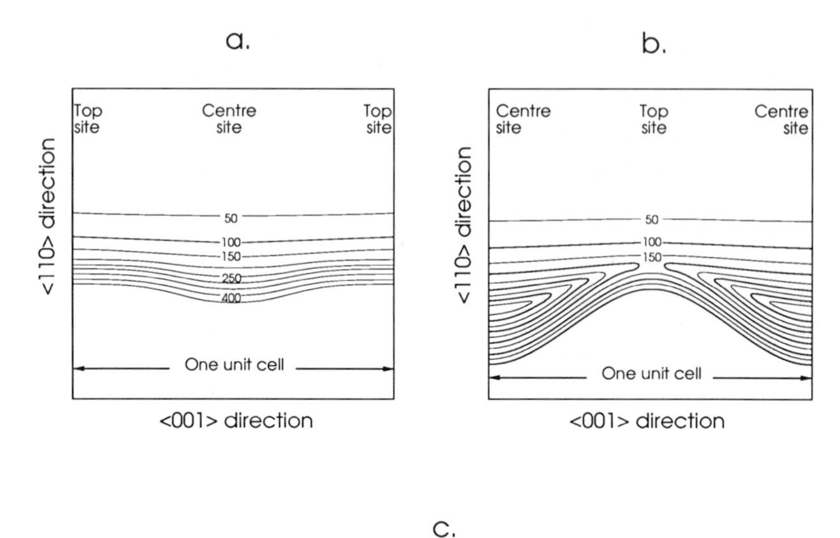

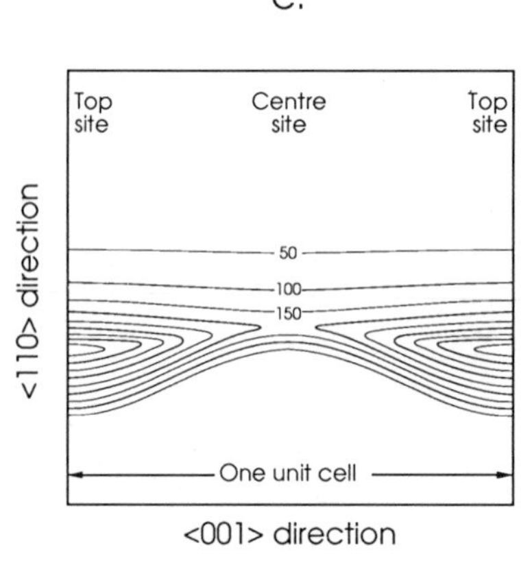

Figure 4.8. Three potentials used to examine the effect of dissociation site within the unit cell of the surface. **(a)** Unreactive surface having no channel to dissociation. **(b)** PES in which the molecule will preferentially dissociate when impacting above a surface atom. **(c)** Dissociation occurs prevalently over the center site.

To create an adiabatic PES, V_{tot}, from these two diabatic states the Landau–Zener coupling construction is employed [44], giving

$$V_{\text{tot}} = \frac{V_{\text{vW}} + V_{\text{rep}} + V_{\text{att}}}{2} - \frac{1}{2}\sqrt{[(V_{\text{vW}} + V_{\text{rep}}) - V_{\text{att}}]^2 + 4\Delta V^2} \tag{4.17}$$

where ΔV is a measure of the barrier reduction caused by hybridization, typically 10–20 meV. A schematic diagram of potentials obtained using this construction for two sites on the surface is shown in Figure 4.7. In this case the minimum barrier would be encountered by a hydrogen molecule impinging on the center site of the copper unit cell, rising to a maximum for impact on the top site.

Table 4.2. Parameters Used in the Construction of the Potential Energy Surface for H_2/Cu(110)[a]

Van der Waals	Repulsive Potential	Attractive Potential			
	$V_0=0.1915$	$V_{diss}=0.165$			
$C_{vW}=0.1665$	$\alpha=1.21$	$\gamma=0.1$			
$z_{vW}=0.563$	$\beta=1.481$	E_{act}©	V_1	g_x	g_y
	$z_0=3.61$				
$k_c=0.4$	$a_x=6.822$	Top	0.2113	−1.7210	0.0
	$a_y=4.823$				
$\Delta V=0.6\times10^{-3}$	$h_x=0.10$	Center	0.1862	1.0290	0.0
	$h_y=0.00$				

[a]Equations 4.12–4.17. All parameters are in au.

To investigate the effect of site dependence of the barrier height within the unit cell, potentials have been selected where unit cell variations in the activation barrier are in the range 200 meV $\leq E_{act} \leq$ 500 meV. This value includes any zero-point energy release on dissociation, as the vibrational coordinate is not explicitly included in this PES. The parameters of the chemisorption potential are then used to investigate the variation in diffraction intensities with the position of the activation barrier minimum within the unit cell.

In all, three different potentials are considered. The first uses only the H_2–surface physisorption potential, Eqs. 4.12–4.14, and is used as a benchmark to aid in identifying changes induced by the presence of the activation barrier. The second potential represents the case when the minimum barrier is located directly over a surface atom, that is, at the on-top site. The third has the minimum lying between surface atoms in the center site. These potentials, shown in Figure 4.8, form the basis for scattering simulations in which the variation of diffraction intensities and sticking probabilities with energy and isotope are probed. The parameters used to construct them are given in Table 4.2.

B. Results

a. Diffraction Data

The case of the nonreactive surface is considered first. No channel to sticking is available and all incoming flux must scatter into the Bragg states. The scattered intensities for the first three diffraction states are shown in Figure 4.9a for H_2 and Figure 4.9b for D_2. On examination of the H_2 scattering data, as the beam energy is raised the initial effect is a redistribution of flux from the [00] specular into higher diffraction states as a result of the increased corrugation felt by the now more deeply penetrating collisions. As the energy increases, the specular continues to fall, becoming less than both the [10] and the [20] beams, until it reaches a sharp minimum at 340 meV, after which it rises again. This effect has been observed in He diffraction [40] and is due to destructive interference between the flux reflecting from the top site and that reflecting from the center site. Within the eikonal

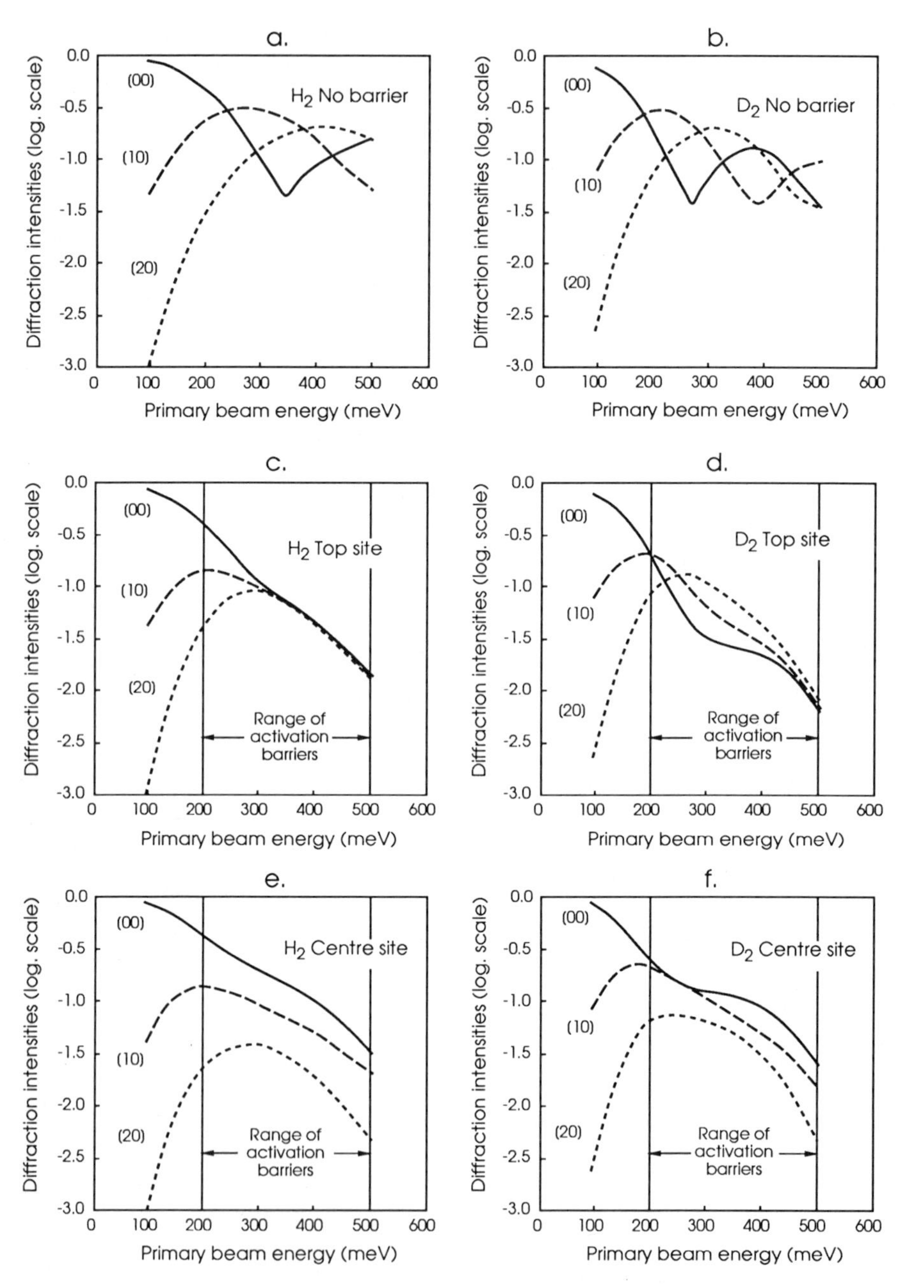

Figure 4.9. Diffraction data for H_2 and D_2 from the unreactive surface **(a,b)** and the reactive surfaces with the minimum barrier over the top **(c,d)** and center **(e,f)** sites. For each potential the results for both H_2 **(a,c,e)** and D_2 **(b,d,f)** are presented.

approximation [45, 46], the energy at which this minimum occurs gives the corrugation parameter directly, as the corrugation amplitude is approximately one quarter of the probe particle wavelength.

A similar trend is seen for deuterium scattering from the same surface (Figure 4.9b), but with the specular minimum falling at a significantly lower energy as the heavier D_2 molecule has a longer wavelength than an H_2 molecule of the same energy and thus encounters the cancellation condition at a lower energy. This feature is also seen in the [10] Bragg state at 390 meV.

Moving now to the reactive potential energy surfaces, a channel to dissociative adsorption is accessible classically at beam energies in excess of 200 meV, and by 500 meV the entire surface will dissociate H_2. The diffraction scam for the case where the minimum activation barrier is positioned directly over a surface atom is shown in Figure 4.9c. Now, at low energies the barrier is not perceived by the molecule and the beam intensities follow the same behavior as those for the unreactive surface. At higher energies all of the beams decay near exponentially as flux is lost to the growing, sticking "hot spot." Similar behavior is seen for the D_2 case (Figure 4.9d), with no minima occurring in any of the beams. This is not unexpected as the cancellation observed is now being hindered by the loss of flux which previously would have reflected from the top site to destructively interfere with flux coming back from the center site.

Intensities calculated for H_2 with the minimum barrier over the center site (Figure 4.9e) show substantially less diffraction occurring than before, with the majority of the flux remaining in the specular beam for all energies. Similar behavior is seen for D_2 (Figure 4.9f), with the residual cancellation effect causing the specular to fall initially but then recover to remain larger than the other beams.

To understand the difference between the scattering from the top and center sites, classical trajectories were run and the population of the parallel momentum states was examined. Four different energies are shown in Figure 4.10 ranging from 180 to 370 meV. With no barrier present these would exhibit a rainbow in momentum space, as the deflection arising from the surface corrugation causes the most probable parallel momentum to also be the largest accessible. This is caused by bunching of the scattering flux from impact parameters close to the point of inflection in the corrugation function. In the lowest-energy case *with* the barrier present, no difference is seen between the two sites, but as the energy increases the rainbow is destroyed only for the center-site PES. This is because in this case the corrugation function is not sinusoidal and has an inflection point closer to the center site, as can be seen in the higher-energy contours of Figure 4.8a. This is removed preferentially when the minimum barrier is above the center site, giving the effect of reduced deflection and consequently less intensity in the higher-order beams. The scale of the momentum in Figure 4.10 is plotted to correspond to the Bragg states, and as the beam energy increases, the population of parallel momentum states spreads out in analogously to the behavior already seen in the quantum scattering calculations.

In all four cases for scattering from reactive surfaces it is important to note that *no* dramatic effects are observed as the activation barrier is reached and surpassed.

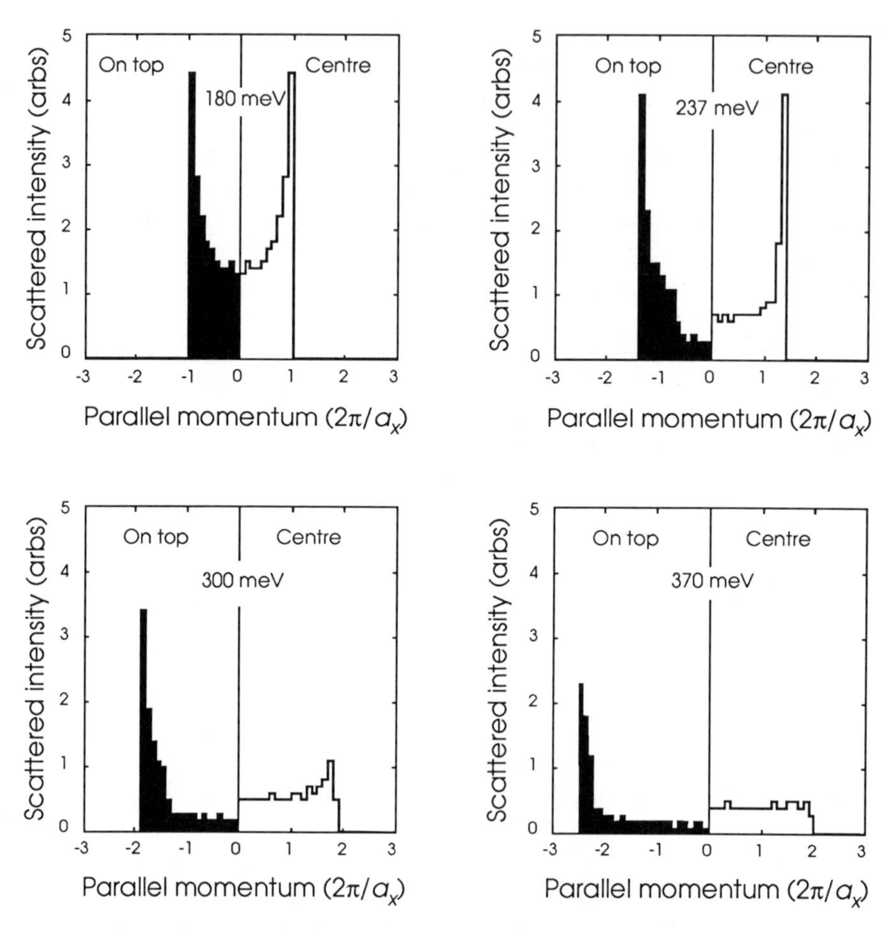

Figure 4.10. Classical simulations showing the parallel momentum distribution after collision with the surface for four different translational energies. On the left half of each figure is the distribution produced with the minimum barrier on top and on the right for the center site PES. The units of momentum are given in reciprocal lattice vectors.

Not only may diffraction and sticking coexist but the diffraction data are so unperturbed by removal of flux that adsorption cannot even be inferred by examination of the diffraction data alone. The only definite signature that may be looked for is the removal of the destructive interference minimum in the specular intensity when comparing D_2 with H_2 data.

b. Sticking Data

The first, and most trivial, estimate of the fraction of the surface that is open to adsorption is achieved by measuring the width of the "hole" from the contour plots shown in Figures 4.8b and 4.8c at a given energy, thus giving a geometric measure of the sticking. As might be expected this value goes from zero to one exactly, within the 300-meV range of the activation barriers. The next measure of the sticking fraction is obtained by running several hundred classical trajectories over one unit cell and counting the number that do not reflect. Finally, as the split operator propagation scheme, used to calculate the diffraction data in the last section, is norm conserving, and the absorbing boundary is constrained to lie well beyond the barrier region, the quantum sticking probability is the complement of the total reflected flux.

Results obtained from these three methods are plotted in Figure 4.11a for H_2 with the minimum barrier over the top site, and in Figure 4.11b for the center site. The most obvious difference in both cases is the extension of finite sticking to energies below the minimum barrier height in the quantum dynamics case. This effect is due entirely to tunneling of the light molecule through the activation barrier. At high energies there are two reasons why a unity dissociation probability is not observed: for the quantum case antitunneling is observed where a particle reflects from a barrier that classically it can surmount; additionally, the particle may be steered so that some normal momentum is deflected, resulting in a lower-energy collision in the reaction coordinate. This steering is seen quite clearly for classical collisions on the PES with the hole on the top site. Here the potential remaining contains the point of inflection and so is more able to steer the molecule before collision. This effect is not seen for the center-site minimum activation barrier PES as the potential remaining at higher energies is very flat and thus the geometric calculation shows good agreement with classical dynamics, implying no prefocusing.

IV. Vibrational Effects and Barrier Topology in H_2 Dissociation

In the four-dimensional calculation presented in Section II, the vibrational coordinate was explicitly included, whereas in Section III it was included as a sleeping coordinate, the assumption being that vibrational motion will evolve adiabatically as an activation barrier is crossed. This issue will now be confronted directly, and in a series of studies on PESs with differing topologies, the coupling between translational and vibrational motion will be investigated [47]. It has long been recognized that the nature of the interaction region of a reactive PES holds the key not only to reaction probabilities, but also to observable excitations in the scattered fragments. In addition, the part played by excitation of internal degrees of freedom depends on the interaction zone topology.

A. Potential Energy Surface

The geometry of the transition state is perhaps the Holy Grail of reactive scattering. There has been considerable effort expended in gas-phase studies [48] to correlate observables to the degree of bond extension necessary to reach the activation barrier to dissociation. In this section we continue in this vein and investigate similar

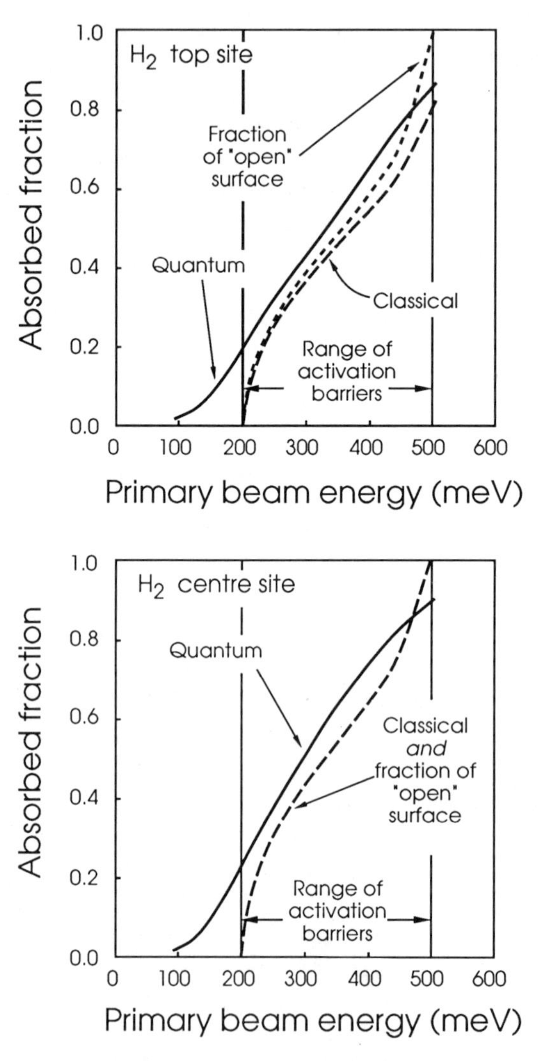

Figure 4.11. Three different methods of measuring the dissociation probability shown for both the top-site and the center-site minimum barrier potentials. The quantum results for both potentials show that tunneling and over-barrier reflection occur, extending the fractional dissociation probability into regions that classically are forbidden. The geometric measurement of the fraction open to dissociation goes from zero to one over the range of the barriers used within the unit cell. In the case of the top-site PES, steering away from the barrier occurs, giving a reduced dissociation probability.

problems in reactive surface scattering. The nature of the interaction potential is simplified to a level at which quantitative statements and comparisons may be made as to the effect of altering the strong interaction region of the PES. The intention is to probe the conditions that can lead to (1) vibrational excitation of the reflected flux, (2) enhanced sticking with internal mode excitation, (3) resonant trapping in dynamic wells, and (4) high-energy dissociated species. These latter, energetic adsorbates are related to the "hot" molecular precursors previously discussed in kinetic studies of water formation [49].

In this section a generalized potential form is set up to simulate a collision of an H_2 molecule with a low corrugation surface with its molecular axis in the plane of the surface. This geometry corresponds to that which will give the lowest-energy configuration for dissociation [30,33,50]. Working in center-of-mass coordinates the two dimensions to be examined are Z, the separation of the center of gravity of the molecule from the surface, and x, the internuclear distance between the atoms of the molecule. To make the energetics, potential construction, and final state projection as simple as possible, while maintaining physical significance, only four components of the interaction region are required and these are now examined in turn.

Region 1. The entrance channel, defined at large Z to be independent of this coordinate and to vary only with x, is constructed to fit the gas-phase potential for H_2 given by the familiar Morse potential having the form

$$V_1(x,Z) = D_1(\exp(-2\alpha_1 x) - 2\exp(-\alpha_1 x)) \tag{4.18}$$

where D_1 is the dissociation energy of the free molecule plus its zero-point energy (in this case giving 4.79 eV) and α_1 (=1.028 au) is the range parameter determined from the vibrational spectra of the molecule. The zero in the x coordinate is set to 1.4 au, the mean internuclear separation of the diatomic including asymmetry effects in the ground state wavefunction. Using this potential it is simple to calculate the vibrational energy levels for the free molecule, using the reduced mass of hydrogen, 0.5 amu or 918 au, the first few energies of which are E_0=269 meV, E_1=783 meV, and E_2=1.265 eV. It is the size of the zero-point energy, 269 meV, that is of particular interest because, depending on whether it is converted into translational energy before or after the barrier is encountered, it must radically alter the outcome of a scattering event. As the molecular vibrational wavefunctions are soluble analytically for a Morse potential, the projection onto these states gives the degree of vibrational excitation for the scattered flux.

Region 2. The exit channel reached when x becomes large and Z is small corresponds to two dissociated atoms in the chemisorption well of the metal, which because of the constraints of the system, are forced to oscillate in phase with each other. The parameterization of this well is set to be identical to that used for the gas-phase diatomic; that is, the depth of each atomic well is $D_1/2$, giving a total depth of D_1 for the dissociated diatomic system. The difference in energies between the molecule and the two adatoms comes purely from the oscillator mass now being 2.0 amu or 3672 au, which, being four times the mass of the gas-phase oscillator, has only half of its zero-point energy. The bond force constants ($2\alpha^2 D$) for both the entrance and exit channels are equal and so the oscillator frequency differs because of the mass disparity in the two dimensions. These rather sweeping approximations do in fact lead to adsorption energies (−2.2 eV per H atom) and bound state frequencies (600 cm^{-1}) that match the experimentally observed values quite closely

[51]. The effect of surface corrugation is neglected partially because a quite smooth crystal face was envisaged but mostly so that the vibrational excitation of the adsorbed atoms could be projected out in a simple manner onto the known bound state functions. Thus, no x dependence is incorporated into this region of the basic potential. The zero in the Z coordinate is arbitrarily set to be the minimum of the adatom Morse potential in the x coordinate.

Region 3. The reaction zone located where both x and Z are small corresponds to the region of space in which the two atoms in the molecule are interacting with one another and the surface. Its design must fulfill several requirements, most fundamental of which is the continuous linking of the entrance with the exit channel. The simplest way of achieving this is to construct an angular dependence whose radial component is a Morse potential and define its center of curvature to lie on the diagonal where the entrance and exit channels are of equal value (i.e., $x = Z$ and $V_1 = V_2$), thus giving a quarter circle forming the mixing region of the PES [52]. The center of this arc is placed at $x = Z = r_0$ and the Morse potential depends on r being defined as $\sqrt{x^2+Z^2}$, giving a potential of the form

$$V_3(r) = D_3(\exp(-2\alpha_3(r-r_0)) - 2\exp(-\alpha_3(r-r_0))) \tag{4.19}$$

The well depth and curvature parameters are defined to vary smoothly from entrance to exit, but as $D_1 = D_2$ and $\alpha_1 = \alpha_2$, D_3 and α_3 are both fixed to the values of the initial and final states and remain constant throughout the reaction region. The result is a smooth transition from the bound states in the entrance channel to the corresponding states in the exit channel of half the original energy (due to the effective mass transform). The basic form of this potential is shown in Figure 4.12a with the center of curvature for the reaction zone, r_0, being set at 1 au from the minimum of both the entrance and exit channels.

Region 4. The barrier to dissociation must now be introduced as the first three components describe an interaction that is classically thermoneutral and quantum mechanically exothermic throughout the reaction. The path of the "valley floor" for this basic potential is simply defined throughout the three regions of entrance, mixing reaction zone, and exit channel. A new coordinate is now used, s, to track the path of this valley, and its origin is set as the entrance to region 3:

$$s = -Z \text{ in region 1} \tag{4.20}$$

$$s = r\theta \text{ in region 3 where } \theta = \tan^{-1}\left(\frac{z-r_0}{x-r_0}\right) \tag{4.21}$$

$$s = \frac{\pi r_0}{2} + x \text{ in region 2} \tag{4.22}$$

Using this new coordinate it is trivial to define a barrier whose position and extent depend on s. The simplest functional form for this is used:

$$V_4 = V_{\text{bar}} \exp\left(-2\beta(s-s_0)^2\right) \tag{4.23}$$

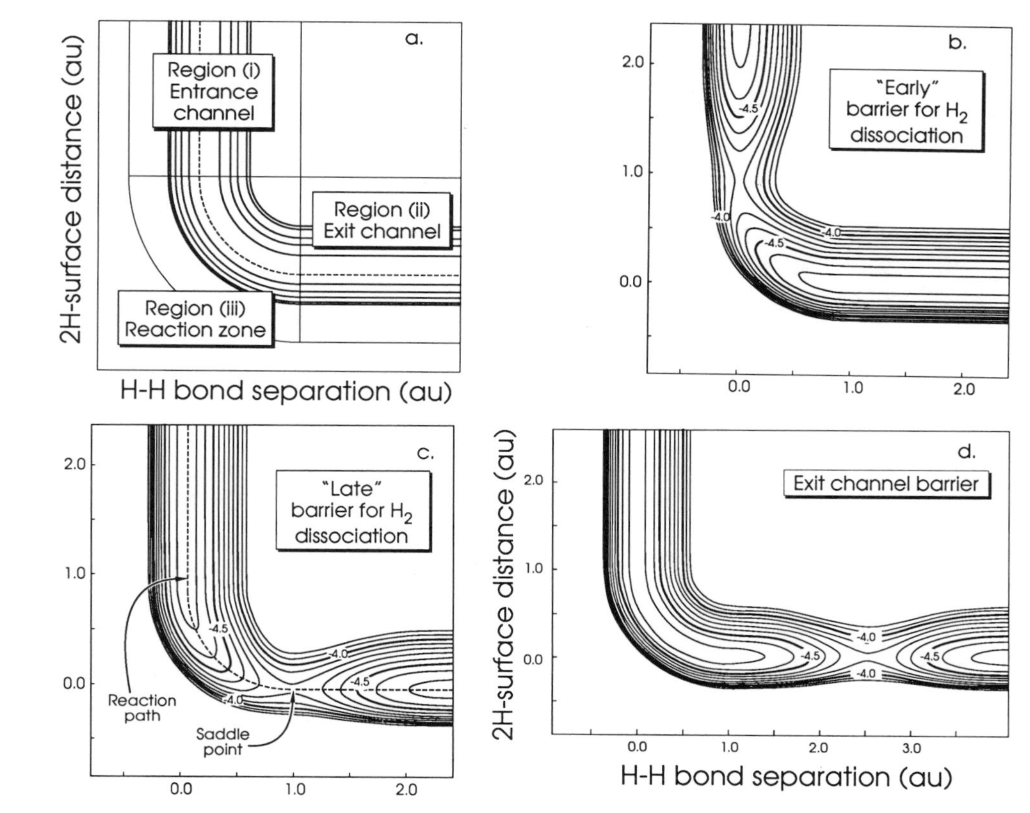

Figure 4.12. The basic unactivated "elbow" potential is shown in **(a)**. To this is added a barrier whose position is varied from the entrance channel in **(b)** to the end of the corner region in **(c)** and finally placed beyond the corner region and into the exit channel, as shown in **(d)**. All contour energies are in electron volts and set relative to the gas-phase energy of two dissociated H atoms.

Depending on the interplay of these states both the height and the position of the barrier will vary. It is the phenomenological effects that the barrier position induces that are to be investigated to introduce the same intuition afforded by the analogous Polanyi rules for gas-phase dynamics [53] to surface reactive scattering. In all, three barrier configurations are examined and these are categorized as follows:

The early barrier consists of a barrier positioned at the boundary between the entrance and reaction zone, $s_0=0$. This simulates a system in which the activated state corresponds to a molecule with its gas-phase equilibrium internuclear separation. This is observed for adsorption on a simple or noble metal [54].

The late barrier represents the case of large perturbation in bond length before the molecular state becomes less favorable than the chemisorbed atomic state. This is achieved by setting the barrier maximum to occur at the beginning of the exit channel, $s_0=1.57$ au.

The exit channel barrier is configured to simulate a system in which the diatomic is able to access a molecular state, with the dissociation barrier being sampled by molecules only after they enter the exit channel. This is performed by setting the barrier maximum to lie 1.5 au into the exit channel at $s_0=3.07$ au.

These three different configurations are depicted in Figures 12b to 12d, respectively. In all cases the width parameter of the Gaussian barriers is set to 1 au with a height (538 meV) equal to twice the zero-point energy of H_2. It should be noted that the early and late barrier potentials are identical if the coordinates were switched. The crucial difference is the mass transform between the two coordinates, an effect that may be removed if the x coordinate is contracted by a factor of 2. To visualize the changes that occur as the corner region is traversed by a molecule, it is useful to define a "reaction path." This closely follows the s coordinate but is defined as the path of steepest *descent* (in mass weighted coordinates) from the barrier maximum or saddle point and will be referred to as the coordinate S. Its properties will now be examined.

B. The Reaction Path: An Effective Potential

The method used to find the reaction path relies on initiation of the search at the saddle point which is the only point, within the range of the barrier, having a zero potential gradient in all directions. For the parameterized potential described above, the saddle point is known analytically and the reaction path is found from

$$\frac{dr'}{dS} = -\frac{\nabla V(r')}{|\nabla V(r')|} \tag{4.24}$$

where $r' = (x', Z)$ and S is the arc length along the path. To calculate S, the variable transformation $x' = x/2$ is employed which makes the mass a scalar quantity. This is equivalent to making a skewed axis representation [55]. It is vital that the S coordinate not be confused with any dynamic trajectory or even an energy profile for the system, as *no* dynamics are included in its construction. Following Marcus [56] it is convenient to define a coordinate ρ that is locally orthogonal to S. The PES $V(x',Z)$

can then, quite generally, be expressed as $V(\rho,S)$, with the classical kinetic energy in these new variables being

$$T = \frac{1}{2}\mu(1 + \kappa\rho)^2\left(\frac{dS}{dt}\right)^2 + \frac{1}{2}\mu\left(\frac{d\rho}{dt}\right)^2 \tag{4.25}$$

where $\kappa(S)$ is the local curvature and μ is the reduced mass. The Hamiltonian in these curvilinear coordinates is

$$\mathcal{H} = \left[-\frac{k^2}{\eta}\frac{\partial}{\partial S}\left(\frac{1}{\eta}\frac{\partial}{\partial S}\right) - \frac{k^2}{\eta}\frac{\partial}{\partial\rho}\left(\eta\frac{\partial}{\partial\rho}\right)\right] + V(\rho,S) \tag{4.26}$$

where $k^2 = 1/2\mu$ and

$$\eta(S) = 1 + \kappa(S)\rho \tag{4.27}$$

With the potential written as

$$V(\rho,S) = V_{\mathrm{RP}}(S) + V_{\mathrm{fit}}(\rho,S) \tag{4.28}$$

the conventional way to solve the quantum dynamics is to calculate $\psi(\rho,S)$ in the entrance and exit channels, where the Schrödinger equation separates, and then match to solutions in the reaction zone [56]. For strongly coupled potentials, $\psi(\rho,S)$ is written in terms of a basis set expansion in the vibrational states of $V_{\mathrm{fit}}(\rho,S)$, and the numerical problem is transformed into a standard coupled-channels problem. This method has recently been employed by Brenig and Kasai [52] for a surface scattering problem analogous to that treated here but for a rectangular potential barrier situated in the reaction zone.

At each point along the reaction path the bound state eigenvalues of $V_{\mathrm{fit}}(\rho,S)$ can be obtained, $E_n(S)$, and it is then possible to define a set of one-dimensional vibrationally adiabatic potentials:

$$V_{\mathrm{eff}}^n(S) = V_{\mathrm{RP}}(S) + E_n(S) \tag{4.29}$$

It is sufficient to fit these oscillator states to Morse potentials which have been defined using an extra parameter, allowing it to tend asymptotically to a best-fit parameter value V_{asym} at large ρ:

$$V_{\mathrm{fit}}(\rho) = V_{\mathrm{dis}} + V_{\mathrm{asym}}(\exp[-2\alpha\rho] - 2\exp[-\alpha\rho]) \tag{4.30}$$

This will, by definition, have its well minimum at $\rho = 0$. This fit is easily achieved by examining the potential at two points straddling the reaction path at a distance $\Delta\rho$ perpendicular to S, giving the asymptotic energy V_{asym}, dissociation energy V_{dis}, and range parameter α as

$$V_{\mathrm{dis}} = \frac{(V_3 - V_2)}{(C-1)^2} \tag{4.31}$$

$$V_{\mathrm{asym}} = V_2 + V_{\mathrm{dis}} \tag{4.32}$$

$$\alpha = -\frac{\ln(C)}{\Delta\rho} \tag{4.33}$$

where $C = \sqrt{(V_2 - V_3)/(V_2 - V_1)}$ in which V_1,V_2, and V_3 are $V_{\text{fit}}(-\Delta\rho)$,$V_{\text{fit}}(0)$, and $V_{\text{fit}}(+\Delta\rho)$, respectively.

The zero-point energy of a given set of Morse parameters may then be found, in atomic units, from the equation

$$E_n = V_{\text{dis}}\left(1 - \left(1 - \frac{\alpha(1+2n)}{\sqrt{8\mu V_{\text{dis}}}}\right)^2\right) \tag{4.34}$$

when $n = 0$ for an oscillator of mass μ. This is then added to the potential found at each point along the S coordinate to give an effective ground state energy that includes the influences of bond relaxation and thus energy release in traversing the reaction zone. Higher oscillator states are similarly calculated by adding the oscillator state energy, from Eq. 4.34, to the reaction path potential energy.

The linearization of the real potential by this technique into S and ρ gives a unique effective potential. It is not, however, possible to perform the reverse process as no information about the curvature of the reaction path is explicitly included. The effective potentials for the four PESs shown in Figure 4.12 are now examined.

No Barrier. This basic form exhibits only the release of the oscillator energy, as the reaction path is of constant energy as depicted in Figure 4.12a. The effective potentials for the first three oscillator states (ground, first, and second) are shown in Figure 4.13a. It should be noted that the release of energy for the first vibrational state is nearly twice as large as that released by the ground state as it traverses the reaction zone. The interplay and efficiency of use of this energy release is examined with the movement of barrier position and shape in the next three potential surfaces. Also of interest is the greater density of vibrational states in the exit channel, giving rise to the possibility of vibrationally excited chemisorbed products for initial translational energies below the $0 \rightarrow 1$ threshold for gas-phase vibrational excitation.

Early Barrier. The *early barrier* shown in Figure 4.13b allows little, if any, of the zero-point vibrational energy release to aid molecular dissociation. This gives an effective barrier height (0.538 eV) and width equal to those encountered when no vibrational effects are included. The higher vibrational states perceive the same barrier height, leading to the prediction that vibrational energy population in the initial entrance channel state will not assist dissociation. Once the barrier has been surmounted, however, the probability that transmitted flux will populate higher vibrational levels is expected to be large, giving vibrationally excited atoms on the surface.

Late Barrier. The *late barrier* simulates the case of near-total release of available vibrational energy before the barrier to dissociation is encountered. In this case there is a greatly reduced barrier of only 0.405 eV and a width half that observed for the early barrier case. It is expected that the vibrational energy population will be most efficiently used in this barrier configuration as the higher vibrational states encounter an even lower barrier as shown in Figure 4.12c.

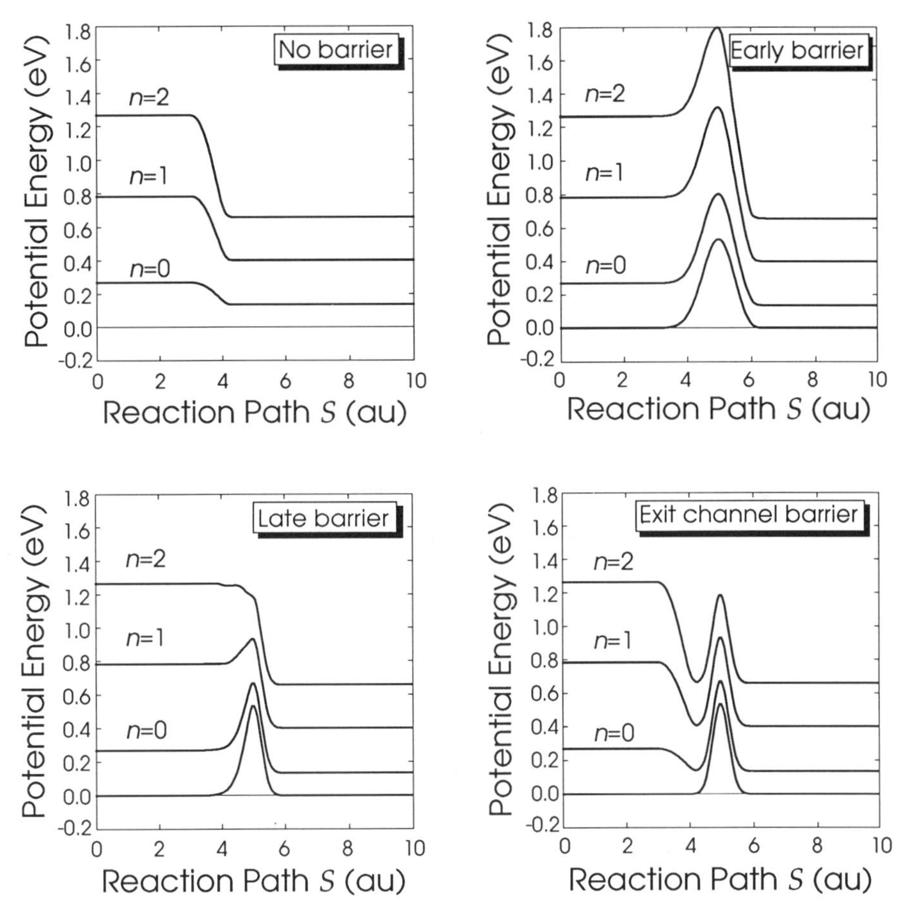

Figure 4.13. One-dimensional effective potentials for the first three vibrational states corresponding to the full two-dimensional potentials presented in Fig. 4.12. It can be seen that the later the barrier is, the thinner and lower is the effective barrier perceived by the dissociating molecule.

Exit Channel Barrier. The *exit channel barrier* has roughly the same height and width as the late barrier case and similar results might therefore be expected. The existence of a well in the first vibrational state effective potential (Figure 4.12d) could introduce dynamic effects, the complexity of which may not be accurately modeled by the ground state effective potential.

The scattering simulation is achieved in a fashion identical to that outlined in the four-dimensional simulation, with an initial state having a Gaussian wavefunction in Z and a pure Morse eigenfunction in x. As the potential asymptotes to a Morse function as $Z \rightarrow \infty$ or $x \rightarrow \infty$, the final state may be projected into the eigenfunctions for either of the scattering channels, thus giving state-to-state excitation probabilities. In addition, for comparison, one-dimensional calculations have been run using the effective potentials.

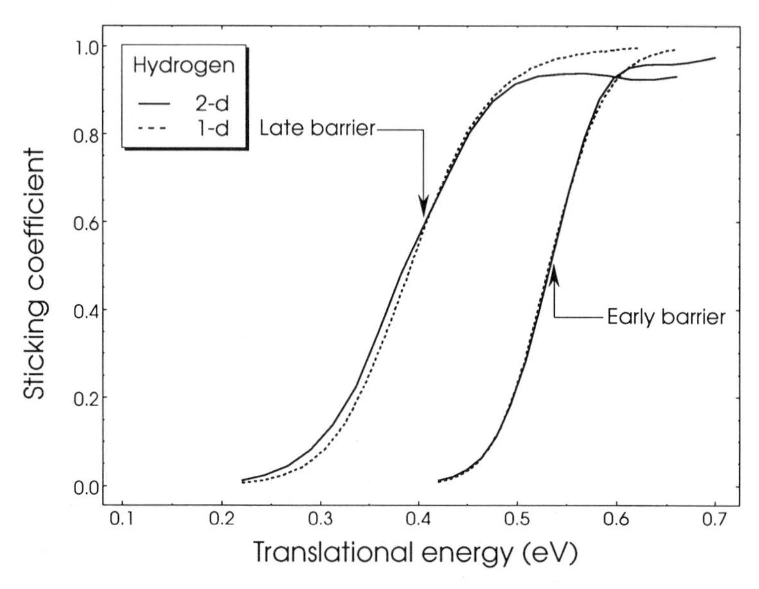

Figure 4.14. Sticking data for the early and late barriers as a function of initial translational energy. The results for both the one- and two-dimensional simulations are shown together with the effective barrier height measured from the one-dimensional potentials (arrows).

C. Results

a. Validity of the One-Dimensional Effective Potential

For all but the exit channel barrier, excellent agreement was observed between the one- and two-dimensional total dissociation probabilities. The major effect excluded from the effective potential is the possibility of excitation into higher vibrational states during the interaction [57]. This is an example of the previously documented bobsled effect [58] in which centrifugal effects cause the incoming wavefunction to swirl up the back wall of the potential, thereby encountering a higher barrier as the minimum energy pathway is not strictly followed. The corollary is that the population of these higher vibrational states in the reaction zone results in a reduction of the available translational energy directed along the reaction coordinate. Figure 4.14 shows the transmitted fraction as a function of initial translational energy for a hydrogen molecule initially in the ground vibrational state for the early and late barrier PES. As can be seen, good agreement is obtained for all but the highest energies, for which the two-dimensional results show a consistently lower sticking coefficient than found using the effective potential. The reason for these deviations may be traced to the population of higher vibrational states and is examined more closely in the next section.

The ordering of these results is best understood by examining the one-dimensional effective potentials (Figures 4.13b and 4.13c). The late barrier is much lower than the early one because all of the released zero-point energy is converted into translational energy, whereas for the early barrier almost none of the zero-point energy is

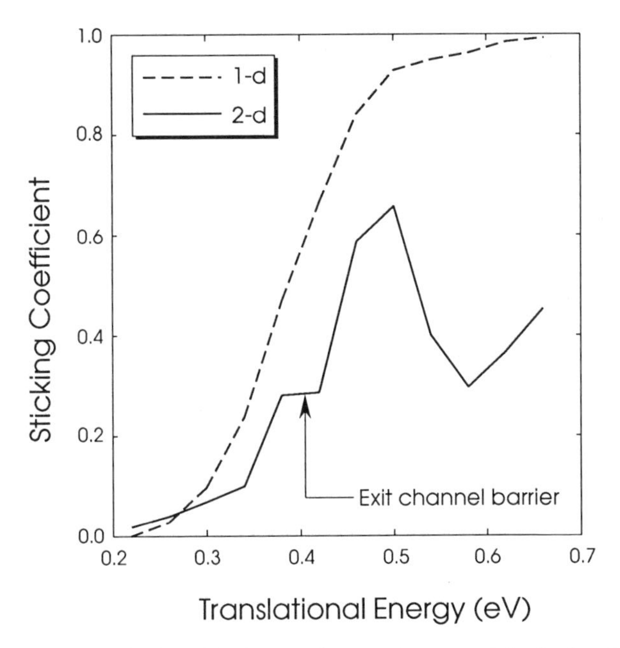

Figure 4.15. Sticking coefficient for the very late barrier potential calculated using the full two-dimensional simulation and the one-dimensional effective potential. In this case the effective potential gives poor agreement with the two-dimensional data because it is unable to simulate the dynamic metastable states observed in the two-dimensional potential. The arrow corresponds to the height of the one-dimensional effective potential barrier.

available along the reaction coordinate. The difference in steepness of the curves is understood if the thickness of the effective potentials is examined. In the case of the early barrier the mass attempting to tunnel through is that of the molecule ($2M$), thus hindering the tunneling process. For the late barrier the tunneling inertia is the reduced mass ($M/2$) which, having a greater de Broglie wavelength, sees a more transparent barrier, leading to finite tunneling over a wider energy range.

For the case of the exit channel barrier the one-dimensional results differ little from those obtained for the late barrier as its height and width are almost identical. In the two-dimensional case, however, radically different results are obtained (Figure 4.15). As a function of translational energy, the sticking fraction shows a large deviation from the usual sigmoidal form. The causes of this may again be found in the excitation of higher vibrational levels, the nature of which is examined later.

The use of deuterium scattering to complement the hydrogen data provides a method of investigating the importance of two key quantum effects, tunneling and zero-point energy. For tunneling, the increased mass of deuterium, a factor of 2 over hydrogen, results in more classical behavior; that is, the S-curve of the transmission coefficient should be more steep, tending toward the step function obtained with classical mechanics. In addition, the greater mass of deuterium results in a $\sqrt{2}$ reduction in the zero-point energy and an increase in the density of the higher vibrational states, again tending toward the classical case of a continuum of energy levels.

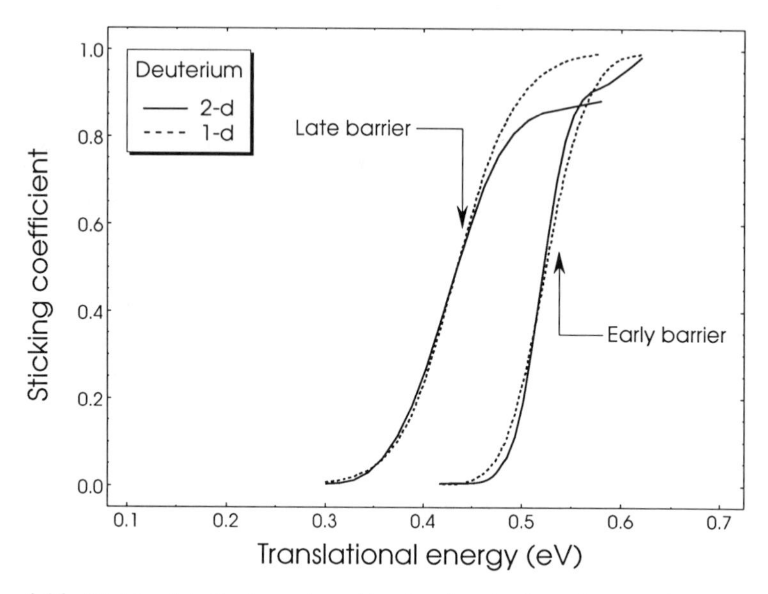

Figure 4.16. Sticking data for the early and late barriers for deuterium as a function of initial translational energy. The results for both one- and two-dimensional simulations are shown together with the effective barrier height measured from the one-dimensional potentials (arrows). This corresponds to Figure 4.14 but the heavier isotope causes the curves to be steeper and more closely spaced.

This reduction in the zero-point energy results in a concomitant reduction in energy release as the reaction zone is traversed. The results for the scattering from the early and late barriers in both one and two dimensions show these effects (Figure 4.16). Again, the agreement between one and two dimensions is good, with the most obvious deviation occurring at high energies in the late barrier case. When these results are compared with those in Figure 4.14, the most obvious difference is that the curves are much closer together. This is due primarily to the lower differential reduction in the barrier for the two potentials, resulting in a reduction in the release of zero-point energy. In addition, the curves are more steep as a result of the decrease in efficiency of tunneling for deuterium because of its greater mass. The ordering of the curves and the trends in the curve steepness are the same as those observed for hydrogen.

b. State-to-State Scattering

In this section the initial state wavefunction is always defined as being the ground state of the gas-phase hydrogen molecule. The one-dimensional results presented in the previous section required the wavefunction to remain in its ground vibrational state; no such constraint exists for the following results.

Early Barrier. The Polanyi rules would say that an entrance channel barrier would give rise to vibrationally excited products. As there is no x–Z coupling before the barrier, the reflected flux would be expected to lie in its ground state. This is indeed

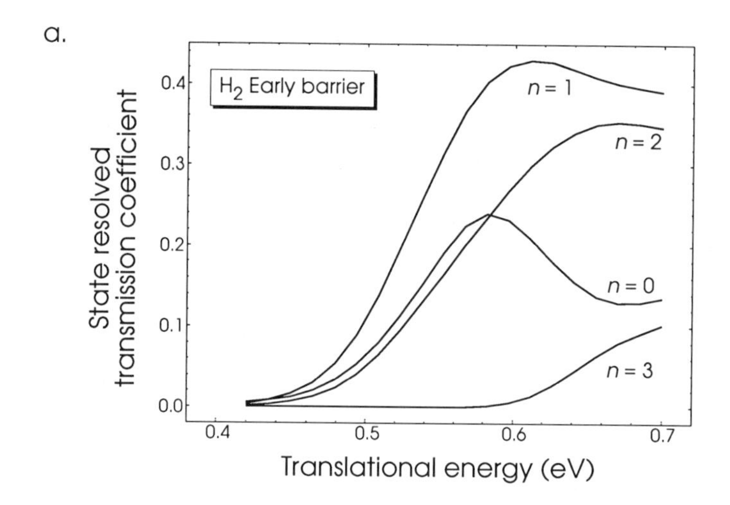

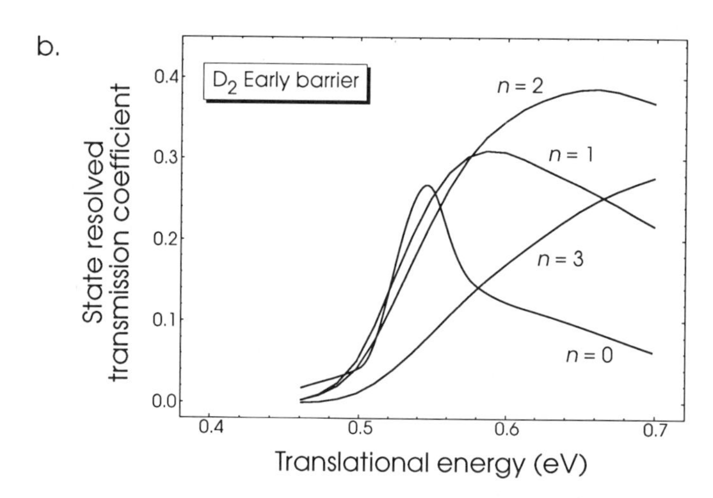

Figure 4.17. Vibrational state-resolved sticking for the two-dimensional early barrier for H_2 **(a)** and D_2 **(b)**. In each case the transmitted flux in the exit channel is projected onto the Morse states and integrated in x. An inversion of intensities is observed resulting from post-barrier curvature in the potential.

the case with less than 10^{-4} of the reflected wavefunction having been excited into the first vibrational state, even at translational energies well in excess of the threshold required for excitation (for H_2 this is 0.514 eV). Any flux that does surmount the barrier, however, encounters a region of strong curvature (bobsled effect), giving rise to substantial excitation in the final transmitted (dissociated) state. The state-resolved scattering intensities for this interaction are depicted in Figure 4.17a for the $0 \rightarrow n$ transitions. The threshold translational energies required to access the states in the exit channel are $0 \rightarrow 1 = 0.122$ eV, $0 \rightarrow 2 = 0.363$ eV, and $0 \rightarrow 3 = 0.590$ eV, with the $0 \rightarrow 0$ pathway being exoergic by 0.135 eV. It is expected therefore that the $0 \rightarrow 3$ transition would be prohibited below 590 meV translational energy. This is indeed the case, with a large excitation probability found post-threshold. The nature of this state is interesting in that it has very little energy in the x coordinate; almost all of its energy is in vibrational excitation of the dissociated atoms, thus giving a highly excited species that moves slowly away from the site of initial surface impact. It is interesting to note that the $0 \rightarrow 1$ transition (and above 0.6 eV the $0 \rightarrow 2$) shows a greater occupation probability than the vibrationally adiabatic $0 \rightarrow 0$ transition. The reason for this is the greater overlap of the initial ground state with higher final vibrational states, in accord with the Franck-Condon principle. Such behavior has been observed for the H_2+F gas-phase reaction [59].

The D_2 results, (Figure 4.17b) exhibit similar excitation probabilities but with slightly different orderings because of the closer spacing of the energy levels and the lower energy release from the zero-point energy. It should also be noted that the $0 \rightarrow 3$ transition may now be accessed at a lower energy. The onset of these excitations is more steep purely because of the greater mass inhibiting access to the exit channel via a tunneling mechanism.

Late Barrier. The form of this potential is such that all of the available zero-point energy is released before the barrier is encountered, resulting in an apparently lower barrier to dissociation. In addition, an incoming molecule can undergo a transition to the first vibrational state before striking the barrier because of the mixing in the PES. Any molecule doing this will almost certainly reflect as its translational energy now lies well below the n=1 activation barrier height (see Figure 4.13c). It is expected therefore that much of the higher-energy reflected flux will appear in the first excited state. This is shown in Figure 4.18a, with the reflected state population favoring the first excited state over the ground state by a factor of 2 for energies in excess of 560 meV. The transmitted flux enters the exit channel purely in the ground state, with less than 0.1% existing in higher vibrational states. This occurs because flux attempting to traverse the barrier after being excited cannot surmount the $n = 1$ barrier and any flux that transmits has no means of exciting oscillator states in the exit channel.

The D_2 results (Figure 4.18b) show an enhanced population of the first excited state probability at lower energies because they are energetically accessible for this heavier molecule. It is also interesting to note that some population of the first excited state in the exit channel now exists at higher energies. This is due to the fact that flux that populates the first excited state before striking the barrier can now surmount the barrier because of the lower overhead of the initial excitation.

Exit Channel Barrier, Resonant Trapping. As mentioned above, the nature of the interaction encountered in the two-dimensional scattering event on this PES is fundamentally different from that predicted by examining the ground state effective reaction path potential. An examination of the first excited state in the effective potential depicted in Figure 4.13d shows a well of ~390 meV, with one side formed by the increasing zero-point energy of the molecule and the other arising from the activation barrier. It is known from the late barrier study that flux encountering the reaction zone will occupy the higher oscillator state and, therefore, must populate

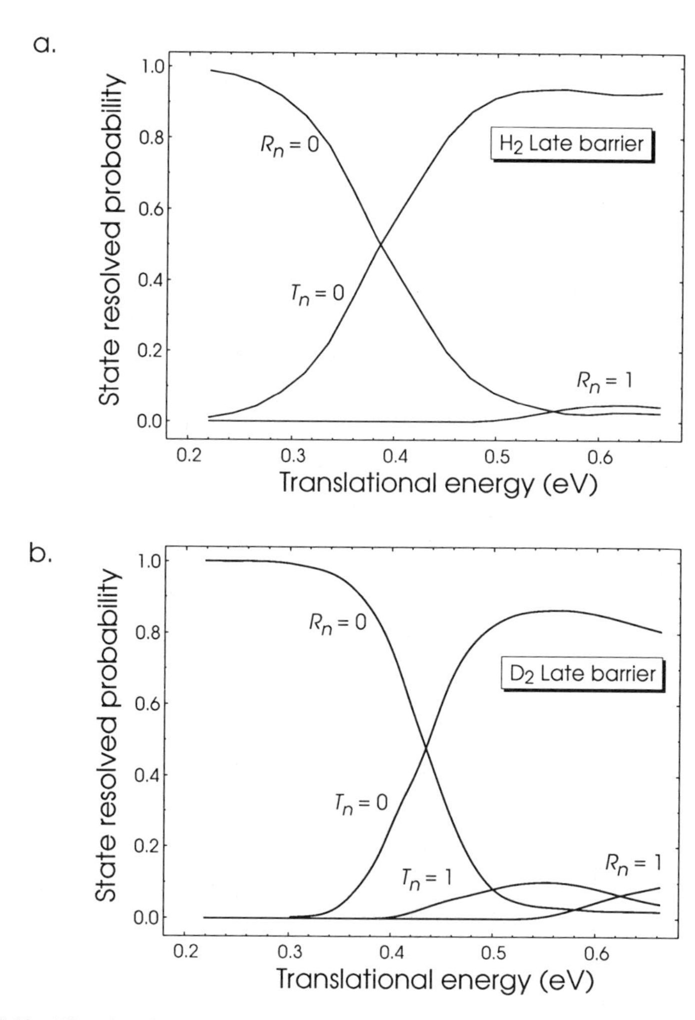

Figure 4.18. Vibrational state-resolved sticking for the two-dimensional late barrier for H_2 **(a)** and D_2 **(b)**. In each case the transmitted flux in the exit channel is projected onto the Morse states and integrated in Z for the reflection coefficient and in x for the transmission coefficient.

this "dynamic well." Although its form is similar to that of a Morse potential, any flux that scatters into it will have only a finite lifetime, and thus the eigenstates in the well will not be sharply defined. Ground state molecules having a translational energy in the range 130 to 500 meV can access this well but then must either deexcite to the ground state or tunnel through the dissociation barrier to the exit channel. To obtain some estimate of the eigenstates in this metastable well, the Morse fitting procedure outlined previously was employed. There are at least five possible states of estimated energies 60, 160, 245, 320, and 370 meV relative to the bottom of the well (to find the approximate translational energy needed to trap into any of these states, from the ground state effective potential, ~ 140 meV must be added).

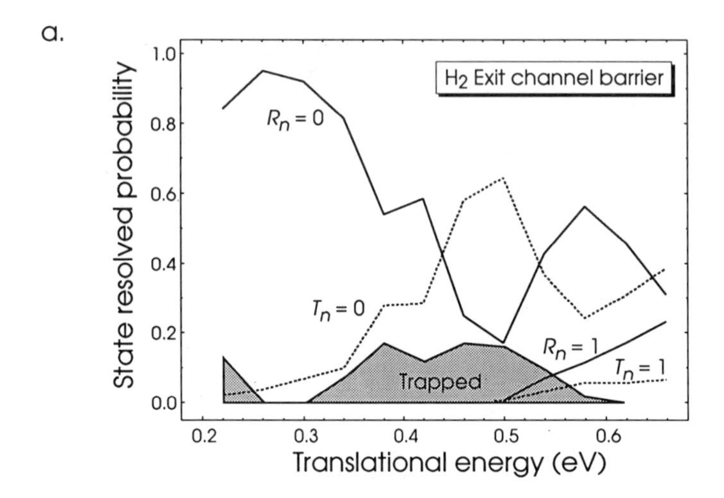

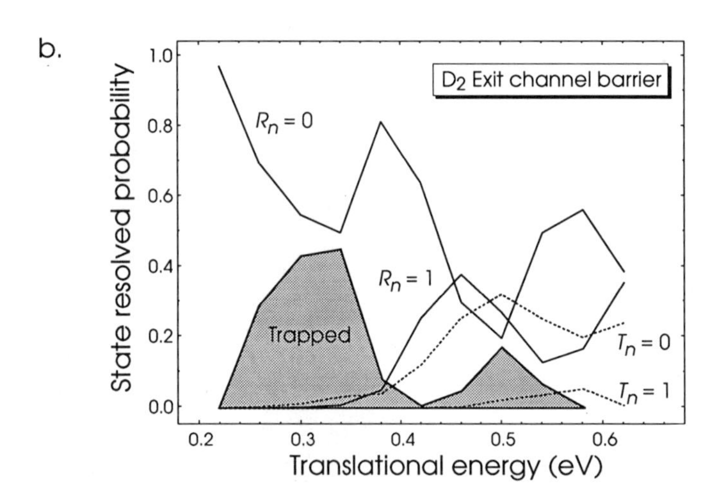

Figure 4.19. Vibrational state-resolved reflection and transmission for the two-dimensional exit channel barrier for H_2 **(a)** and D_2 **(b)**. Also shown (shaded) is the flux still trapped in the corner region. This resonance, which will eventually decay, is shown here at a time equal to twice that used to resolve the other potentials.

The one-dimensional effective potential results for transmission proved to be almost identical to those obtained for the late barrier case, which is to be expected as they exhibit the same barrier height and roughly the same width. For the two-dimensional simulation, the scattering probabilities are shown in Figure 4.19a for the reflected, transmitted, and trapped components (shaded region). The definition of trapping used here is the amount of flux in the prebarrier zone after twice the time used to resolve the late barrier interaction has elapsed. This is in the range 0.3 to 0.5 ps depending on the initial translational energy. As this state is not stable, it decays after a finite length of time with flux either (1) deexciting and then transmitting or

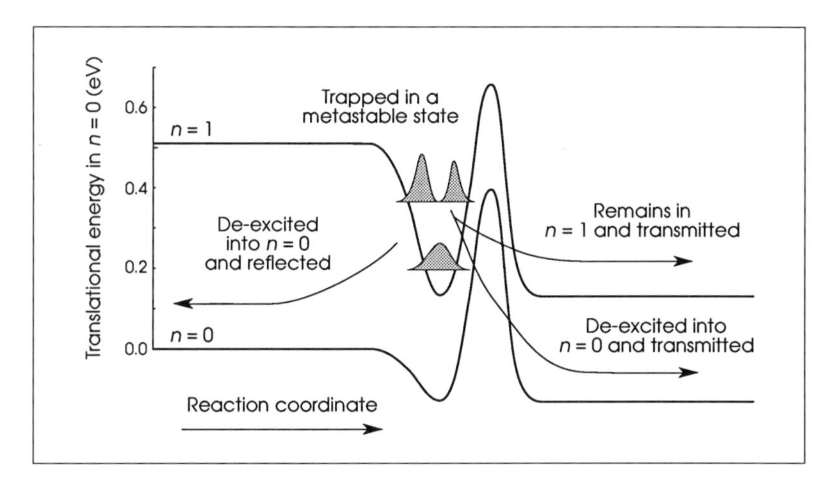

Figure 4.20. Schematic representation for the trapping phenomena observed in the exit channel PES where the well exists in the $n = 1$ effective potential. A molecule is prepared in the $n = 0$ state with a translational energy degenerate with one of the states in the $n = 1$ dynamic well. Its energy is not sufficient to allow gas-phase vibrational excitation and consequently it has three possible fates: (1) it can deexcite into the $n = 0$ state and transmit, (2) it can deexcite and reflect, or (3) it can tunnel through the $n = 1$ barrier to dissociate on the surface.

reflecting at the ground state barrier, or (2) tunneling through the first excited state barrier and dissociating. These states are closely related to Feshbach resonances and the decay schemes are sketched in Figure 4.20.

On examination of the scattering data, substantial trapping is seen at a translational energy of 220 meV. The wavefunction in the reaction zone at the end of the interaction, for a translational energy of 220 meV, is shown in Figure 4.21. The major features are a single node in the ρ direction, implying an $n = 1$ state, and the absence of a node along the reaction coordinate, indicating a population in its ground state. For energies between 260 and 300 meV no flux trapped into the quasibound states, as these energies lie in the band gap of the dynamic well. At energies above 300 meV, trapping is observed continuously up to $\sim$600 meV.

Figure 4.22 shows the time development of one such scattering event where the initial translational energy is 380 meV. The wave packet approaches the reaction zone and overshoots, climbing the back wall of the PES ($t = 3666$ au), again demonstrating the bobsled effect [58]. At time 6333 au the majority of the packet is beginning its journey back to the gas phase and will emerge as a promptly scattered molecule in the $n = 0$ vibrational state. Part of the packet has, however, remained in the bowl-shaped region before the barrier ($t = 10{,}000$ au), and the nodal structure identifies this fraction to be in the first excited state of the $n = 1$ dynamic well. The dominant channel for decay of this state is to scatter back into the gas phase rather than tunnel through either the $n = 0$ or $n = 1$ barrier. This would be in line with a Franck–Condon description of the deexcitation process.

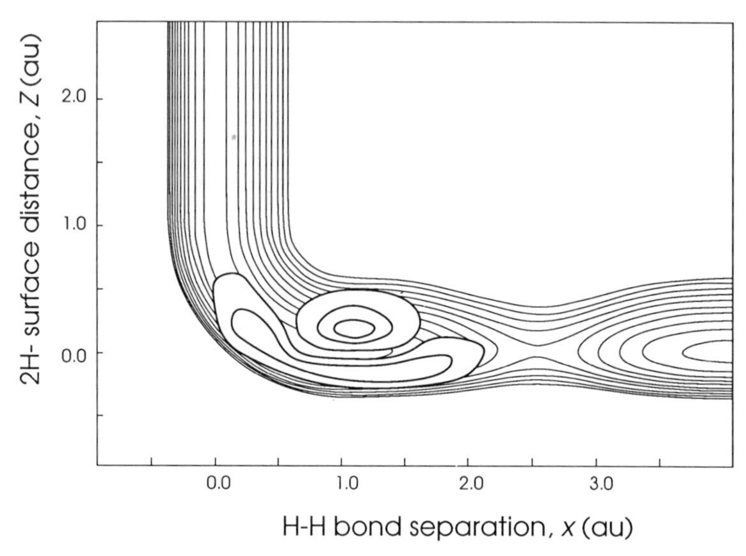

Figure 4.21. Wavefunction remaining in the interaction zone of the exit channel barrier potential for an initial translational energy of 220 meV H_2. The single node along the reaction coordinate seen in the wavefunction shows it to lie in the ground state of the $n = 1$ effective potential.

The high-energy reflection data in Figure 4.19a show that as soon as the threshold for $n = 1$ scattered molecules is exceeded, the trapping process is replaced by reflection into this state and by transmission into the first level of the exit channel. Accompanying this is a sharp rise in reflection into the $n = 0$ reflected state, showing again that following deexcitation from the dynamic well, the preferred method of escape from the interaction region is reflection from the barrier rather than transmission through to the exit channel.

The corresponding D_2 results (Figure 4.19b) indicate a similar behavior but at lower energies because of the heavier mass of D_2. The excitation of the $n = 1$ state in the exit channel is accessible at lower energies, but is populated to roughly the same low extent as in the H_2 case, showing that little of the excited prebarrier flux transmits through to the exit channel. The excitation of the reflected D_2 molecules is much greater, showing almost 40% population at 460 meV.

It is important to remember that the "trapping" described here contains *no energy loss process* and should not be confused with trapping induced by collision with a damped oscillator. Here it is purely a dynamic effect of excitation into a higher-energy resonance from which escape is possible only by slow processes, giving rise to a long residence time.

c. *Stimulated Dissociation by State Selection*

To investigate the possibility of internal energy aiding the dissociation process, the initial wavefunction is now prepared in the $n = 1$ vibrational state. Results for the

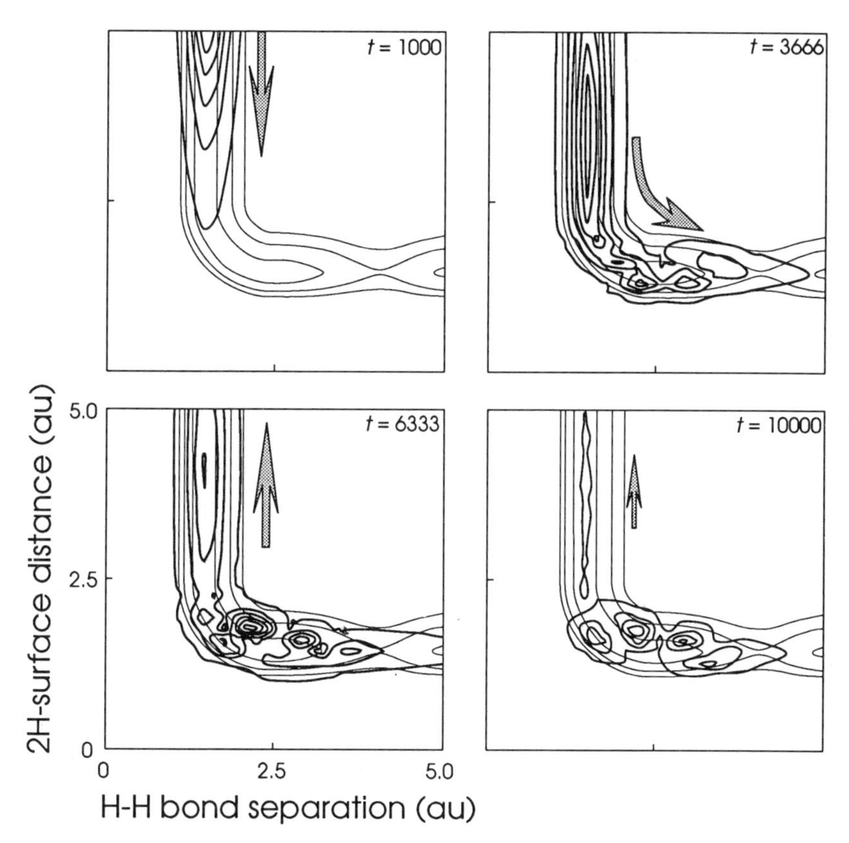

Figure 4.22. Time lapse sequence of the interaction of H_2 with the exit channel PES. A molecule, initially in its ground vibrational state with 0.38 meV translational energy, excites into the Feshbach resonance shown in Figure 4.20. From the nodal structure of the wavefunction, it is possible to characterize the state by two quantum numbers, $m = 1$ and $n = 1$, referring to bound states along and perpendicular to the reaction path, respectively. The arrows indicate the predominant direction of flux. In the final frame there is a small component of the wavefunction in the entrance channel which is the steady leaking out of the bound state back into the vacuum. The time is shown in atomic time units (1000 atu = 24.2 fs). The PES contour energies can be obtained by referring to Figure 4.12.

dissociation probability as a function of translational energy for the early barrier are shown in Figure 4.23a together with the $n = 0$ data. It is clear that the extra 514 meV of vibrational energy has resulted in a negligible alteration of the position of the S-curve, the only effect being a slight broadening of the energy range of the curve. This illustrates the effects of a "sleeping coordinate," in this case the molecular vibration whose energy cannot be converted into the reaction coordinate despite the fact that the available energy is 1.5 times the height of the barrier. The ground state population of the reflected flux remained at less than 0.01% of the total flux for all energies, showing that almost no deexcitation occurs in the barrier region, confirming that the scattering is vibrationally adiabatic.

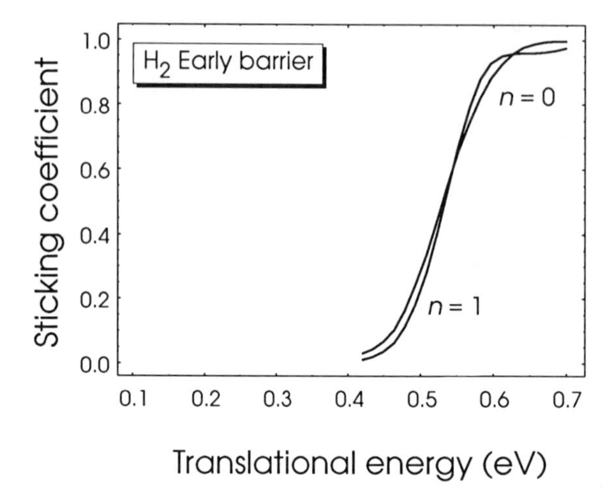

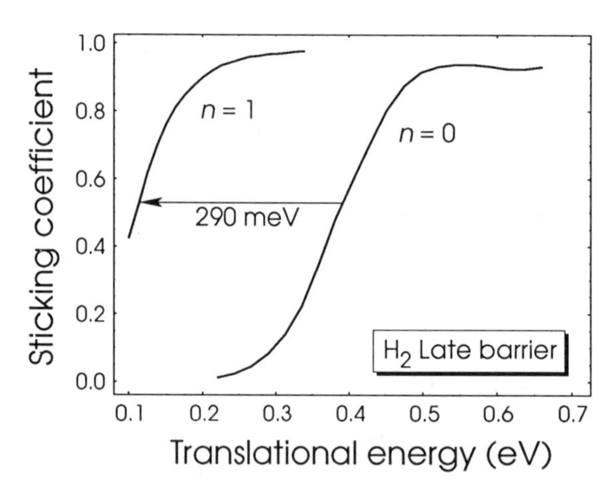

Figure 4.23. Plots showing the effect of vibrational excitation on the sticking coefficient depending on the position of the barrier along the reaction coordinate. As the barrier moves into the exit channel the $n = 1$ curve shifts much more than the $n = 0$ curve.

In the case of the late barrier, the efficacy of vibrational energy is greatly increased as seen in Figure 4.23b. Now there is a shift in the curve of ~290 meV to lower energies when compared with the $n = 0$ case. It is interesting to note that if all of the vibrational energy were available for barrier traversal, then there would be 640 meV of energy available (783/2 meV from the change in mass in the $n = 1$ state in the curvature region and an additional 257 meV if it then deexcited into the ground state of the exit channel), making the system unactivated! This is simply *not* the case and the observation of a finite but much reduced barrier is best understood by examining the $n = 1$ effective potential for the late barrier shown in Figure 4.13c. The height of this barrier is 146 meV relative to a zero-kinetic-energy n=1 molecule, whereas if the diatomic were vibrationally deexcited *before* striking the barrier, its total energy would lie well above that of the ground state barrier as can be seen in Figure 4.13c. This again indicates the inefficiency of the deexcitation process before the collision with the barrier; the population of scattered molecules in the ground state never rises above 5% even in this late barrier system.

The conclusion from these studies is that it is the release of vibrational energy from the first excited state in the entrance channel to the first excited state in the exit channel, $1 \rightarrow 1$ "vibrational adiabaticity," that is the dominant process, rather than deexcitation onto the ground state before the barrier is struck.

V. Summary

In this chapter we have given some examples of the recent progress made in studies of the quantum dynamics of gas–surface processes. Although quantum phenomena have hitherto been regarded as perhaps an interesting but esoteric diversion, a wide range of new experimental results are beginning to demonstrate that a quantum description of surface processes is required in many instances. It has been shown that the sticking of Ne on cold surfaces necessitates a quantum description of the lattice dynamics [60]. This also holds for the trapping of H_2 on Cu surfaces [61]. Diffraction has been observed for moderately heavy species such as Ar and N_2 [62,63]. The strong isotope effect observed in the dissociation of CH_4 indicates that tunneling may be occurring [64,65], while the enhanced dissociation of H_2 found for vibrationally excited species also has a quantum signature [19]. To theoretically model these sorts of events, a tractable scheme for solving the motion of a quantum wave packet is desirable. Although stationary state methods can, in some instances, be used, generally speaking, such methods do not deliver any of the insight afforded by time-dependent methods.

In all probability, the neatest method for solving the dynamics of a reactive encounter (e.g., dissociative adsorption) with a finite-temperature surface is to treat the incoming molecule fully quantum-mechanically and the surface semiclassically. This has the advantage that the detailed motion of the adsorbate is treated exactly whereas that of the surface is treated in a more statistical fashion. It is rather unnecessary, in most cases, to know the detailed motion of the ith atom in the jth layer! To embark on this road, an efficient method for solving the Schrödinger equation is required along with a coupling scheme to the lattice. The first of these problems may be achieved by using any of the recently developed grid methods for propagation of a quantum wave packet [66]. We have chosen to use the split operator method [27] in the examples presented here as it is efficient and provides insight into the dynamics during the collision process. It is important to stress that this is not the only method available and for certain classes of Hamiltonians it cannot be used at all [66]. It may be possible to treat the semiclassical subsystem by using the method of semiclassical Gaussian wave packets [67] coupled to the projectile by mean field methods [69], but this must await further development.

We have examined, in some detail, the interaction of a hydrogen molecule with a variety of potential energy surfaces specifically constructed to mimic physically interesting systems. The basis for these potentials is electronic total energy calculations. As mentioned in the Introduction, we have not attempted to fit PES parameters to satisfy any particular experimental results; instead, we have presented the dynamic consequences of particular classes of PES topologies.

Initially we examined the interplay between translational, vibrational, and rotational energy on the dissociative adsorption of H_2 [34]. The PES had an activation barrier of ~0.2 eV which, using conventional terminology, was located in the entrance channel. By examination of the results from two-, three-, and four-dimensional quantum simulations, it was possible to determine that the vibrational coordinate couples significantly with the rotational degree of freedom, only weakly with the translational motion across the unit cell, but strongly with that perpendicular to the surface. These effects were then studied in turn.

The strong coupling with rotational energy was examined in the form of a model calculation for the dissociative adsorption based on a simple steric model [37]. It is well known that because of its sphericity, rotations in H_2 couple only weakly to translational degrees of freedom; it seems clear, therefore, that the strongest isotropies in the scattered rotational distributions will arise from an angular selectivity in the dissociation. Calculations for the dissociation barrier height as a function of angle confirm intuition to the extent that molecules lying flat experience a lower barrier. Therefore there will be a significant θ dependence in the interaction potential which, in turn, results in a high degree of rotational excitation. This is particularly interesting in that measurements of the rotational distributions *and particularly their vector properties* should give direct information on the transition state geometry [69].

The strong coupling found between dissociation and translational energy normal to the surface has been well known from the time of Lennard-Jones [70] and has formed the basis for much of the hand-waving explanations often found in experimental studies. The decision was made to probe the dependence on translational energy a little deeper and potentials were generated having barrier heights with an identical variation across the unit cell but minima at either an on-top or a center site [31]. It was found that the diffraction intensities carry certain signatures as to the exact location of the col in the PES. Again, this information could be usefully employed in creating a detailed map of the transition state geometry.

Finally, the consequences wrought on the dissociation probability by the exact location of the activation barrier have been discussed [47]. This particular topic is inextricably tied to the efficacy of vibrational energy in surmounting activation barriers. A series of potentials were constructed having the barrier located in either the entrance or the exit channel. It was found that the propensity rules developed by Polanyi [53] were substantiated: for early barriers, translational energy aids dissociation and products are vibrationally excited; for late barriers, vibrational energy aids dissociation and products are translationally excited. The ramifications for surface dynamics have been presented and the usefulness of a reaction path analysis has been discussed. For the case when a barrier occurs very late in the exit channel, structure is observed in the scattering probabilities which arises from the formation of dynamic wells in the reaction path PES. Molecules remain trapped in these resonances for comparatively long periods which in real systems would enable secondary processes, such as phonon scattering and ultimately trapping, to occur. Estimates of the activation barrier height for H_2/Cu have been discussed in the light of results obtained for vibrationally enhanced dissociative adsorption.

Acknowledgments

We acknowledge Bret Jackson and Jens Nørskov for stimulating discussions. We thank Neville Richardson for his constant interest and a critical reading of the manuscript.

References

1. S. L. Bernasek, *Adv. Chem. Phys.* **41** (1980), 477.
2. J. A. Barker, and D. J. Auerbach, *Surf. Sci. Rep.* **4** (1985), 1.
3. D. W. Goodman, *Acc. Chem. Res.* **17** (1984), 194.
4. G. Comsa, and R. David, *Surf. Sci. Rep.* **5** (1985), 145.
5. M. J. Cardillo, *Langmuir* **1** (1985), 4.
6. S. T. Ceyer, *Annu. Rev. Phys. Chem.* **39** (1988), 479.
7. C. T. Rettner, H. Stein, and E. K. Schweizer, *J. Chem. Phys.* **89** (1988), 3337.
8. J. C. Tully, *Acc. Chem. Res.* **14** (1981), 188.
9. R. B. Gerber, *Chem. Rev.* **87** (1987), 29.
10. J. W. Gadzuk, *Annu. Rev. Phys. Chem.* **39** (1988), 395.
11. G. Ehrlich, in *Activated Chemisorption* (R. Vanselow, and R. Howe, Eds.), pp. 1–64, Springer-Verlag, Berlin, 1988.
12. A. E. DePristo and A. Kara, *Adv. Chem. Phys.*, **77** (1990), 163.
13. J. C. Lin, N. Shamir, Y. B. Zhao, and R. Gomer, *Surf. Sci.* **231** (1990), 333.
14. A. Kara, and A. E. DePristo, *J. Chem. Phys.* **88** (1988), 2033.
15. C. T. Rettner, E. K. Schweizer, and H. Stein, *J. Chem. Phys.* **93** (1990), 1442.
16. P. Avouris, D. Schmeisser, and J. E. Demuth, *Phys. Rev. Lett.* **48** (1982), 199.
17. S. Andersson and J. Harris, *Phys. Rev. Lett.* **48** (1982), 545.
18. H. J. Robota, W. Vielhaber, M. C. Lin, J. Segner, and G. Ertl, *Surf. Sci.* **155** (1985), 101.
19. K. D. Rendulic, A. Winkler, and H. Karner, *J. Vac. Sci. Technol. A* **5** (1987), 488.
20. J. Lapujoulade, and J. Perreau, *Phys. Scr.* **T4** (1983), 138.
21. B. E. Hayden, and C. L. A. Lamont, *Phys. Rev. Lett.* **63** (1989), 1823.
22. G. Anger, A. Winkler, and K. D. Redulic, *Surf. Sci.* **220** (1989), 1.
23. J. Harris, *Appl. Phys. A* **47** (1988), 63.
24. J. Harris, and S. Andersson, *Phys. Rev. Lett.* **55** (1985), 1583.
25. J. K. Nørskov, *J. Chem. Phys.* **90** (1989), 7461.
26. J. A. Fleck, J. R. Morris, and M. D. Feit, *Appl. Phys.* **10** (1976), 129.
27. M. D. Feit, J. J.A. Fleck, and A. Steiger, *J. Comput Phys.* **47** (1982), 412.
28. A. Goldberg, H. M. Schey, and J. L. Schwartz, *Am. J. Phys.* **35** (1967), 177.
29. D. Kosloff, and R. Kosloff, *J. Comput. Phys.* **52** (1983), 35.
30. B. Jackson, and H. Metiu, *J. Chem. Phys.* **86** (1986), 1026.

31. D. Halstead, and S. Holloway, *J. Chem. Phys.* **88** (1988), 7197.
32. D. Halstead, and S. Holloway, *Chem. Phys. Lett.* **154** (1989), 181.
33. M. R. Hand, and S. Holloway, *J. Chem. Phys.* **91** (1989), 7209.
34. U. Nielsen, D. Halstead, S. Holloway, and J. K. Nørskov, *J. Chem. Phys.* **93** (1990), 2879.
35. G. O. Sitz, A. C. Kummel and R. N. Zare, *J. Vac. Sci. Technol. A* **5** (1987), 513.
36. R. C. Mowrey, Y. Sun, and D. J. Kouri, *J. Chem. Phys.* **91** (1989), 6519.
37. S. Holloway, and B. Jackson, *Chem. Phys. Lett.* **172** (1990), 40.
38. P. K. Johansson, *Surf. Sci.* **104** (1981), 510.
39. A. J. Cruz, and B. Jackson, *J. Chem. Phys.* **94** (1991), 5715.
40. C. Capozzi, S. M. Francis, D. Roscoe, N. V. Richardson, and S. Holloway, *J. Vac. Sci. Technol. A* **5** (1987), 1049.
41. J. Harris, and A. Liebsch, *Phys. Rev. Lett.* **49** (1982), 341.
42. P. Nordlander, C. Holmberg, and J. Harris, *Surf. Sci.* **152** (1985), 702.
43. N. D. Lang, in *Theory of the Inhomogeneous Electron Gas* (S. Lundqvist, and N. H. March, Eds.), p. 189, Plenum Press, New York, 1983.
44. E. E. Nikitin, *Theory of Elementary Atomic and Molecular Processes in Gases*, Clarendon, Oxford, 1974.
45. T. Engel, and K. H. Rieder, *Springer Tracts in Modern Physics*, Springer-Verlag, Berlin, 1982.
46. G. Boato, P. Cantini, U. Garibaldi, A. C. Levi, L. Mattera, R. Spadaccini, and G. E. Tommei, *J. Phys. C* **15** (1973), L394.
47. D. Halstead, and S. Holloway, *J. Chem. Phys.* **93** (1990), 2859.
48. J. C. Polanyi, *Science* **236** (1987), 680.
49. J. Harris, and B. Kasemo, *Surf. Sci.* **105** (1981), L281.
50. M. Asscher, O. M. Becker, G. Haase, and R. Kosloff, *Surf. Sci.* **206** (1988), L880.
51. P. Nordlander, S. Holloway, and J. K. Norskov, *Surf. Sci.* **136** (1984), 59.
52. W. Brenig, and H. Kasai, *Surf. Sci.* **213** (1989), 179.
53. J. C. Polanyi, *Acc. Chem. Res.* **5** (1972), 161.
54. J. Harris, S. Andersson, C. Holmberg, and P. Nordlander, *Phys. Scr.* **T13** (1986), 155.
55. R. B. Bernstein, *Chemical Dynamics via Molecular Beam and Laser Techniques*, Oxford Univ. Press, Oxford, 1982.
56. R. A. Marcus, *J. Chem. Phys.* **45** (1966), 4493.
57. E. A. McCullough, and R. E. Wyatt, *J. Chem. Phys.* **54** (1970), 3578.
58. R. D. Levine, and R. B. Bernstein, *Molecular Reaction Dynamics and Chemical Reactivity*, Oxford Univ. Press, Oxford, 1987.
59. S. Latham, J. F. McNutt, and R. E. Wyatt, *J. Chem. Phys.* **69** (1978), 3746.
60. H. Schlichting, D. Menzel, T. Brunner, W. Brenig, and J. C. Tully, *Phys. Rev. Lett.* **60** (1988), 2515.
61. S. Andersson, L. Wilzén, M. Persson, and J. Harris, *Phys. Rev. B* **40** (1989), 8146.
62. K. H. Rieder and W. Stocker, *Phys. Rev. B* **31** (1985), 3392.
63. E. K. Schweizer, and C. T. Rettner, *Phys. Rev. Lett.* **62** (1989), 3085.
64. C. T. Rettner, H. E. Pfnür, and D. J. Auerbach, *Phys. Rev. Lett.* **54** (1985), 2716.
65. C. T. Rettner, F. Fabre, J. Kimman, and D. J. Auerbach, *Phys. Rev. Lett.* **55** (1985), 1904.

66. R. Kosloff, *J. Phys. Chem.* **92** (1988), 2087.
67. E. J. Heller, *J. Chem. Phys.* **62** (1975), 1544.
68. E. J. Heller, *J. Chem. Phys.* **64** (1976), 63.
69. S. Holloway, and X. Y. Chang, *Trans. Faraday Soc.*, to be published.
70. J. E. Lennard-Jones, *Trans. Faraday Soc.* **28** (1932), 333.

CHAPTER 5

Relationship of Reaction Energetics to the Mechanism and Kinetics of Heterogeneously Catalyzed Reactions

Alexis T. Bell

I. Introduction

The transformation of reactants to products on the surface of a catalyst can be envisioned to proceed via a sequence of elementary steps, commonly referred to as the reaction mechanism. As mechanisms cannot be deduced from first principles, they are usually postulated on the basis of physicochemical knowledge and intuition. Depending on how much information is available, the nature of the adsorption sites and the structure of the adsorbed reactants, intermediates, and products may be specified either very crudely or in some detail.

More often than not, more than one reaction mechanism can be proposed for a given stoichiometric reaction. It is also conceivable that if two or more competing pathways occur, one pathway may be preferred under one set of reaction conditions, and another may predominate under another set of conditions. The challenge then is to select the set of elementary processes that provides the best representation of the observables.

Assessment or evaluation of a postulated mechanism is possible, in principle, if the rate coefficient for each elementary step can be specified, together with its dependence on temperature and adsorbate coverage. With this information in hand, the kinetics of the overall reaction could be described in terms of the dynamics of each elementary step. Subsequent analysis would then show which processes predominate and what factors control the overall catalyst activity and selectivity. Similar calculations carried out for a series of metals would show whether the same mechanism prevails for all metals or whether changes occur as a consequence of differences in the properties of individual metals.

To carry out the procedure described in the preceding paragraph it would be necessary to have a reliable theory for predicting rate coefficients on the basis of the properties of the metal surface and the relevant surface species. A critical element in such a theory would be the potential energy of the interactions between adsorbates and the metal surface. If this potential hypersurface could be defined, it would be possible to predict the preferred sites for adsorption, the equilibrium structure of the adsorbate, the strength of adsorbate binding, and the height of the barriers for transforming one adspecies into another. Using transition state theory, one could then determine the rate coefficients for elementary process occurring on the surface.

This chapter discusses the use of reaction energetics to delineate energetically preferred reaction pathways and the effects of metal composition on catalytic activity and selectivity. The influence of adsorbate coverage on reaction energetics and, hence, on the rate coefficients for elementary reactions is also examined.

II. Calculations of Reaction Energetics

The energy hypersurface governing the progress of heterogeneously catalyzed reactions is a complex function depending on the position of the adsorbate (or adsorbates) relative to the metal surface, as well as the structure of the adsorbate itself. In principle, adsorbate–surface potentials should be obtained by accurate *ab initio* quantum mechanical calculations. In Chapter 1 of this book, Siegbahn and Wahlgren review the current status of quantum chemical calculations of the interactions of atoms and small molecules with metal surfaces. They discuss the progress and difficulties of *ab initio* approaches and demonstrate that only a few models are promising for calculating reaction energetics. In particular, they show that by use of the bond-prepared cluster model, calculations can be made of the heats of adsorption of H, O, N, and CO on Ni and Cu that are in good quantitative agreement with experimental measurements. Such calculations also appear to be successful for determining the gas–surface potential for H_2, O_2, and CH_4 dissociation on Ni surfaces. Although these results are impressive, it must be recognized that satisfactory *ab initio* calculations are so expensive that they can be carried out only for relatively small systems. Moreover, full geometric relaxation and calculation of the complete potential hypersurface for most systems of chemical interest are not possible at this time. (The discussion of dissociation of adsorbates on metal surfaces by Halstead and Holloway in Chapter 4 provides further examples of these limitations.)

Naturally, from early on, attempts have been made to use simpler, semiempirical quantum chemical techniques (see Chapter 1), particularly the extended Hückel method (EHM) [1] and its refined version, the atomic superposition-electron delocalization-molecular orbital (ASED-MO) method [2]. So far, however, calculations of adsorption energies based on such approaches produce large errors relative to experiment, usually of tens of kilocalories per mole [3]. As the sources of the errors are not controllable, these semiempirical quantum chemical methods do not seem promising as a means for obtaining quantitatively acceptable estimates of reaction energetics.

An alternative to quantum chemical modeling is phenomenological modeling. Here the prevailing approach has been to use linear correlations (Polanyi–Semenov, Bronsted, Hammond, Balandin, etc.) between the activation energy and the enthalpy of reaction [4,5]. Arguably, such relationships are restricted to a comparison within a given class of reactions (e.g., X + AB → XA + B, where reactant X is varied while reactant AB is kept fixed). Even if the reaction enthalpies are known accurately, which is usually not the case, the parameters in the relationship must be determined empirically.

Clearly, to be of heuristic value, phenomenological modeling should not postulate energetic interrelations but rigorously deduce them from a set of general principles. A model of this type, the bond order conservation-Morse potential (BOC-MP) method, has recently been developed by Shustorovich [3,6]. This analytic model is based on a small number of well-defined assumptions and requires very few parameters, all of which are obtainable from experimental observations. Extensive experience with the BOC-MP approach has shown that it yields heats of adsorption and activation energies for various elementary reactions that are in close agreement with experiment [3,6]. The model is also capable of describing the effects of adsorbate coverage on both heats of adsorption and activation energies [3,6a], something that none of the other models can do. In view of the accuracy of the BOC-MP method and its broad utility, we discuss it in more detail. An overview of the method is given in Section III and examples of its application to surface reactions are presented in Sections IV and V of this chapter. Some aspects of the BOC-MP method have already been discussed in Chapters 2 and 3.

III. Bond Order Conservation-Morse Potential Approach

The BOC-MP approach is based on two principal assumptions. The first is that each two-center interaction between an adsorbate atom A and a surface metal atom M is described by a Morse potential, and the total heat of adsorption is given by the sum of all A–M two-center interactions. The second assumption of the BOC-MP approach is that during the interaction of a molecular or atomic species with a metal surface, the total bond order, x, is conserved and normalized to unity (see Chapters 2 and 3). As the BOC-MP formalism is described extensively in Refs. [3,6], and touched upon in Chapters 2 and 3, we will only recall the relevant formulas needed to calculate the heat of adsorption for molecular adsorbates, Q, and the activation barriers for dissociation, recombination, and disproportionation reactions, ΔE^*.

For atomic chemisorption where an adatom A interacts with n metal atoms M, the M_n–A bond energy Q_n increases monotonically as the value of n increases. The magnitude of Q_n is given by

$$Q_n = Q_A = Q_{0A}\ (2\text{-}1/n) \tag{5.1}$$

where Q_{0A} is the maximum M–A two-center bond energy. The value of Q_n reaches the absolute maximum in the hollow n-fold site, so that the observed heat of atomic chemisorption Q_A (the atomic binding energy) can be identified with Q_n.

For molecular chemisorption, one should distinguish between weak and strong M_n–AB bonding. The weakly bound AB molecules typically have a closed electronic shell (for example, H_2, N_2, CO, NH_3, and H_2O) or unpaired electrons occupying substantially delocalized molecular orbitals (e.g., NO and O_2). The strongly bound AB molecules have unpaired electrons, which retain atomic characteristics. Examples of such species are CH, CH_2, NH, NH_2, OH, and OCH_3.

The simplest case corresponds to AB adsorbed perpendicular to a surface with the A end down. For such $\eta^1\mu_n$ coordination (one adatom A interacting with n metal M atoms), the heat of chemisorption for weak M_n–AB bonding is

$$Q_{AB,n} \approx \frac{Q_{0A}^2}{(Q_{0A}/n) + D_{AB}} \text{ for } D_{AB} > \frac{n-1}{n} Q_{0A} \tag{5.2}$$

where D_{AB} is the gas-phase A–B bond energy. If AB is coordinated via both A and B interacting with two metal atoms ($\eta^2\mu_2$ coordination), the bonding energy becomes

$$Q_{AB} = \frac{ab(a+b) + D_{AB}(a-b)^2}{ab + D_{AB}(a+b)} \tag{5.3}$$

where

$$a = Q_{0A}^2(Q_{0A} + 2Q_{0B})/(Q_{0A} + Q_{0B})^2$$

and

$$b = Q_{0B}^2(Q_{0B} + 2Q_{0A})/(Q_{0A} + Q_{0B})^2$$

For homonuclear A_2 ($a = b = \frac{3}{4} Q_{0A}$), Eq. 5.3 reduces to

$$Q_{A_2} = \frac{\frac{9}{2} Q_{0A}^2}{3Q_{0A} + 8D_{AB}} \tag{5.4}$$

For strong M_n–AB bonding with $\eta^1\mu_n$ coordination through the A atom, the heat of adsorption can be expressed as

$$Q_{AB} = \frac{Q_A^2}{Q_A + D_{AB}} \tag{5.5}$$

Equation 5.5 is an analog of Eq. 5.2 (for $n = 1$) where Q_{0A} is substituted for Q_A. In addition to weak and strong bonding, one can imagine the intermediate case which may be described by interpolation between the two extremes. In particular, for $\eta^1\mu_n$ coordination, one can simply average Eq. 5.2 and Eq. 5.5 as

$$Q_{AB} = \frac{1}{2}\left[\frac{Q_{0A}^2}{(Q_{0A}/n) + D_{AB}} + \frac{Q_A^2}{Q_A + D_{AB}}\right] \tag{5.6}$$

The intermediate M_n–AB bond strength is expected for monovalent radicals of tetravalent carbon, such as, for example, CH_3 and HCO.

A formula for the heat of adsorption of a chelated radical A–X–A (such as HCOO) has also been developed. In this case,

$$Q_{A_2(X)} = \frac{\frac{3}{2} Q_A^2}{Q_A + D_{AX}} \tag{5.7}$$

If AB approaches a surface from the gas phase, the activation barrier $\Delta E^*_{AB,g}$ for the dissociation $AB_g \rightarrow A_s + B_s$ depends explicitly on the chemisorption energies of all adsorbates; namely, the barrier can be approximated as

$$\Delta E^*_{AB,g} = \frac{1}{2}\left(D_{AB} + \frac{Q_A Q_B}{Q_A + Q_B} - Q_{AB} - Q_A - Q_B \right) \tag{5.8}$$

The dissociation barrier $\Delta E^*_{AB,g}$ from a chemisorbed state will be larger than $\Delta E^*_{AB,g}$ just by the amount of the molecular heat of chemisorption Q_{AB}:

$$\Delta E^*_{AB,s} = \Delta E^*_{AB,g} + Q_{AB} \tag{5.9}$$

Substituting Eq. 5.8 into Eq. 5.9, we obtain

$$\Delta E^*_{AB,s} = \frac{1}{2}\left(D_{AB} + \frac{Q_A Q_B}{Q_A + Q_B} + Q_{AB} - Q_A - Q_B \right) \tag{5.10}$$

In general, for the reverse reaction of recombination, the activation barriers can be calculated from the relevant thermodynamic cycles. Specifically, for the recombination of chemisorbed A_s and B_s to chemisorbed AB_s or gas-phase AB_g, the activation barriers $\Delta E^*_{A-B,s}$ and $\Delta E^*_{A-B,g}$ may be the same or different, depending on the sign of the dissociation barrier $\Delta E^*_{AB,g}$ from the gas phase; namely,

$$\Delta E^*_{A-B,s} = \Delta E^*_{A-B,g} = Q_A + Q_B - D_{AB} + \Delta E^*_{AB,g} \qquad \text{if } \Delta E^*_{AB,g} > 0 \tag{5.11}$$

or

$$\Delta E^*_{A-B,g} = \Delta E^*_{A-B,s} - \Delta E^*_{AB,g} = Q_A + Q_B - D_{AB} \qquad \text{if } \Delta E^*_{AB,g} < 0 \tag{5.12}$$

If one defines D_{AB} as the difference between total bond energies of AB_g and A_g plus B_g, the definition is valid for A and B being either atomic or molecular adsorbates. Thus, the BOC-MP formalism treats the reaction energetics in a uniform way for both diatomic and polyatomic molecules.

The BOC-MP approach can also handle disproportionation reactions $X + Y \rightleftharpoons Z + F$. Here, X_s and Y_s are taken as a chemisorbed quasimolecule XY_s with the gas-phase bond energy $D_{XY} = D_X + D_Y$ and the heat of chemisorption $Q_{XY} = Q_X + Q_Y$. If X, Y, Z, and F are arranged such that

$$D_{XY} = D_X + D_Y > D_Z + D_F, \tag{5.13}$$

the dissociation bond energy, D, of the quasimolecule XY into the fragments Z and F is

$$D = D_X + D_Y - D_Z - D_F > 0 \tag{5.14}$$

Then, for the forward direction of the disproportionation reaction, the activation barrier can be calculated from Eq. 5.10 as

$$\Delta E^*_{XY,s} = \frac{1}{2}\left(D_X + D_Y - D_Z - D_F + \frac{Q_Z Q_F}{Q_Z + Q_F} + Q_X + Q_Y - Q_Z - Q_F \right) \tag{5.15}$$

For the reverse direction, the barrier can be calculated as the recombination barrier from the relevant analogs of Eq. 5.11 or 5.12 combined with Eq. 5.15.

The preceding formulas for Q and ΔE^* are derived for low adsorbate coverages, $\theta \leq 1/n$, when the surface metal atoms of a unit cell, M_n–A or M_n–AB, interact with only one adspecies (in other words, when there are no adspecies in the adjacent unit cells). For medium and high coverages, $\theta > 1/n$, the coverage and coadsorption effects should be taken into account. The effects of lateral interactions can also be formulated within the context of the BOC-MP approach. When only through-metal interactions influence the heat of adsorption, the heat of adsorption for atoms can be written as

$$Q^*_{A,n} = \sum_{i=1}^{n} \frac{Q_{A,n}}{nm_i} (2 - 1/m_i) \tag{5.16}$$

and for molecules as

$$Q^*_{AB,n} = \sum_{i=1}^{n} \frac{Q_{AB,n}}{nm_i} (2 - 1/m_i) \tag{5.17}$$

where $Q_{A,n}$ ($Q_{AB,n}$) is the heat of adsorption for atom A (molecule AB) at zero coverage, and m_i is the number of adsorbates bonded to the ith metal atom. More elaborate equations can be written if direct adsorbate–adsorbate (A–A) interactions are important in addition to through-metal (M–A) interactions [6a,7]. The dependence of the activation energy on local site occupancy for various processes can be calculated by substituting the values of $Q^*_{A,n}$ and/or $Q^*_{AB,n}$ into Eqs. 5.8–5.12 and 5.15.

IV. Use of Reaction Energetics in Defining Reaction Mechanisms

Knowledge of the thermochemistry of a reaction can be used to determine the most energetically favorable pathways by which a reaction might proceed. From an evaluation of the activation barriers associated with each elementary step it is possible to identify processes for which the barrier height is so large that the probabilities of such processes occurring are negligible. Such analyses can identify which of several exit channels is most likely to occur and, hence, which elementary processes control product selectivity. Comparison of the energetics for different metals makes it possible to project the effects of metal composition on both activity and selectivity. Below, we illustrate several examples of the use of reaction energetics to interpret reaction mechanisms and the effects of metal composition. In each of these illustrations, the BOC-MP approach has been used to determine the heats of adsorption of all adspecies and the activation barriers for all relevant steps. For the most part, these calculations have been performed for empty surfaces, and, as a consequence, the mechanistic deductions drawn are strictly valid in the limit of zero adsorbate coverage. It will be shown that nevertheless such deductions are consistent with experimental observations made under conditions of finite coverage.

Table 5.1. Heats of Chemisorption (Q) and Total Bond Energies in the Gas Phase(D) and in Chemisorbed States ($D + Q$) for Species Involved in the Decomposition of Formic Acid on Ag(111), Ni(111), and Fe/W(110)[a,b]

Species	D^c	Coord.	Ag Q	Ag $D+Q$	Ni Q	Ni $D+Q$	Fe/W Q	Fe/W $D+Q$
C	–		<120	<120	171	171	200	200
H	–		<52	<52	63	63	66	66
O	–		80	80	115	115	125	125
CO	257		6	263	27	284	35	292
OH	102	$\eta^1\mu_3$	35	137	61	163	69	171
CO_2	384	$\eta^2\mu_2$	3	387	6	390	8	392
HCO	274	$\eta^1\mu_1$	<27	301	50	324	65	339
HCOO	384[d]	$\eta^1\mu_2$	34	418	59	443	66	450
		$\eta^2\mu_2$	39	423	71	455	81	465
HCOOH	481	$\eta^2\mu_2$	14	495	26	507	31	512

[a]As the value of Q_C is known experimentally only for Ni(111), the extrapolated values of Q_C were used for the other surfaces [3]. To rectify errors introduced by this extrapolation, the experimental values of Q_{CO} were used. All energies are in kcal/mol.
[b]Experimental values of D, Q_A, and Q_{CO} are taken from Table XIV of Ref. [3].
[c]Experimental values taken from Ref. [9].
[d]Reference [11].

A. Formic Acid Decomposition

Studies of formic acid decomposition on transition and posttransition metals have shown that adsorbed formic acid readily dissociates at or below room temperature to form formate species and that these species then decompose to release either CO_2 or CO. The BOC-MP approach has recently been used by Shustorovich and Bell [3,8] to analyze the periodic regularities of this reaction on Ag(111), Ni(111), and Fe/W(110) [Fe/W(110) represents a model surface for which adsorbate binding energies are intermediate between those on Fe and W, which are closely similar]. The purpose of these calculations was to address the following questions: (1) Why is the formate species the prevailing intermediate? (2) What is the preferred coordination of the formate species? (3) How does the formate species decompose into CO_2 and CO?

Table 5.1 lists the values of Q and $D + Q$ for each of the chemisorbed species thought to be involved in the decomposition of formic acid. As the value of Q_C is known experimentally only for Ni(111), the extrapolated values of Q_C were used for the other surfaces [3]. To rectify the errors introduced by this extrapolation, the experimental values of Q_{CO} were used. The activation barriers for elementary reactions calculated using the BOC-MP approach are summarized in Table 5.2. The values of Q, $D + Q$, and ΔE^* listed in Tables 5.1 and 5.2 are taken from Ref. [3] and are considered to be more accurate than those given in Ref. [8], as the former are based on more accurate estimates of the heat of adsorption of carbon on Ag and the heat of formation of formate in the gas phase [3].

It is apparent that the decomposition of formic acid on Ni and Fe/W should differ from that occurring on Ag. For the first two metals, the value of ΔE^*_g for the

Table 5.2. Calculated Activation Barriers for Forward (ΔE^*_f) and Reverse (ΔE^*_r) Elementary Reactions Involved in Formic Acid Decomposition on Ag(111), Ni(111), and Fe/W(110)[a]

Reaction		ΔE^*_f			ΔE^*_r	
	Fe/W	Ni	Ag	Fe/W	Ni	Ag
$HCOOH_g \leftrightarrows HCOO_s + H_s$	−22	−15	6	50	37	0
$\leftrightarrows HCO_s + OH_s$	−13	−2	43	29	6	0
$HCOOH_s \leftrightarrows HCOO_s + H_s$	9	11	20	28	22	0
$HCOO_s \leftrightarrows HCO_s + O_s$	22	25	42	21	9	0
$\leftrightarrows CO_s + OH_s$	12	13	23	11	5	0
$\leftrightarrows H_s + CO_{2,s}$	(8	5	0	1	2	16)[b]
$CO_{2,s} \leftrightarrows CO_s + O_s$	1	6	44	27	15	0
$HCO_s \leftrightarrows H_s + CO_s$	2	0	0	22	23	14

[a]The values of D and Q are from Table 5.1. All energies are in kcal/mol.
[b]As this reaction in the gas phase is practically thermoneutral (see Table 5.1), the formal value of D_{CH} is equal to zero, so that the values of ΔE^*_f appear to be underestimated [3].

dissociative adsorption of formic acid from the gas phase is negative. It indicates that if formic acid is adsorbed molecularly at low temperatures, it will dissociate preferentially to desorbing on elevation of the temperature. The products of such dissociation may be either $HCOO_s$ and H_s or HCO_s and OH_s, the activation barrier for dissociation into $HCOO_s$ being lower over Ni and Fe/W. Of the two carbon-containing products, only $HCOO_s$ may be observable as HCO_s will decompose spontaneously to CO_s and H_s.

By contrast to Ni and Fe/W, dissociative adsorption of formic acid on Ag exhibits a modest activation barrier for dissociation from the gas phase. When dissociation takes place, it will occur preferentially to $HCOO_s$, because the activation barrier for this process is much smaller (by 37 kcal/mol) than that for the formation of HCO_s.

Formate species can exhibit either mono- or dicoordination. As shown by Table 5.1, the heat of formate adsorption for the η^2 (di-) coordination is larger than that for the η^1 (mono-) coordiantion, the difference in the heats of adsorption increasing in the order Ag < Ni < Fe/W from 5 to 15 kcal/mol. Consistently, the experimental literature indicates that η^2 coordination is observed overwhelmingly [11–23], although on a low-activity surface such as Cu(100) [24], η^1- and η^2- coordinated formate species appear to coexist.

The decomposition of $HCOO_s$ groups can be perceived to occur via two alternative paths:

$$HCOO_s \rightarrow CO_{2,s} + H_s$$

$$HCOO_s \rightarrow HCO_s + O_s$$

The results presented in Table 5.2 indicate the dependence of these paths on metal composition. Thus, the decomposition of $HCOO_s$ into $CO_{2,s}$ and H_s is strongly preferred over Ag but the decomposition into HCO_s and O_s is preferred over Ni and especially over Fe/W. HCO_s is projected to be unstable; that is, once formed this species is expected to decompose spontaneously to CO_s and H_s.

The formic acid decomposition pathways projected by the BOC-MP approach are in good agreement with experimental observations. On surfaces of Mo [11], Fe [12], Ru [13], Rh [14], Ni [15,25] Pd [16], Pt [17], Cu [18], Ag [19,26], and Au [27], formate species have been observed as the first stable intermediate of formic acid decomposition and no evidence has been found for formyl species [19,28]. The high activation barrier for HCOOH dissociation on Ag and Au is demonstrated by the fact that here formate formation can be observed only in the presence of preadsorbed oxygen [19,26,27], which significantly decreases the activation energy barrier for HCOOH dissociation [3]. The results presented in Table 5.2 also agree with the observation that CO_2 is the sole product of formic acid decomposition over Ag(110) [19], Au(110) [27], Cu(100) [19], and Pt(111) [19], whereas CO is the major product formed over Ru(001) [19], Ni(110) [19], Ni(111) [25], and Fe(100) [12], and only CO is formed over Mo(100) [19]. It is noted further that while BOC-MP calculations indicate that CO_2 is formed by $HCOO_s$ decomposition on all of the metals studied, the calculated barrier for CO_2 dissociation to CO_s and O_s rapidly decreases from 44 kcal/mol for Ag to 1 kcal/mol for Fe/W. Thus, if CO_2 is formed during HCOOH decomposition on Ni and Fe/W, it will decompose into CO_s and O_s.

B. Methanol Synthesis from CO and CO_2

The mechanisms by which CO and CO_2 undergo hydrogenation to form methanol have been the subject of extensive experimental investigation. Based on infrared [29–33] and trapping [34–36] studies carried out with Cu catalysts, it has been suggested that HCO groups are the key intermediates in CO hydrogenation, whereas

Table 5.3. Heats of Chemisorption (Q) and Total Bond Energies in the Gas-Phase (D) and Chemisorption ($Q + D$) States for Species Involved in Methanol Formation and the Water-Gas-Shift Reaction on Cu(111) and Pd(111)[a]

		Cu		Pd	
Species	D^b	Q	$D + Q$	Q	$D + Q$
H	–	56	56	62	62
O	–	103	103	87	87
C	–	120	120	160	160
CO	257	12	269	34	291
CO_2	384	5	389	4	388
HCO	274	27	301	44	318
H_2CO	361	16	377	12	373
H_3CO	383	55	438	43	426
CH_3OH	487	15	502	11	498
HCOO	384[c]	59	443	46	430
OH	102	52	154	40	142
H_2O	220	14	234	10	230
H_2	104	5	109	7	111

[a]See footnote *a* in Table 5.1. All energies are in kcal/mol.
[b]Experimental values [9].
[c]Reference [10].

Table 5.4. Calculated Enthalpies of Reaction (ΔH_f) and Activation Barriers for Forward (ΔE^*_f) and Reverse (ΔE^*_r) Elementary Reactions Involved in Methanol Formation and the Water Gas Shift Reaction on Cu(111) and Pd(111)[a]

Reaction	M	ΔH_f	ΔE^*_f	ΔE^*_r
$CO_s \leftrightarrows C_s + O_s$	Cu	46	51	5
	Pd	44	50	6
$CO_s \leftrightarrows CO_g$	Cu	12	12	0
	Pd	34	34	0
$CO_s + H_s \leftrightarrows HCO_s$	Cu	24	24	0
	Pd	35	35	0
$HCO_s + H_s \leftrightarrows H_2CO_s$	Cu	−20	0	20
	Pd	7	16	9
$H_2CO_s + H_s \leftrightarrows H_3CO_s$	Cu	−5	4	9
	Pd	9	10	1
$H_3CO_s \leftrightarrows CH_{3,s} + O_s$	Cu	16	18	2
	Pd	4	16	12
$H_3CO_s + H_s \leftrightarrows CH_3OH_s$	Cu	−8	10	18
	Pd	−10	7	17
$CH_3OH_s \leftrightarrows CH_3OH_g$	Cu	15	15	0
	Pd	11	11	0
$CO_s + OH_s \leftrightarrows HCOO_s$	Cu	−20	0	20
	Pd	3	11	8
$CO_s + OH_s \leftrightarrows CO_{2,s} + H_s$	Cu	−22	0	22
	Pd	−17	1	18
$HCOO_s \leftrightarrows HCO_s + O_s$	Cu	39	39	0
	Pd	25	27	2
$HCOO_s + H_s \leftrightarrows H_2CO_s + O_s$	Cu	19	19	0
	Pd	32	32	0
$HCOO_s \leftrightarrows CO_{2,s} + H_s$	Cu	−2	2	4
	Pd	−20	0	20
$H_2O_s \leftrightarrows OH_s + H_s$	Cu	24	26	2
	Pd	26	26	0
$OH_s \leftrightarrows O_s + H_s$	Cu	−5	16	21
	Pd	−7	15	22
$CO_{2,s} \leftrightarrows CO_s + O_s$	Cu	17	28	11
	Pd	10	34	24
$CO_{2,s} \leftrightarrows CO_{2,g}$	Cu	5	5	0
	Pd	4	4	0
$H_{2,s} \leftrightarrows H_{2,g}$	Cu	5	5	0
	Pd	7	7	0
$H_{2,s} \leftrightarrows 2H_s$	Cu	−3	12	15
	Pd	−13	9	22
$H_{2,g} \leftrightarrows 2H_s$	Cu	−8	7	15
	Pd	−20	2	22
$HCOO_s + 4H_s \rightarrow CH_3OH_s + OH_s$	Cu	11		
	Pd	38		
$HCOO_s + 5H_s \rightarrow CH_3OH_s + H_2O_s$	Cu	−13		
	Pd	12		
$CO_s + H_2O_s \leftrightarrows CO_{2,s} + H_{2,s}$	Cu	5		
	Pd	22		

[a] All energies are in kcal/mol.

Table 5.5. Comparison of Calculated Heats of Adsorption (Q) with Experimental Values

Adspecies	Surface	Q (kcal/mol)		
		Calc.	Exp.	Ref.
H_2O	Cu(111)	14	14^a	[47]
			$11–13^a$	[48]
	Pd(111)	10	10^b	[49]
CH_3OH	Cu(111)	15	17^a	[47]
CO_2	Cu(111)	5	4–5	[50]

[a]For Cu(110).
[b]For Pd(100).

HCOO groups are the key intermediates in CO_2 hydrogenation. Formyl intermediates have also been proposed in the synthesis of methanol from CO and H_2 over Pd [37]. Experimental studies have shown as well that the rate of methanol synthesis from CO is very small on unpromoted Cu but proceeds much more rapidly over unpromoted Pd [37–45]. By contrast, the hydrogenation of CO_2 to methanol proceeds readily over Cu, whereas over Pd, CO_2 hydrogenation yields only CO and H_2O via the reverse water gas shift (WGS) reaction. To investigate the origins of the observed differences between Cu and Pd, Shustorovich and Bell [46] have carried out theoretical analyses of the hydrogenation of CO and CO_2 over Cu(111) and Pd(111) surfaces.

Table 5.3 lists the experimental values of D_{AB}, Q_A, and Q_{CO} as well as the calculated values of Q_{AB} and the total bond energies $D_{AB} + Q_{AB}$ for the relevant chemisorbed species formed on Cu(111) and Pd(111). Table 5.4 lists the calculated values of ΔE^* and ΔH for the elementary processes believed to be involved in the synthesis of methanol from CO and CO_2, and in the WGS reaction.

Verification of the calculations of Q and ΔE^* can be made only in the limited number of cases where experimental data are available. The data for Q are presented in Table 5.5 and those for ΔE^* are presented in Table 5.6. As can be seen, the differences between calculated and experimentally observed values of Q and ΔE^* are typically 1–3 kcal/mol.

We turn next to an assessment of the energetics associated with alternative pathways for the synthesis of methanol from CO and CO_2 and for the WGS reaction. In

Table 5.6. Comparison of Calculated Activation Barriers (ΔE^*) with Experimental Values

Reaction	Surface	ΔE^* (kcal/mol)		
		Calc.	Exp.	Ref.
$H_{2,g} \rightarrow 2H_s$	Cu(111)	7	5	[3]
$H_2O_s \rightarrow H_s + OH_s$	Cu(111)	26	27	[51]
$O_s + H_s \rightarrow OH_s$	Cu(111)	21	22	[52]
$CO_s + O_s \rightarrow CO_2$	Pd(111)	24	25	[3]
$CH_3O_s \rightarrow H_2CO_g + H_s$	Cu(111)	25	24	[53]
$OH_s + CO_s \rightarrow HCOO_s$	Rh(100)	10	8	[54]

the absence of O_s or OH_s, the formation of methanol from CO can be envisioned to proceed via the following sequence of steps:

1. $H_s + CO_s \rightarrow HCO_s$
2. $H_s + HCO_s \rightarrow H_2CO_s$
3. $H_s + H_2CO_s \rightarrow H_3CO_s$
4. $H_s + H_3CO_s \rightarrow H_3COH_s$
5. $CH_3OH_s \rightarrow CH_3OH_g$

As the first intermediate is HCO_s, this mechanism is referred to as the formyl mechanism. Table 5.4 shows that the largest activation energy, ΔE_1^*, is for reaction 1. The magnitude of this barrier is 24 kcal/mol for Cu and 35 kcal/mol for Pd. Comparison of these values with the activation energy for desorption of adsorbed CO indicates that the activation barrier $\Delta E_1^*$for CO hydrogenation to formyl, reaction 1, is significantly larger than the activation barrier for CO desorption for Cu (12 kcal/mol), but of comparable magnitude for Pd (34 kcal/mol). As a consequence, on a relative basis, the formyl route to methanol should be more efficient on Pd(111) than on Cu(111).

If OH_s groups are present on the catalyst surface, these species can react with adsorbed CO to form formate groups. The formate species can then serve as precursors to methanol. The sequence of elementary steps involved in the formate mechanism can be written as follows:

6. $OH_s + CO_s \rightarrow HCOO_s$
7. $H_s + HCOO_s \rightarrow H_2CO_s + O_s$
3. $H_s + H_2CO_s \rightarrow H_3CO_s$
4. $H_s + H_3CO_s \rightarrow H_3COH_s$
5. $H_3COH_s \rightarrow H_3COH_g$

Table 5.4 shows that the formation of formate species is projected to be strongly exothermic on Cu ($\Delta H = -20$ kcal/mol) but weakly endothermic ($\Delta H = 3$ kcal/mol) on Pd. The activation barrier for this process is calculated to be zero for Cu and 11 kcal/mol for Pd. Once formed, the formate species can either decompose to HCO_s and O_s or undergo hydrogenolysis to form H_2CO_s and O_s (reaction 7). The calculated activation barriers presented in Table 5.4 indicate that decomposition should be preferred on Pd, but hydrogenolysis on Cu. The subsequent steps in the formation of methanol are identical to those appearing in the formyl mechanism presented above and, judging by the activation barriers, the hydrogenolysis of formate to H_2CO_s appears to be the rate-determining step.

The hydrogenation of CO_2 can result in either CO_s and OH_s, or $HCOO_s$. Table 5.4 shows that for Cu the activation barrier for formate formation is significantly smaller than that for disproportionation into CO_s and OH_s, whereas for Pd, the activation barriers for the two processes are almost identical. Furthermore, the activation barrier for $CO_{2,s}$ hydrogenation to $HCOO_s$ is close to the activation barrier for $CO_{2,s}$ desorption from Cu(111), but is significantly greater than the activation barrier for $CO_{2,s}$ desorption from Pd(111). As a consequence, the ease of formation of $HCOO_s$ should be substantially greater on Cu than on Pd. Judging by the activation barriers, we also conclude that, other conditions being equal, on Cu the hydrogenation of CO_2 to methanol should proceed more readily than the hydrogenation of CO to methanol.

The WGS reaction can occur in parallel with the synthesis of methanol; hence, it is of interest to examine the energetics for this reaction. A plausible mechanism for the WGS reaction can be written as follows:

$$
\begin{array}{rrcl}
8. & H_2O_s & = & OH_s + H_s \\
9. & CO_s + OH_s & = & CO_{2,s} + H_s \\
6. & CO_s + OH_s & = & HCOO_s \\
10. & HCOO_s & = & CO_{2,s} + H_s \\
11. & 2H_s & = & H_{2,s} \\
12. & OH_s & = & O_s + H_s \\
13. & CO_s + O_s & = & CO_{2,s}
\end{array}
$$

Table 5.4 indicates that the activation barrier for H_2O_s dissociation (reaction 8), which is 26 kcal/mol for both Cu and Pd, is the highest of all the barriers, and, consequently, the dissociative adsorption of water is expected to be a rate-determining step. Once OH_s groups are formed, they can react with adsorbed CO_s to produce $CO_{2,s}$ and H_s. Two pathways can be envisioned for this to occur. The first is disproportionation via reaction 9, whereas the second is via the formation and subsequent decomposition of $HCOO_s$, reactions 6 and 10. Inspection of Table 5.4 indicates that based on energetic considerations, disproportionation will be preferred on both Cu and Pd. Thus, formate species are not necessary intermediates in the mechanism of WGS reaction, although on Cu, some CO_2 may also be formed via the formate route, because as noted earlier, the activation barrier to produce formate groups from CO_s and OH_s is calculated to be zero, and the activation barrier for decomposition of $HCOO_s$ to $CO_{2,s}$ and H_s is only 2 kcal/mol.

It should be noted that the direct oxidation of CO_s via the reaction $CO_s + O_s \rightarrow CO_{2,s}$ (reaction 13) is not expected to be significant, as the activation barriers for reactions 12 and 13 (16 and 11 kcal/mol, respectively, for Cu, and 15 and 24 kcal/mol, respectively, for Pd) are much higher than those for CO_2 formation from CO_s and OH_s (reaction 9).

The reverse WGS reaction can occur in parallel with the hydrogenation of CO_2. The mechanism for this process is the same as that shown above for the WGS reaction, as a consequence of the principle of microscopic reversibility. The activation barrier for the reverse of reaction 10 is calculated to be 4 kcal/mol for Cu and 20 kcal/mol for Pd, whereas the calculated desorption activation barriers for $CO_{2,s}$ are 5 and 4 kcal/mol, respectively. As a consequence, the reverse WGS reaction is expected to proceed very effectively over Cu but not over Pd. It is also of note that on Cu the activation barrier for the decomposition of $HCOO_s$ to form CO_s and OH_s is only 1 kcal/mol greater than that for the hydrogenolysis of $HCOO_s$ to form H_2CO_s, a rate-limiting step along the path to methanol. This suggests that the formation of methanol from CO_2 can compete effectively with the reverse WGS reaction on Cu.

The above analysis of methanol synthesis from CO suggests that in the absence of adsorbed hydroxyl groups, Cu should be relatively inactive compared with Pd as a consequence of the relationship between the activation energies for the formation of formyl groups and those for the desorption of adsorbed CO. This projection is qualitatively consistent with experimental observations for both unsupported and silica-supported Cu and Pd. Unpromoted Cu powder shows negligible activity for methanol formation [39,40], whereas under comparable conditions, Pd(110) shows noticeable activity [45]. Similarly, for unpromoted Cu/SiO_2 and Pd/SiO_2 under identical reaction conditions (T = 573 K, H_2/CO = 2, P = 10.6 atm), the turnover frequency for methanol synthesis is more than an order of magnitude lower for Cu/SiO_2 than for Pd/SiO_2 [37,45].

If OH_s groups can be produced on the surface of Cu, then the reaction of CO to methanol is projected to occur via the formate mechanism, which is more favorable energetically than the formyl mechanism. This projection appears to be inconsistent with the findings of Au and Roberts [57], who failed to observe reaction 6 on Cu(111); however, as discussed below, there might be reasons other than energetics that determine the rate of reaction 6. In any event, the projection that in the formate route to methanol the hydrogenolysis of formate groups (reaction 7) is the rate-determining step is fully consistent with experiment [51,55].

The BOC-MP calculations clearly project that on Cu, CO_2 should undergo hydrogenation to methanol much more readily than CO, because of the low activation barrier to form $HCOO_s$ and the fact that the pathway from this intermediate to H_2CO_s is energetically more favorable than that involving the hydrogenation of CO_s. The formation of methanol should be accompanied by some CO formation by the reverse WGS reaction, as the activation barrier for the decomposition of $HCOO_s$ to CO_s and OH_s (20 kcal/mol) is comparable to that for the hydrogenolysis of this species to H_2CO_s and O_s (19 kcal/mol). These projections of the BOC-MP model are completely consistent with experimental observation. For example, it has recently been reported [56] that for unpromoted Cu/SiO_2, the rate of methanol formation from CO_2 is 50-fold faster than that from CO, under comparable conditions. Also consistent with the projection of the model is the observation that comparable amounts of CO and CH_3OH are produced during CO_2 hydrogenation [56], indicating that the reverse WGS reaction competes effectively with the reaction of CO_2 to CH_3OH.

Unlike Cu, the hydrogenation of CO_2 over Pd is not projected to produce methanol because both the activation barrier for hydrogenolysis of CO_2 to CO_s and OH_s (18 kcal/mol) and that for the hydrogenation of $CO_{2,s}$ to $HCOO_s$ (20 kcal/mol) are much higher than the activation barrier for desorption of $CO_{2,s}$ (4 kcal/mol). Even if formate species are formed, the activation barrier for the decomposition of $HCOO_s$ to CO_s and OH_s is much smaller than that for hydrogenolysis of $HCOO_s$ to H_2CO_s.

Finally, one can add that for H_2O_s dissociation (reaction 8) on Cu(111), the activation barrier is calculated to be 26 kcal/mol, in very close agreement with the experimental estimate of 27 kcal/mol given by Campbell and Daube [51]. Judging by the activation barriers, this dissociation step is projected to be the rate-determining step of the WGS reaction, in agreement with experiment [50,51].

Although the qualitative projections concerning the mechanisms of methanol synthesis and the WGS reaction deduced from the BOC-MP calculations are generally consistent with experimental observation, two important features require further discussion [46]. The first is the uncertainty concerning the formation and transformations of formate species, particularly reactions 6 and 10. It has been shown [3] that the BOC-MP method underestimates the activation barriers of reactions 6 and 10 and, therefore overemphasizes the attractiveness of the formate route to methanol. This overemphasis does not affect the qualitative conclusions for Pd(111), where it is projected that the formate route is not significant in the hydrogenation of either CO or CO_2. In the case of Cu(111), however, the projection that CO hydrogenation occurs preferentially via the formate route requires some qualifications. Mention has already been made of the findings of Au and Roberts [57] that at 295 K on Cu(111), CO_s does not interact with OH_s. Of course, there is a possibility that the activation barrier is too high for significant reaction to occur, and we simply underestimated the activation barrier, but this is not thought to be the case. Indeed, the barrier for reaction 6 has been measured by Ho on Rh(100) and was found to be as low as 8 kcal/mol [54], in close agreement with the BOC-MP calculated value of 10 kcal/mol [3]. Because on Rh(100) the value of Q_{CO} = 32 kcal/mol is much higher than Q_{CO} = 12 kcal/mol on Cu(111), but the values of Q_O are practically the same for the two surfaces (102 and 103 kcal/mol, respectively [58]), the activation barrier on Cu should be smaller than on Rh, and we can set 8 kcal/mol as the upper bound for the activation barrier on Cu. Consequently there should be another reason for reaction 6 to be slow. The clue might be an observation by Ho [54] that for Rh(100), reaction 6 could not be observed for OH_s coverages below 0.5. If this condition of high OH_s coverage also holds for Cu surfaces, it could explain the findings of Au and Roberts [57]. Moreover, if Chinchen *et al.* [38] are correct in their claim that the surface coverage of Cu by OH_s is very small under methanol synthesis conditions, the discussion of whether or not reaction 6 occurs on unpromoted metallic Cu may be purely academic.

The second point requiring further discussion is the mechanism of the WGS reaction. It has been projected (see above) that this reaction should proceed via reactions 6 plus 8–12, in which OH_s is the active oxidizing agent. This conclusion is to be compared with the previous suggestions that the WGS reaction proceeds via the sequence [50]

$$H_2O_s \rightarrow 2H_s + O_s$$

$$CO_s + O_s \rightarrow CO_{2,s}$$

The choice of OH_s or O_s as the principal oxidizing agent in the WGS reaction may depend on the concentrations of these groups under reaction conditions. One can add that the activation barrier for the reaction $CO_s + O_s \rightarrow CO_{2,s}$ on Cu(111) has been measured by Habraken *et al.* [59] to be 18–20 kcal/mol. This value is much larger than the heat of CO adsorption, 12 kcal/mol. By contrast, in all other reported cases of CO_s oxidation to CO_2, including surfaces of Pt, Pd, Rh, Ru, and Ag, the activation barrier for CO oxidation has invariably been found to be smaller than the heat of CO adsorption [3]. The latter relationships is also predicted by the BOC-MP model [3], which makes the experimental value of the activation energy reported for Cu(111) [59] seem puzzling.

C. Hydrocarbon Synthesis from CO and H_2

The hydrogenation of CO can produce methane and/or a variety of C_2+ paraffins and olefins. C_2+ aldehydes and alcohols are also observed, albeit to a lesser extent, unless the catalyst is specifically promoted to produce these products. Experimental investigations of the mechanism of CO hydrogenation to hydrocarbons [60] suggest that the C–O bond of adsorbed CO is broken at an early stage of reaction and that CH_x (x = 1–3) species serve as precursors to the formation of methane and C_2+ hydrocarbons. Aldehydes and alcohols are believed to be derived from acyl groups formed by the insertion of CO_s into the carbon–metal bond of alkyl species. The plausibility of this scenario has recently been investigated by Shustorovich and Bell using the BOC-MP approach [61,62].

Because the BOC-MP approach explicitly treats only mono- and dicoordinated species, the analysis was limited to C_1 and C_2 hydrocarbons and oxygenates. The

Table 5.7. Heats of Chemisorption (Q) and Total Bond Energies in the Gas-Phase (D) and Chemisorption States ($D + Q$) for Species Involved in Fischer–Tropsch Synthesis on Cu(111), Pt(111), Ni(111), and Fe/W(110)[a]

		Cu		Pt		Ni		Fe/W	
Adsorbate	D^b	Q	$D+Q$	Q	$D+Q$	Q	$D+Q$	Q	$D+Q$
H	–	56	56	61	61	63	63	66	66
O	–	103	103	85	85	115	115	125	125
C	–	120	120	150	150	171	171	200	200
CH	81	72	153	97	178	116	197	142	223
CH_2	183	48	231	68	251	83	266	104	287
CH_3	293	26	319	38	331	48	341	62	355
CH_4	398	5	403	6	404	6	404	7	405
OH	102	52	154	39	141	61	163	69	171
H_2O	220	14	234	10	230	17	237	19	239
CO	257	12	269	32	289	27	284	36	293
CH_3CO	565	27	592	40	605	50	615	66	631
CH_3CHO	651	16	667	11	662	19	670	22	673
CH_3CH_2O	668	54	722	41	709	64	732	72	740
CH_3CH_2OH	771	14	785	11	782	18	789	21	792
CH_3C	376	71	447	97	473	115	491	141	517
CHCH	392	9	401	14	406	18	410	25	417
CH_2CH	421	40	461	44	465	55	476	71	492
CH_2CH_2	538	8	546	12	550	15	553	20	558
CH_3CH	466	49	515	70	536	85	551	107	573
CH_3CH_2	576	26	202	39	615	49	625	64	640
CH_3CH_3	674	4	678	5	679	5	679	6	680

[a]The values of Q_A (A = H, O, C) and Q_{CO} are from Ref. [3]. All energies are in kcal/mol.
[b]From Ref. [9]. (Also see Table 5.10.)

Table 5.8. Calculated Activation Barriers for Forward (ΔE_f^*) and Reverse (ΔE_r^*) Directions of Elementary Reactions Involved in Fischer–Tropsch Synthesis on Cu(111), Pt(111), Ni(111), and Fe/W(110)[a]

	Cu		Pt		Ni		Fe/W	
Reaction	ΔE_f^*	ΔE_r^*	ΔE_f^*	ΔE_r^*	ΔE_f^*	ΔE_r^*	ΔE_f^*	ΔE_r^*
$CO_g \leftrightharpoons CO_s$	0	12	0	32	0	27	0	36
$\leftrightharpoons C_s + O_s$	39	5	18	6	6	35	−14	68
$CO_s \leftrightharpoons C_s + O_s$	51	5	50	6	33	35	22	54
$CO_s + H_s \leftrightharpoons C_s + OH_s$	51	0	59	0	29	16	20	32
$C_s + H_s \leftrightharpoons CH_s$	31	8	38	5	42	5	46	3
$CH_s + H_s \leftrightharpoons CH_{2,s}$	5	27	13	25	17	23	24	22
$CH_{2,s} + H_s \leftrightharpoons CH_{3,s}$	0	32	7	26	16	24	19	21
$CH_{3,s} + H_s \leftrightharpoons CH_{4,s}$	0	28	6	18	14	14	24	8
$\leftrightharpoons CH_{4,g}$	0	23	6	12	14	8	24	1
$CH_{3,s} + C_s \leftrightharpoons CH_3C_s$	7	15	19	11	29	8	43	5
$CH_3C_s + H_s \leftrightharpoons CH_3CH_s$	10	22	18	20	22	19	27	17
$CH_{3,s} + CH_s \leftrightharpoons CH_3CH_s$	0	43	0	27	10	23	24	19
$CH_3CH_s + H_s \leftrightharpoons CH_3CH_{2,s}$	0	31	7	25	13	24	20	21
$CH_{3,s} + CH_{2,s} \leftrightharpoons CH_3CH_{2,s}$	0	52	0	33	6	24	20	18
$\leftrightharpoons CH_2CH_{2,s} + H_s$	0	52	0	29	11	20	28	10
$CH_3CH_{2,s} \leftrightharpoons CH_2CH_{2,s} + H_s$	4	4	4	0	11	2	16	0
$CH_{2,s} + CH_{2,s} \leftrightharpoons CH_{4,s}$	0	84	0	48	11	32	34	32
$CH_{2,s} + CH_s \leftrightharpoons CH_2CH_s$	0	77	2	38	18	31	39	21
$CH_2CH_s + H_s \leftrightharpoons C_2H_{4,g}$	0	21	1	13	8	7	20	−3
$CH_3CH_{2,s} + H_s \leftrightharpoons CH_3CH_{3,s}$	0	20	10	13	19	10	29	3
$\leftrightharpoons CH_3CH_{3,g}$	0	16	10	8	19	5	32	−3
$CH_{3,s} + CO_s \leftrightharpoons CH_3CO_s$	2	6	16	1	14	4	20	3
$CH_3CO_s \leftrightharpoons CH_3C_s + O_s$	42	0	47	0	33	24	28	39
$CH_3CO_s + H_s \leftrightharpoons CH_3C_s + OH_s$	47	0	52	0	32	8	28	19
$\leftrightharpoons CH_3CHO_s$	0	19	14	10	18	10	29	5
$\leftrightharpoons CH_3CHO_g$	0	3	15	−1	27	−9	46	−17
$CH_3CHO_s + H_s \leftrightharpoons CH_3CH_2O_s$	7	6	14	0	8	7	8	9
$CH_3CH_2O_s \leftrightharpoons CH_3CH_{2,s} + O_s$	19	2	18	9	13	21	9	34
$CH_3CH_2O_s + H_s \leftrightharpoons CH_3CH_2OH_s$	10	18	18	30	19	13	24	10
$\leftrightharpoons CH_3CH_2OH_g$	10	3	18	19	24	−5	35	−11

[a]The values of D and Q are from Table 5.7. All energies are in kcal/mol.

calculated values of Q and $D + Q$ for appropriate C_1 and C_2 adsorbates are given in Table 5.7. Also listed in this table are atomic heats of chemisorption, Q_A, and the total gas-phase bond energies, D_{AB}, which are the experimental parameters used to calculate Q_{AB}.

The activation barriers, ΔE^*, for the elementary steps involved in Fischer–Tropsch (F–T) synthesis are collected in Table 5.8. The results presented in Table 5.8 provide a basis for identifying the energetically preferred path (or paths) for producing C_s and $CH_{x,s}$ ($x = 1$–3) in the periodic series Fe/W(110), Ni(111), Pt(111), and Cu(111). C_s can be produced either by direct dissociation of CO ($CO_{g,s} \rightarrow C_s + O_s$) or by hydrogen-assisted dissociation of adsorbed CO ($H_s + CO_s \rightarrow C_s + OH_s$). As the activation barrier for dissociation of CO from the gas phase on Fe/W is negative ($\Delta E^*_{CO,g} = -14$ kcal/mol), this process will occur spontaneously. For Cu, the value of $\Delta E^*_{CO,g}$ rises to 39 kcal/mol and, hence, is large enough to preclude dissociative adsorption. The activation barriers on Ni and Pt are intermediate between those on Fe/W and Cu. Table 5.8 shows that the activation barrier for hydrogen-assisted dissociation of CO_s to produce C_s is smaller than that for direct dissociation of CO_s on Fe/W and Ni, but for Cu and Pt the reverse is true. Because for Cu and Pt, the activation barriers for CO_s dissociation, both direct and hydrogen-assisted, are significantly larger than the barrier for CO_s desorption, the calculations project that the dissociation of CO_s will not occur appreciably on these metals. For Ni, the barrier height for desorption of CO_s is practically the same as that for hydrogen-assisted CO_s dissociation, whereas for Fe/W, the latter barrier is much smaller (by 16 kcal/mol) than the former. As a consequence, it is projected that under conditions of F-T synthesis, dissociation of adsorbed CO should occur readily on Ni and with particular ease on Fe/W.

$CH_{x,s}$ species appear to be formed typically via stepwise hydrogenation of C_s or the dissociation of CH_xO_s (viz., $CH_xO_s \rightarrow CH_{x,s} + O_s$). As seen from Table 5.8, only on Fe/W may the formation of CH_s occur differently, namely, via hydrogen-assisted CO_s dissociation, $CO_s + H_s \rightarrow CH_s + O_s$. This process has a smaller barrier than that for direct hydrogenation of carbidic carbon, $C_s + H_s \rightarrow CH_s$ (39 and 46 kcal/mol, respectively). For Ni, however, the principal path to CH_s and further to $CH_{x,s}$ species seems to be the stepwise hydrogenation of C_s, whereas the dissociation of CH_3O_s is the principal path to $CH_{3,s}$ on Pt and Cu (cf. the similar findings in the BOC-MP analysis of methanation and methanol synthesis over Ni, Pd, Pt [63], and Cu [46]).

Table 5.8 shows very clearly that the hydrogenation of C_s to CH_s is strongly endothermic on all metal surfaces. Subsequent addition of hydrogen by and large remains endothermic on Fe/W but occurs exothermically on Ni, Pt, and Cu. Accordingly, the activation barrier for the addition of the first atom of hydrogen is much larger than that for the addition of subsequent hydrogen atoms, the value of ΔE^* usually decreasing as more hydrogen is added. The calculations also project that the activation barrier for each of the hydrogenation steps decreases in the order Fe/W > Ni > Pt > Cu.

Let us next examine the processes for C–C chain growth. One means by which chain propagation can occur is the insertion of $CH_{x',s}$ groups into the metal–carbon bond of a $CH_{x,s}$ group, that is, $CH_{x,s} + CH_{x',s} \rightarrow CH_xCH_{x',s}$ ($x = 1,2,3$; $x' = 1,2$). Table 5.8 shows that the value of ΔE^* decreases as x and x' increase; hence, the lowest activation barrier is for $x = 3$ and $x' = 2$. An alternative process for the formation of C–C bonds is CO insertion into the metal–carbon bond of $CH_{x,s}$ groups. The barrier for such insertions decreases as x increases, becoming smallest for $CH_{3,s}$. Table 5.8 shows that for a given $CH_{x,s}$ group the value of ΔE^* for CO_s insertion is larger than that for $CH_{2,s}$ insertion on Cu, Pt, and Ni. It is also projected that for any given elementary process, the value of ΔE^* decreases in the order Fe/W $\gg$ Ni > Pt $\approx$ Cu.

C_2 hydrocarbons can be formed from $C_2H_{5,s}$ species by either β-hydrogen elimination to form C_2H_4 or by α-hydrogen addition to form C_2H_6. It is evident from Table 5.8 that the activation barrier for olefin formation is lower than that for paraffin formation on Fe/W, Ni, and Pt.

The BOC-MP calculations project that the formation of acyl groups by CO insertion into the metal–carbon bond of $CH_{3,s}$ groups is exothermic only on Cu but endothermic on other metals. The activation barriers for this process are 2, 16, 14, and 20 kcal/mol for Cu, Pt, Ni, and Fe/W, respectively. The acyl groups formed can either decompose (by cleaving the C–O bond) to CH_3C_s and O_s or undergo hydrogenation to CH_3CHO_s. With the exception of Fe/W, the activation barrier for the first of these processes is substantially larger than that for the second. As the formation of the acyl group is a necessary first step in the production of C_2-oxygenates, the calculations suggest that the synthesis of such products would be most favorable over Cu and least probable on Fe/W. Table 5.8 also shows that on Pt the activation barrier for hydrogenation of CH_3CHO_s to $CH_3CH_2O_s$ is larger than that for the desorption of acetaldehyde, but on other metals the reverse is true. The ethoxide groups, $CH_3CH_2O_s$, can either decompose to $CH_3CH_{2,s}$ and O_s or undergo hydrogenation to form ethanol. The barrier for decomposition is smaller on Fe/W and Ni, whereas the barrier for hydrogenation is smaller on Cu; on Pt both barriers are practically the same. Thus, in summary, one would not expect to see significant formation of C_2-oxygenates on Fe/W and Ni, because the intermediate species, particularly $CH_3CH_2O_s$, are prone to cleave the C–O bond. By contrast, one would expect to see ethanol formed over Cu and a mixture of ethanol and acetaldehyde might be formed over Pt.

As has already been explained, the dissociative adsorption of gas-phase CO to produce carbidic carbon should be easy on Fe/W(110) and Ni(111) but difficult on Pt(111) and practically impossible on Cu(111). On the other hand, the activation barriers for hydrogenation of C_s to $CH_{x,s}$ and the barriers for chain propagation via $CH_{2,s}$ addition rapidly increase in the order Cu $<$ Pt $<$ Ni $<$ Fe/W. This suggests that bimetallic catalysts containing Fe and Pt or Cu might be more effective for producing C_2 hydrocarbons than those containing Fe alone.

The nonactivated dissociation of gas-phase CO on Fe/W inevitably leads to a high coverage of carbidic carbon. To assess the effects of carburizing an Fe surface, calculations were carried out of the heat of adsorption of CO and the barrier for dissociation of CO from the gas phase on Fe(100)-$c(2\times2)$C,O. This structure is formed by spontaneous dissociation of CO on a clean Fe(100) surface [64] and may be representative of the active surface of real F-T iron catalysts. As seen from Table 5.9, such partial carburization of an Fe(100) surface results in a reduction of the heat of adsorption of CO from 34 to 20 kcal/mol and in an increase in the activation barrier for the dissociation of gas-phase CO from -11 to 16 kcal/mol. As follows from

Table 5.9. Dissociation of CO on Fe(100): Coadsorption Effects on Q and ΔE^{*a}

Surface	$\theta_{C,O}$	Q_C	Q_O	Q_{CO}	$\Delta E^*_{CO,g}$
bcc Fe(100)	0	200^b	120^b	34^b	-11^c
Fe(100)$-c(2\times2)$C,O	0.5	178^c	82^c	20^c	16^c

[a] All energies are in kcal/mol.
[b] Experimental estimates [3].
[c] BOC-MP calculated values for $D_{CO} = 257$ kcal/mol.

Tables 5.7 and 5.8, the values of Q_C, Q_O, Q_{CO}, and ΔE^*_{CO} on Fe(100)-c(2×2)C,O appear to be intermediate between those for metallic Ni(111) and Pt(111) surfaces.

Comparisons between the BOC-MP projections concerning the mechanism of F–T synthesis and experimental observation should be made with caution, as the BOC-MP calculations were carried out for well-defined metal surfaces, whereas most experiments were carried out on supported metal catalysts. As a consequence, we consider only those experimental data that appear to be relevant to the model analysis.

The projection that the dissociative adsorption of gas-phase CO should occur spontaneously on Fe/W is in good agreement with numerous studies [60a,64–66]. In particular, Vink *et al.* [64] found that CO dissociates without activation on Fe(100), forming the Fe(100)-c(2×2)C,O structure on which CO chemisorbs molecularly with Q_{CO} = 20–24 kcal/mol, depending on coverage. For comparison, the BOC-MP calculations for Fe(100)-c(2×2)C,O (see Table 5.9) give Q_{CO} = 20 kcal/mol and the activation barrier $\Delta E^*_{CO,g}$ = 16 kcal/mol, from which it can be concluded that CO does not chemisorb dissociatively. The projection that the presence of hydrogen facilitates CO dissociation on Fe/W and Ni is consistent with measurements of the rates of direct versus hydrogen-assisted CO dissociation in Ni [67] and Ru [68] surfaces. The concomitant conclusion that the iron surface active for hydrocarbon formation should not be metallic but distinctly carbided also agrees with numerous observations [60a,64–66]. Particularly relevant to the present discussion is the observation that during F–T synthesis metallic Fe is rapidly carburized to form FeC_x [60a]. Moreover, in a model study of the hydrogenation of CO over polycrystalline iron, Krebs and co-workers [66], using Auger electron spectroscopy (AES) and x-ray photoelectron spectroscopy (XPS) techniques, found that the maximum turnover frequency for methane is not representative of a clean Fe surface but rather of a surface covered by a large amount of carbidic carbon. The BOC-MP model also projects that the dissociative adsorption of CO should be activated on Ni, Pt, and Cu, with the height of the activation barrier rapidly rising in the order Ni $\ll$ Pt $\ll$ Cu. Indeed, various Ni surfaces, including the single-crystal Ni(111) and Ni(100) as well as the polycrystalline Ni/Al_2O_3, are known to dissociate CO, producing a moderate amount of carbidic carbon (e.g., θ_C = 0.1–0.2 on Ni(100) [67]), which can be readily hydrogenated to methane, the methanation kinetics being structure insensitive [67]. At the same time, CO can hardly dissociate on Pt surface, where only the kinked Pt areas appear to be active [69], and no direct CO dissociation has been observed on Cu surfaces.

The calculations presented in Table 5.8 lead to the conclusion that chain propagation should occur via the insertion of $CH_{2,s}$ groups into the metal–carbon bond of alkyl species. The activation barrier for this process is significantly smaller than that for the insertion of CO_s. These projections are in strong agreement with experiment. In particular, in the isotopic tracer studies carried out by Brady and Pettit [70,71], the isotopic distribution of ^{13}C in the hydrocarbons produced from a mixture of ^{13}CO, $^{12}CH_2N_2$, and H_2 over a Co catalyst could only be explained by a chain propagation process involving the addition of $^{12}CH_2$ groups, while adsorbed ^{13}CO did not participate directly in chain propagation. Brady and Pettit [71] note further that the observation of a Schultz–Flory distribution of products characterized by a single value of α, the probability of chain growth, is indicative of a single mode of stepwise chain growth. The overwhelming experimental evidence supports the view that this single mode is described by $CH_{2,s}$ addition to adsorbed alkyl groups, $C_nH_{2n+1,s} + CH_{2,s} \rightarrow C_{n+1}H_{2n+3,s}$ [60a], in full agreement with the BOC-MP projection.

As metallic Fe is unstable during F–T synthesis and is rapidly carburized to form FeC_x [60a], under steady-state conditions, the C–C chain growth and termination effectively occur over an iron carbide surface. As shown above in Table 5.9, the effects of carburization of Fe are to move the heats of adsorption and activation energies toward those of a metal intermediate between Ni and Pt, making the recombination and desorption barriers much smaller than those on Fe. This is a very important conclusion because it explains why iron-based F–T catalysts can be efficient in the first place. Indeed, as Table 5.8 clearly shows, no products, neither hydrocarbons nor oxygenates, can be thermally desorbed without decomposition from clean Fe surfaces, in full agreement with various surface science studies [72]. Finally, the BOC-MP prediction that α-olefins should be primary products of hydrocarbon formation on Fe/W, Ni, and Pt is consistent with the observed F–T product distributions [60a].

D. Hydrocarbon Decomposition

The decomposition of hydrocarbons on transition metals is a problem of considerable interest as it relates to the formation of coke, a common cause of catalyst deactivation. The energetics of ethylene and acetylene decomposition on Fe/W, Ni, and Pt have also been examined by Shustorovich and Bell using the BOC-MP approach [72]. The results of these calculations are presented in Table 5.10 and 5.11. A number of interesting conclusions can be drawn from these calculations.

We begin by considering the energetics of C_2H_4 dehydrogenation to ethylidyne. This process is projected to be strongly exothermic on Fe/W ($\Delta H = -25$ kcal/mol), practically thermoneutral on Ni ($\Delta H = -1$ kcal/mol), and moderately endothermic on Pt ($\Delta H = 16$ kcal/mol). The formation of ethylidyne from ethylene can occur via either CH_3CH_s or CH_2CH_s as shown:

$$CH_2CH_{2,s} \xrightarrow{1} CH_3CH_s \xrightarrow{2} CH_3C_s$$
$$CH_2CH_{2,s} \xrightarrow{1'} CH_2CH_s \xrightarrow{2'} CH_3C_s$$

Because the activation barriers for isomerization (steps 1 and 2′) cannot be determined within the BOC-MP framework [3], it is not possible to project which path might be preferred. Recent experimental evidence reported by Zaera [73] suggests that, on Pt(111) and probably other metal surfaces, ethylidyne formation occurs preferentially via a vinyl intermediate, that is, steps 1′ and 2′.

Once formed, the fate of CH_3C_s appears to be sensitive to metal composition. On Pt(111), the BOC-MP calculations predict CH_3C_s to be relatively stable, as the

Table 5.10. Heats of Chemisorption (Q) and Total Bond Energies in the Gas-Phase (D) and Chemisorption ($D + Q$) States for Species Involved in the Decomposition of C_2 Hydrocarbons on Fe/W(110), Ni(111), and Pt(111)[a]

		$Q_{C_2H_x}$			$D_{C_2H_x} + Q_{C_2H_x} + (8-x)Q_H$[c]		
C_2H_x	$D_{C_2H_x}$[b]	Fe/W	Ni	Pt	Fe/W	Ni	Pt
H_3C-CH_3	674	6	5	5	812	805	801
H_3C-CH_2	576	64	49	39	838	814	798
H_3C-CH	466	107	85	70	837	803	780
$H_2C=CH_2$	538	20	15	12	822	805	794
H_3C-C	376	141	115	97	847	806	778
$H_2C=CH$	421	71	55	44	822	791	770
$H_2C=C$	348	110	87	71	854	813	785
$HC\equiv CH$	392	25	18	14	813	788	772
$HC\equiv C$	259	106	84	69	827	784	755
$CH_3 + CH_3$	586	124	96	76	842	808	784
$CH_3 + CH_2$	476	166	131	106	840	796	765
$CH_3 + CH$	374	204	164	135	842	790	753
$CH_3 + C$	293	262	219	188	885	827	786
$CH_2 + CH_2$	366	208	166	136	838	784	746
$CH_2 + CH$	264	246	199	165	840	778	734
$CH_2 + C$	183	304	254	218	883	815	767
$CH + CH$	162	284	232	194	842	772	722
$CH + C$	81	342	287	247	885	809	755
$C + C$	0	400	342	300	928	846	788
$CH_4 + CH_4$	796	14	12	12	810	808	808

[a]Values of Q_A are from Table 5.7. All energies are in kcal/mol.
[b]From Refs. [3,9].
[c]Normalized for the stoichiometry C_2H_8. For C_2H_x (or $CH_y + CH_{x-y}$) the remaining $8 - x$ atoms H are assumed to be atomically adsorbed.

calculated C$-$C bond scission barrier is 11 kcal/mol. This barrier decreases to 8 kcal/mol for Ni(111) and to 5 kcal/mol for Fe/W(110). Thus, it is projected that the stability of CH_3C to C$-$C bond cleavage decreases in the order Pt > Ni > Fe/W, in agreement with experiment [74–76]. Of further note is the fact that on Pt(111), the activation barrier for $CH_3C_s \rightarrow CH_{3,s} + C_s$ (11 kcal/mol) is smaller than that for $CH_3C_s \rightarrow CH_2C_s + H_s$ (13 kcal/mol). This indicates that dehydrogenation of CH_3C_s should be preceded by C$-$C bond cleavage, in full agreement with ^{13}C nuclear magnetic resonance (NMR) observations [77,78].

The hydrogen atoms released during the dehydrogenation of ethylene can react with part of the adsorbed ethylene to produce ethyl groups, $CH_3CH_{2,s}$. As the activation barrier for this process is very small (see Table 5.10), the formation of $CH_3CH_{2,s}$ groups should occur readily. The calculations presented in Table 5.10 suggest that the activation barrier for hydrogen elimination from the β carbon of the ethyl group to form $CH_2CH_{2,s}$ is much smaller than that for hydrogen elimination from the α carbon to form CH_3CH_s, the difference being particularly significant for Pt. These projections are consistent with the recent experimental findings on the decomposition of C_2H_x species on Pt(111) reported by Zaera [79].

Table 5.11. Calculated Activation Barriers for Forward (ΔE_f^*) and Reverse (ΔE_r^*) Reactions of Species Involved in the Decomposition of C_2 Hydrocarbons on Fe/W(110), Ni(111), and Pt(111)[a]

C_2H_x			D^b_{CX}	ΔE_f^*			ΔE_r^*		
				Fe/W	Ni	Pt	Fe/W	Ni	Pt
C_2H_6	$CH_3CH_{3,g}$	$\leftrightarrows CH_3CH_{2,s} + H_s$	98	−3	5	8	32	19	10
	$CH_3CH_{3,s}$	$\leftrightarrows CH_3CH_{2,s} + H_s$	98	3	10	13	29	19	10
C_2H_5	$CH_3CH_{2,s}$	$\leftrightarrows CH_{3,s} + CH_{2,s}$	100	18	24	33	20	6	0
		$\leftrightarrows CH_2CH_{2,s} + H_s$	38	16	11	4	0	2	0
		$\leftrightarrows CH_3CH_s + H_s$	110	21	24	25	20	13	7
C_2H_4	$CH_2CH_{2,g}$	$\leftrightarrows CH_{2,s} + CH_{2,s}$	172	−2	17	36	36	11	0
		$\leftrightarrows CH_2CH_s + H_s$	117	−3	7	13	20	8	1
	$CH_2CH_{2,s}$	$\leftrightarrows CH_{2,s} + CH_{2,s}$	172	18	32	48	34	11	0
		$\leftrightarrows CH_2CH_s + H_s$	117	17	22	25	17	8	1
	CH_3CH_s	$\leftrightarrows CH_{3,s} + CH_s$	92	19	23	27	24	10	0
		$\leftrightarrows CH_2CH_s + CH_s$	45	25	21	18	10	9	8
		$\leftrightarrows CH_3C_s + H_s$	90	17	19	20	27	22	18
C_2H_3	CH_2CH_s	$\leftrightarrows CH_{2,s} + CH_s$	157	21	31	38	39	18	2
		$\leftrightarrows CHCH_s + H_s$	29	14	9	5	5	6	7
		$\leftrightarrows CH_2C_s + H_s$	73	5	7	9	37	29	24
	CH_3C_s	$\leftrightarrows CH_{3,s} + C_s$	83	5	8	11	43	29	19
		$\leftrightarrows CH_2C_s + H_s$	28	2	15	13	9	22	20
C_2H_2	$CHCH_g$	$\leftrightarrows CH_s + CH_s$	230	−4	19	36	54	21	0
		$\leftrightarrows CHC_s + H_s$	133	−12	1	11	39	18	8
C_2H_2	$CHCH_s$	$\leftrightarrows CH_s + CH_s$	230	21	37	50	50	21	0
		$\leftrightarrows CHC_s + H_s$	133	13	19	25	27	18	8
	CH_2C_s	$\leftrightarrows CH_{2,s} + C_s$	165	20	27	32	49	29	14
		$\leftrightarrows CHC_s + H_s$	89	34	32	31	4	3	1
C_2H	CHC_s	$\leftrightarrows CH_s + C_s$	178	13	22	29	71	47	29

[a]The values of D and Q are from Table 5.10. All energies are in kcal/mol.
[b]The difference between the gas-phase total bond energies of the reactant and products.

The BOC-MP calculations show that acetylene chemisorbs more strongly than ethylene. The results in Table 5.10 suggest that dissociative adsorption to form C_2H_s and H_s should occur readily on Ni and spontaneously on Fe/W. On Pt, where the activation barrier for this process is 11 kcal/mol, the first surface reaction may be isomerization $CHCH_s \rightarrow CH_2C_s$ for which $\Delta H = -13$ kcal/mol. While the activation barrier for this process cannot be determined within the BOC-MP framework, the projection that the process is exothermic is consistent with the observation of CH_2C_s species on Pt(111) by Ibach and Lehwald [80] using EELS and on alumina-supported Pt by Wang *et al.* [78,81] using ^{13}C NMR spectroscopy. It should be noted though that the isomerization of acetylene to vinylidene may not be a major route for C–C bond scission as the activation barrier for the process $CH_2C_s \rightarrow CH_{2,s} + C_s$ is 32 kcal/mol. Dehydrogenation $CH_2C_s \rightarrow CHC_s + H_s$, for which the activation barrier is 31 kcal/mol, followed by C–C bond scission to form CH_s and C_s, for which the activation barrier is 29 kcal/mol, may be slightly preferred. Thus, some loss of hydrogen is expected before C–C bond rupture, in agreement with ^{13}C NMR data reported by Wang *et al.* [78,81] for C_2H_2 interacting with supported Pt. Another alternative appears to be disproportionation, $2C_2H_{2,s} \rightarrow CH_3C_s + CHC_s$, for which the BOC-MP method projects on Pt(111) an activation energy of 26 kcal/mol [3]. Consistent with this, Avery [82] has found that C_2H_2 does not desorb from Pt(111) and undergoes disproportionation above 370 K. Using Avery's data, Carter and Koel [83] have estimated the activation energy for disproportionation to be 23 kcal/mol.

For C_2H_4 and C_2H_2 chemisorbed on Fe/W(110), BOC-MP calculations indicate that the desorption energies are larger than the activation barriers for both C–H and C–C dissociation. In other words, it is predicted that not only the C–H but also C–C bond cleavage is nonactivated from the gas phase. Thus, on Fe surfaces one can expect ethylene to decompose rapidly to acetylene and further into CH_x fragments. This projection is consistent with the fact that under heating of chemisorbed CHCH (obtained either by adsorption of gas-phase C_2H_2 or by decomposition of adsorbed C_2H_4), only CH_x intermediates have been observed on various Fe surfaces by Erley *et al.* [84] and Seip *et al.* [85]. For metals between Fe/W and Pt, the decomposition products may be a variety of C_2H_x and CH_x species, depending on metal composition and reaction conditions. In particular, on Ni(111), the BOC-MP calculations project that chemisorbed ethylene dehydrogenates to vinyl species with an activation barrier of 22 kcal/mol, and that vinyl species further decompose via the sequence $CH_2CH_s \rightarrow CHCH_s + H_s \rightarrow CHC_s + 2H_s \rightarrow CH_s + C_s + 2H_s$, with the activation barriers of 9, 19, and 22 kcal/mol, respectively. Although vinylidene formation from vinyl via the reaction $CH_2CH_s \rightarrow CH_2C_s + H_s$ appears to be easy, the further decomposition of CH_2C_s to form $CH_{2,s}$ plus C_s or CHC_s plus H_s requires much larger activation barriers (27 and 32 kcal/mol, respectively). These BOC-MP projections are consistent with the recent observations of Zhu *et al.* [86,87] for C_2H_4 and C_2H_2 decomposition on Ni(111) and Ni(100). Similarly, rapid decomposition of $CHCH_s$ to HCC_s and $CH_{x,s}$ species on Ni surfaces has been reported by Lehwald and Ibach [88] and Stroscio *et al.* [89]. On the other hand, on Ru(001), the activity of which is intermediate between that of Pt(111) and Ni(111), H_xCC_s species with $x = 1$–3 (resulting from isomerization $CHCH_s \rightarrow CH_2C_s$, dehydrogenation, $CHCH_s \rightarrow CHC_s + H_s$, and rehydrogenation, $CHCH_s + H_s \rightarrow [CH_2CH_s] \rightarrow CH_3C_s$) have been identified by EELS by Parmeter *et al.* [90] and Jacob *et al.* [91].

V. Relationship of Reaction Energetics to Reaction Dynamics

In the absence of adsorbate–adsorbate interactions, the rate of an elementary surface reaction can be represented by

$$r = k\theta_i \tag{5.18}$$

or

$$r = k\theta_i\theta_j \tag{5.19}$$

where k is the rate coefficient, and θ_i and θ_j are the surface coverages for species i and j. The rate coefficient obeys an Arrhenius expression of the form

$$k = \nu \exp(-E/k_b T), \tag{5.20}$$

where ν is the preexponential, E is the activation barrier, and k_b is Boltzmann's constant. Many catalyzed reactions occur, however, under conditions of high coverage where adsorbate–adsorbate interactions cannot be neglected. In such cases, k becomes a complex function of θ, which does not obey the Arrhenius form of Eq. 5.20. As discussed in a recent review by Lombardo and Bell [7], adsorbate–adsorbate interactions affect not only the activation energy for a reaction but also the local configuration of adsorbates in the immediate vicinity of a site where reaction occurs.

Lombardo and Bell [7,92–94] have recently demonstrated that the BOC-MP approach can be used in conjunction with a Monte-Carlo technique to represent the effects of adsorbate–adsorbate interactions on the dynamics of surface processes. In the balance of this section, we outline that approach and illustrate its application to the desorption of CO and H_2 from different metal surfaces.

Molecules are assumed to participate in three rate processes: adsorption, diffusion, and desorption. At any instant in time, molecules on the surface occupy well-defined sites on a single-crystal lattice, the coordination of the adsorbate being specific to the metal and the adsorbate. As the local coverage varies with position on the surface, the dynamics of the three rate processes should be site specific. This specificity is taken into account through probabilities of adsorption, diffusion, and desorption which depend on the local surface environment.

The probability of adsorption on a site i in the time interval Δt is defined as [7,92–94]

$$P_i^a = S_0 P a_s / (2\pi m k_b T_g)^{1/2} \, \Delta t = S_0 F \Delta t \tag{5.21}$$

where S_0 is the sticking coefficient at zero coverage, P is the pressure, a_s is the area per site, m is the mass of an adsorbate, and T_g is the temperature of the gas phase. The factor F on the far right-hand side of Eq. 5.21 is the flux of adsorbates per site. The total rate of adsorption, expressed as a turnover frequency, r_a, based on the number of surface metal atoms, N_s, is given by

$$\begin{aligned} r_a &= \sum_i N_{a,i} \,/\, (N_s \, \Delta t) \\ &= N_a \,/\, (N_s \, \Delta t) \end{aligned} \tag{5.22}$$

The probability of desorption from site i in the time interval Δt is represented as [7,92–94]

$$P_i^d = \nu_i \exp[-E_{d,i}/k_b T] \, \Delta t \tag{5.23}$$

where ν_i is the frequency factor for desorption and $E_{d,i}$ is the activation energy for desorption. The value of $E_{d,i}$ is determined by means of the BOC-MP approach. The total rate of desorption, r_d, from all local environments is simply

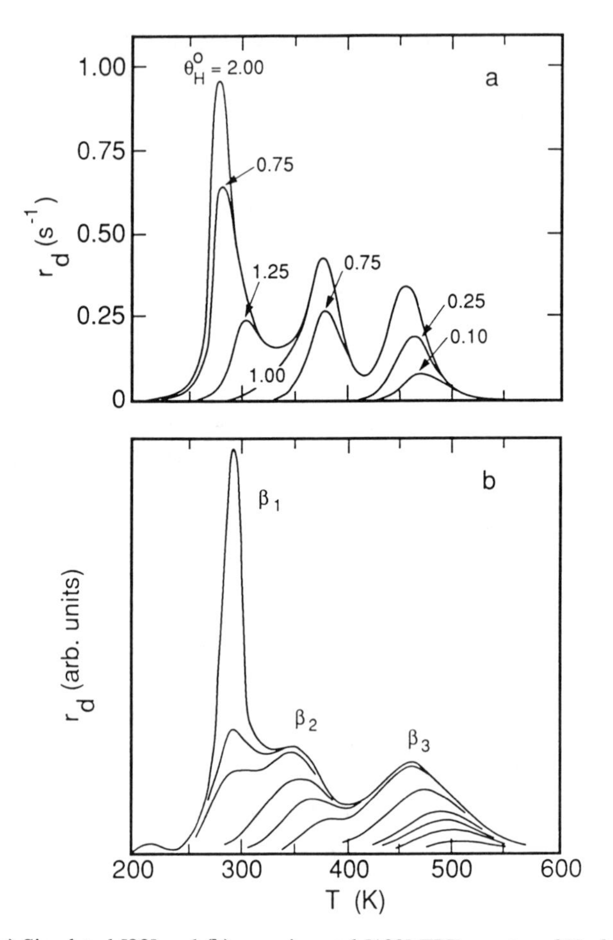

Figure 5.1. **(a)** Simulated [92] and **(b)** experimental [102] TPD spectra of H_2 from Mo(100).

$$r_d = \sum_i N_{d,i} / (N_s \, \Delta t) \\ = N_d / (N_s \, \Delta t) \tag{5.24}$$

As the activation energy for surface diffusion is typically 10–15% of that for desorption (see Chapter 3), whereas the frequency factors for the two processes are comparable [7,92–94], surface diffusion is expected to be a much more rapid process than desorption. As a consequence, diffusion is not treated as a rate process, but rather, it is assumed that adsorbate atoms or molecules will reposition themselves to achieve an equilibrium distribution. The equilibrium distribution of adsorbates is assumed to form in accordance with Kawasaki statistics [7,92–94], where the probability that an adsorbate on site i moves to an adjacent site j is given by

$$P_{ij} = \frac{\exp\,[-(Q_i - Q_j) / k_b T]}{1 + \exp\,[-(Q_i - Q_j) / k_b T]} \tag{5.25}$$

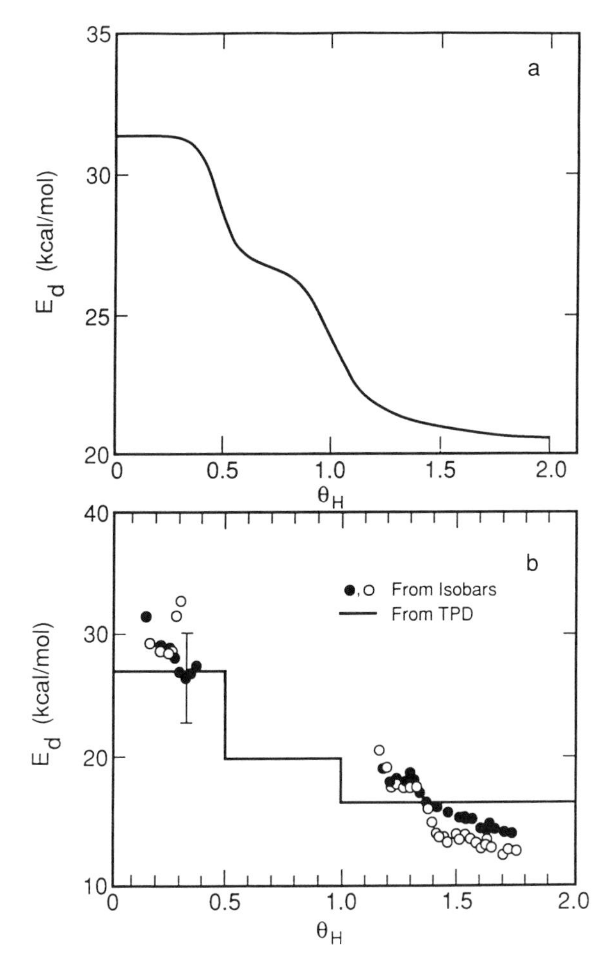

Figure 5.2. **(a)** Calculated [92] and **(b)** experimentally observed [103] variations of E_d with θ_H for H_2 desorption from Mo(100).

where Q_i and Q_j are the heats of adsorption for an adsorbate at sites i and j, respectively. Equation 5.25 weights diffusional jumps according to the magnitude of $Q_i - Q_j$. It should be noted that if $Q_i = Q_j$, then $P_{ij} = 0.5$, and the two sites have equal probability of occupancy. For a sufficiently large number of jumps, application of Eq. 5.25 produces an equilibrium distribution of adsorbates. The values of Q_i and Q_j in Eq. 5.25 are determined using the BOC-MP approach.

Lombardo and Bell [92] have simulated the desorption of H_2 from Mo(100) assuming, in accordance with the experimental observation, that H atoms occupy twofold bridging sites and have a maximum coverage of 2.0 [95–97]. No allowance was made for reconstruction of the Mo(100) surface, as experimental evidence suggest that reconstruction occurs primarily below 300 K and at low adsorbate coverages [96–101].

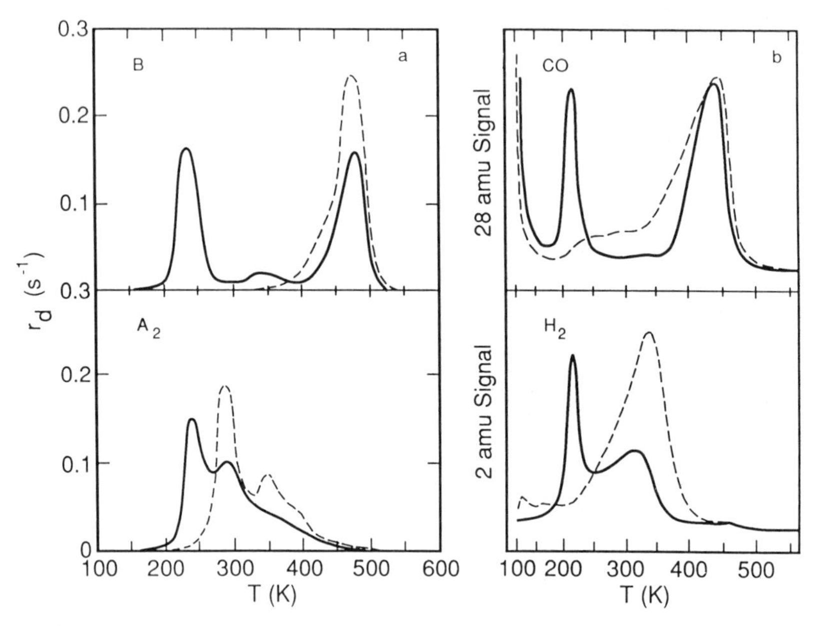

Figure 5.3. **(a)** Simulated TPD spectra of B and A_2 [93]. **(b)** Experimental TPD spectra of CO and H_2 desorption from Ni(100) [104]. The dashed curves represent desorption of each species adsorbed separately, and the solid curves represent desorption of both species coadsorbed.

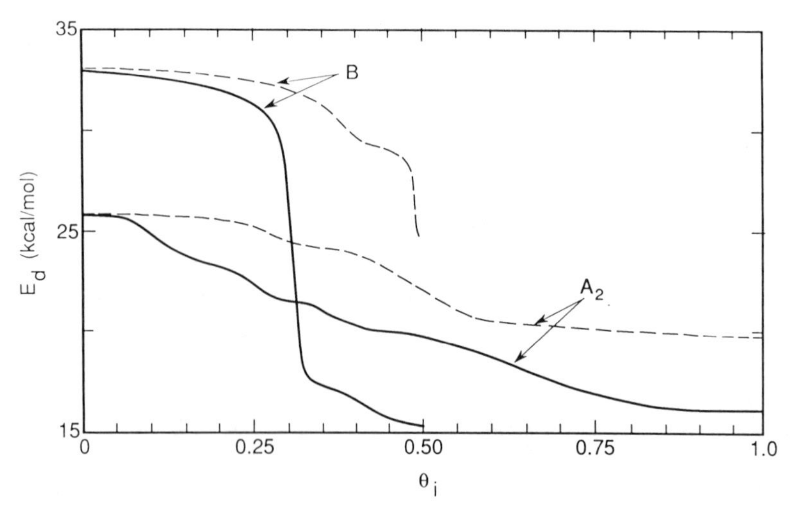

Figure 5.4. Variation in E_d with θ_i for the simulations shown in Figure 5.3a [93]. The dashed curves represent the desorption of each species adsorbed separately, and the solid curves represent the desorption of both species coadsorbed.

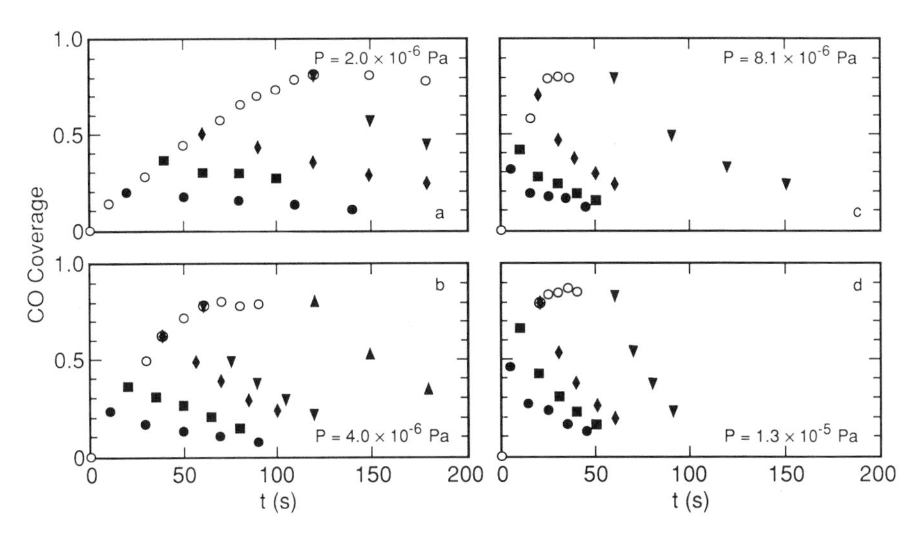

Figure 5.5. Experimentally observed variation in CO coverage versus time for polycrystalline Pd at $T_s = 380$ K [106]. The open circles represent the total CO coverage and the filled symbols represent the decay in the coverage of $C^{18}O$ for $t > t_1$. The total CO pressure for each experiment is as follows: **(a)** $P = 2.0 \times 10^{-6}$ Pa, **(b)** $P = 4.0 \times 10^{-6}$ Pa, **(c)** $P = 8.1 \times 10^{-6}$ Pa, **(d)** $P = 1.3 \times 10^{-5}$ Pa.

As illustrated by Figures 5.1a and 1b, the simulated temperature-programmed desorption (TPD) spectra for H_2 desorption from a Mo(100) surface agree with the experimentally observed spectra [100]. Figure 5.2 shows a comparison of the simulated and experimentally observed profiles of the activation energy as a function of coverage. Both plots exhibit a stepwise decrease in activation energy as the coverage increases. It is concluded that the observed decrease in the activation energy with increasing coverage is caused by a decrease in the heat of adsorption of H atoms when more than one atom is bonded to a given metal atom.

Lombardo and Bell [93] have also simulated TPD spectra for coadsorbed species using the BOC-MP approach to describe the effects of adsorbate coverage on the energetics of desorption. Nearest-neighbor interactions between the adsorbates and the metal surface as well as nearest-neighbor interactions between the coadsorbates were taken into account. The presence of a strongly bound coadsorbate on a bcc(100) surface was shown to shift the associative desorption spectrum of adsorbed atoms to lower temperatures, in qualitative agreement with experimental results for H_2 coadsorbed with strongly bound atomic species such as C, O, and N on Mo(100) and Fe(100) surfaces.

In the same paper Lombardo and Bell [93] have described the simulation of TPD spectra for the concurrent desorption of B molecules and the associative desorption of A atoms from an fcc(100) surface. Two types of behavior were observed. In one case, both species exhibited new low-temperature features not present in the TPD spectra of A and B when each species was adsorbed alone. In the second case, only the more weakly bound species displayed new spectral features. Both types of

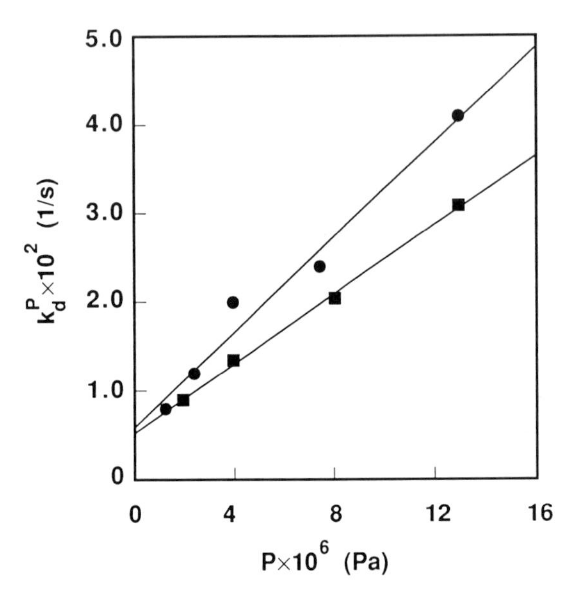

Figure 5.6. Variation of k_d^p versus *P*. ●, Analysis of experimental data presented by Yamada *et al.* [105]; ■, values obtained by Monte-Carlo simulation [94].

behavior have been observed for the codesorption of CO and H_2: the first for a Ni(100) surface, and the second for a Rh(100) surface [104]. Figure 5.3a shows TPD spectra for A_2 and B when each species is adsorbed separately and when both species are coadsorbed. The experimentally observed TPD spectra for CO and H_2 on Ni(100) are shown in Figure 5.3b. A comparison of the two figures demonstrates the qualitative agreement between the model predictions and the experimental observations. In Figure 5.4 are shown the activation energy profiles versus coverage for both the pure component and coadsorption simulations. For the coadsorbed A_2 and B, the activation energies are lower than for the respective pure component activation energies and show larger variation with coverage.

More recently, Lombardo and Bell [94] have extended their Monte-Carlo calculations to describe the effects of adsorbate pressure on the kinetics of CO desorption. Of particular interest was the simulation of results observed by Yamada *et al.* [105,106] who carried out the following experiments. At $t = 0$, $C^{18}O$ at a pressure *P* was adsorbed on a clean Pd surface held at a surface temperature T_s. At $t = t_1$, the gas phase was quickly switched from $C^{18}O$ to $C^{16}O$ while maintaining a constant pressure, and then at $t = t_1 + t_2$, the surface was flashed to high temperature. The TPD spectra resulting from the final step were integrated to determine the coverages of $C^{18}O$ and $C^{16}O$, $\theta(C^{18}O)$, $\theta(C^{16}O)$, and the total coverage, $\theta_T = \theta(C^{18}O) + \theta(C^{16}O)$. Repetition of the above procedure for different values of t_1 and t_2 was used to generate plots of coverage versus time. Figure 5.5 shows examples of the data obtained by Yamada *et al.* [106]. As seen in the figure, the total coverage initially rises with time until it reaches an equilibrium value for a given pressure. As the pressure is increased, higher equilibrium coverages are achieved in shorter periods.

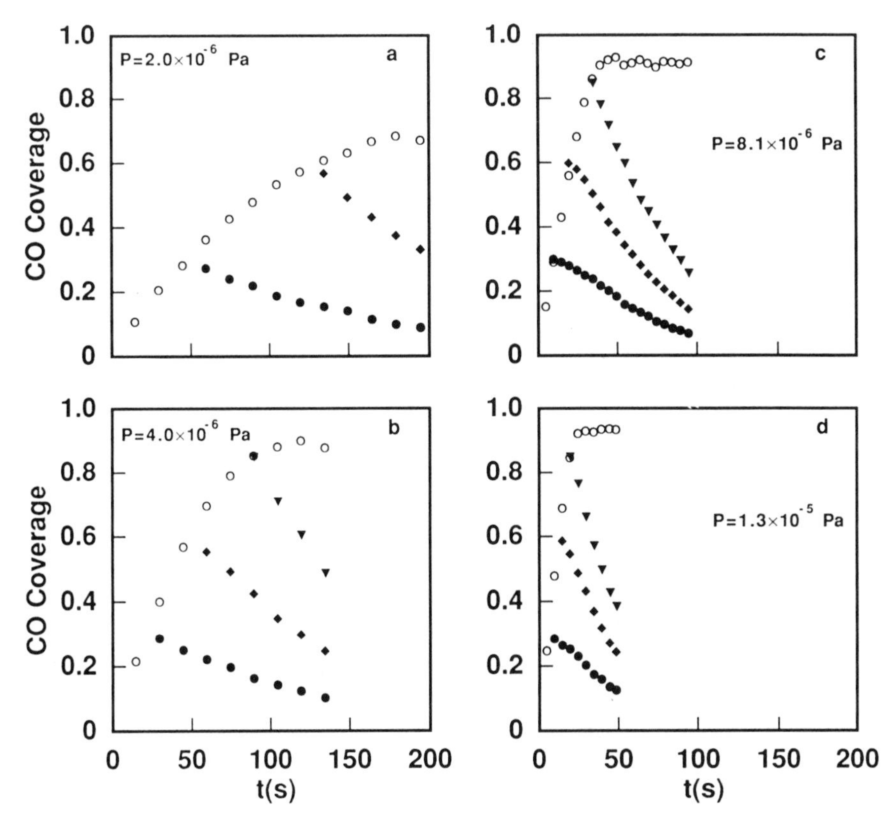

Figure 5.7. Simulated variation in CO coverage versus time for Pd(100) at T_s = 380 K [94]. The open circles represent the total CO coverage and the filled symbols represent the decay in the coverage of $C^{18}O$ for $t > t_1$. The total CO pressure for each simulation is as follows: **(a)** $P = 2.0 \times 10^{-6}$ Pa, **(b)** $P = 4.0 \times 10^{-6}$ Pa, **(c)** $P = 8.1 \times 10^{-6}$ Pa, **(d)** $P = 1.3 \times 10^{-5}$ Pa.

Also shown in Figure 5.5 are the curves of the decay of $\theta(C^{18}O)$ for $t > t_1$. The rates of $\theta(C^{18}O)$ decay are seen to depend on the pressure and the coverage at which the isotope switch is made. Yamada *et al.* [106] obtained rate coefficients for the desorption of $C^{18}O$ by fitting the decay curves of $\theta(C^{18}O)$ to single exponential functions of the form $\theta(C^{18}O) = \theta°(C^{18}O)\exp[-k_d^p(t-t_1)]$, where $\theta°(C^{18}O)$ is the value $\theta(C^{18}O)$ at $t = t_1$ and k_d^p is the apparent rate coefficient for desorption. As seen in Figure 5.6, k_d^p exhibits a linear dependence on the pressure.

Simulations of the results of Yamada *et al.* [106] were carried out for a Pd(100) surface on which CO molecules were placed in both types of bridging sites and a maximum of two molecules was allowed to be bonded to each metal atom. The activation energy for desorption from a totally bare surface was taken as 31.5 kcal/mol and the preexponential factor for desorption was taken as 10^{16} s^{-1}. Based on the experimental results obtained by Yamada *et al.* [105,106], the sticking coefficient for adsorption was written as

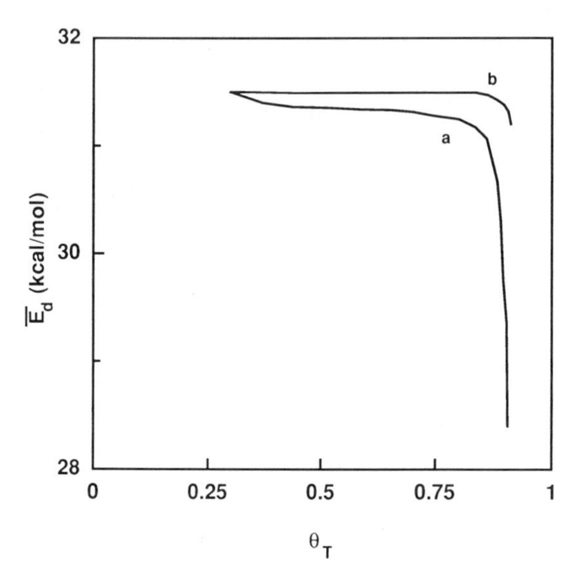

Figure 5.8. Variation in the average value of E_d as a function of total CO coverage [94]. **(a)** $C^{18}O$ molecules that desorb, **(b)** all of the $C^{18}O$ molecules on the surface. The results shown are for the simulation presented in Figure 5.7c for the case where the isotope switch is made at a $C^{18}O$ coverage of 0.30.

$$S(\theta) = 0.88\left(1 + \frac{0.05\theta}{1-\theta}\right)^{-1} \tag{5.26}$$

The results of the simulations are shown in Figure 5.7. The equilibrium coverage is seen to increase with increasing CO pressure whereas the time required to reach the equilibrium coverage decreases. Comparison with Figure 5.5 reveals qualitative resemblance of the general trends. As the simulations were carried out for a Pd(100) surface whereas the experimental results were obtained on polycrystalline Pd, quantitative agreement between the simulations and experiment was not sought. Nevertheless, the level of agreement between experiment and simulation is sufficiently high to suggest that the simulations capture the essential features of the experiments.

The curves of CO coverage versus time presented in Figure 5.7 were analyzed according to the method used by Yamada *et al.* [105]. As shown in Figure 5.6, treating the data in this fashion leads to an apparent dependence of the desorption rate constant on pressure, even though no direct mechanism for adsorption-assisted desorption was included in the model. It should also be noted that the dependence of k_d^p on CO pressure determined from the simulations agrees rather well with the dependence observed by Yamada *et al.* [105].

Before addressing the origin of the apparent pressure dependence of k_d^p, it should be noted that the isotope switch from $C^{18}O$ to $C^{16}O$ is made under two sets of conditions: before the total coverage is at its equilibrium value and after the total coverage has reached its equilibrium value. When the switch is made before the equilibrium coverage is reached, the desorption of $C^{18}O$ occurs while the total coverage is increasing. The rate at which the total coverage increases is proportional to the gas pressure. When the switch is made after the equilibrium coverage is reached, the desorption of $C^{18}O$ occurs at a constant total coverage that is proportional to the pressure. Thus, in both cases, the total coverage of CO is determined by the pressure.

Figure 5.8 shows that the average activation energy of the desorbing $C^{18}O$ molecules decreases from 31.5 kcal/mol at $\theta_T = 0.35$ to 27.5 kcal/mol at $\theta_T = 0.90$. As a consequence, changes in θ_T resulting from the effects of pressure will influence the activation energy for desorption. It therefore follows that the pressure dependence in k_d^p observed in Figure 5.6 can be ascribed to the effects of lateral interactions on the activation energy for CO desorption, which, in turn, can be related back to changes in θ_T with pressure.

VI. Conclusions

Determination of the energetically preferred pathway for transforming reactants to products on a metal surface requires knowledge of the activation barriers for each elementary process. At present, accurate *ab initio* quantum chemical calculations of activation barriers are not feasible and/or too expensive for practically all systems of chemical interest. Semiempirical quantum chemical calculations of realistic systems are affordable but insufficiently accurate to be relied on. Reasonable estimates can be made, though, using phenomenological approaches, particularly the BOC-MP method. This method requires as input only heats of atomic adsorption and gas-phase molecular bond energies. Heats of adsorption for molecular species and molecular fragments and activation barriers for a wide variety of elementary processes can then be calculated using well-defined algebraic relationships. The BOC-MP approach also provides a satisfactory means for identifying the energetically preferred pathways for metal-catalyzed processes, as demonstrated above by its application to an analysis of formic acid decomposition, CO hydrogenation to methanol and hydrocarbons, and hydrocarbon decomposition. Based on BOC-MP calculations, it is possible to project the elementary steps that control catalyst activity and selectivity and the effects of changes in metal composition. The BOC-MP approach can also be used to calculate heats of adsorption and activation energies as a function of surface coverage. Monte-Carlo simulations of desorption from metal surfaces, in which the BOC-MP method is used to determine the local activation energy for desorption, provide a good description of both isothermal and TPD rates.

Acknowledgment

This work was supported by the Director, Office of Basic Energy Sciences, Chemical Sciences Division, of the U.S. Department of Energy under Contract DE-AC03-76SF00098.

References

1. For a review, see R. Hoffmann, *Solids and Surfaces: A Chemist's View of Bonding in Extended Structures*, VCH, New York, 1988.
2. For a recent review, see A. B. Anderson, in *Theoretical Aspects of Heterogeneous Catalysis* (J. B. Moffat, Ed.), Ch. 10, Van Nostrand, New York, 1990.
3. E. Shustorovich, *Adv. Catal.* **37** (1990), 101.

4. For a review, see A. A. Balandin, *Adv. Catal.* **19** (1969), 1.
5. For more recent applications, see, for example, J. Ziolkowski, *J. Catal.* **100** (1986), 45, and references therein.
6. E. Shustorovich, (a) *Surf. Sci. Rep.* **6** (1988), 1; (b) *Acc. Chem. Res.* **21** (1988), 183; (c) *J. Mol. Catal.* **54** (1989), 301.
7. S. J. Lombardo, and A. T. Bell, *Surf. Sci. Rep.*, in press.
8. E. Shustorovich, and A. T. Bell, *Surf. Sci.* **222** (1989), 371.
9. *CRC Handbook of Chemistry and Physics*, pp. 171–190, CRC Press, Boca Raton, Fla., 1984–1985.
10. (a) G. C. Schatz, M. S. Fitzcharles, and L. B. Harding, *Faraday Disc. Chem. Soc.* **84** (1987), 359; (b) B. Ruscis, M. Schwarz, and J. Berkowitz, *J. Chem. Phys.* **91** (1989), 6780.
11. S. L. Miles, S. L. Bernasek, and J. L. Gland, *Surf. Sci.* **127** (1983), 271.
12. J. B. Benziger, and R. J. Madix, *J. Catal.* **74** (1982), 67.
13. N. R. Avery, B. H. Toby, A. B. Anton, and W. H. Weinberg, *Surf. Sci.* **122** (1982), L574.
14. F. Solymosi, J. Kiss, and I. Kovacs, *J. Phys. Chem.* **92** (1988), 796.
15. R. J. Madix, J. B. Gland, G. E. Mitchell, and B. A. Sexton, *Surf. Sci.* **125** (1983), 481.
16. C. Egawa, I. Dai, S. Naito, and K. Tamaru, *Surf. Sci.* **176** (1986), 491.
17. P. Hoffmann, S. R. Bare, N. V. Richardson, and D. A. King, *Surf. Sci.* **133** (1983), 589.
18. (a) B. E. Hayden, K. Prince, D. P. Woodruff, and A. M. Bradshaw, *Surf. Sci.* **133** (1983), 589. (b) E. Iglesia, and M. Boudart, *J. Catal.* **81** (1983), 214.
19. N. D. S. Canning, and R. J. Madix, *J. Phys. Chem.* **88** (1984), 2437.
20. J. E. Crowell, J. G. Chen, and J. T. Yates, Jr., *J. Chem. Phys.* **85** (1986), 3111.
21. R. J. Madix, *Adv. Catal.* **29** (1980), 1.
22. A. Puschmann, J. Haase, M. D. Cropper, C. E. Riley, and D. P. Woodruff, *Phys. Rev. Lett.* **54** (1985), 2250.
23. M. D. Cropper, C. E. Riley, D. P. Woodruff, A. Puschmann, and J. Haase, *Surf. Sci.* **171** (1986), 1.
24. (a) B. A. Sexton, *Surf. Sci.* **88** (1979), 319. (b) L. H. Dubois, T. H. Ellis, B. R. Zegarski, and S. D. Kevan, *Surf. Sci.* **172** (1986), 385.
25. J. B. Benziger, and G. R. Schoofs, *J. Phys. Chem.* **88** (1984), 4439.
26. M. Barteau, M. Bowker, and R. J. Madix, *Surf. Sci.* **94** (1980), 303.
27. D. A. Outka, and R. J. Madix, (a) *Surf. Sci.* **179** (1987), 361; *J. Am. Chem. Soc.* **109** (1987), 1708.
28. J. E. Crowell, J. G. Chen, and J. T. Yates, Jr., *J. Chem. Phys.* **85** (1986), 3111.
29. J. Saussey, J. C. Lavalley, J. Lamotte, and T. Rais, *J. Chem. Soc. Chem. Commun.* (1982), 278.
30. J.-P. Hindermann, E. Schlieffer, H. Idris, and A. Kiennemann, *J. Mol. Catal.* **33** (1985), 133.
31. A. Deluzarche, J.-P. Hindermann, A. Kiennemann, and R. Kieffer, *J. Mol. Catal.* **31** (1985), 225.
32. J. F. Edwards, and G. L. Schrader, *J. Catal.* **94** (1985), 175.
33. Y. Amenomiya, and T. Tagawa, in *Proceedings, 8th International Congress on Catalysis, Berlin*, Vol. II, p. 557, Dechema, Frankfurt-am-Main, 1984.
34. A. Deluzarche, J. Cressley, and P. Kieffer, *J. Chem. Res. (S)* **136** (1979) (M), 1657.
35. A. Deluzarche, J.-P. Hindermann, and R. Kieffer, *J. Chem. Res. (S)* **72** (1981) (M), 934.
36. G. A. Vedage, R. Pitchai, R. G. Herman, and K. Klier, *Am. Chem. Soc. Div. Fuel Chem. Prepr. Pap.* **29**, No. 5 (1984), 196.
37. R. F. Hicks, and A. T. Bell, *J. Catal.* **91** (1985), 104.

38. G. C. Chinchen, M. S. Spencer, K. C. Waugh, and D. A. Whan, *J. Chem. Soc. Faraday Trans.* **83** (1987), 2193.

39. G. R. Sheffer, and T. S. King, *J. Catal.* **115** (1989), 376.

40. G. R. Sheffer, and T. S. King, *J. Catal.* **116** (1989), 488.

41. J. A. Brown Bourzutschky, N. Homs, and A. T. Bell, *J. Catal.* **124** (1990), 52.

42. M. L. Poutsma, L. F. Elek, P. A. Ibarbia, A. P. Risch, and J. A. Rabo, *J. Catal.* **52** (1978), 157.

43. Yu. A. Ryndin, R. F. Hicks, A. T. Bell, and Yu. I. Yermakov, *J. Catal.* **70** (1981), 287.

44. F. Fajula, R. G. Anthony, and J. H. Lunsford, *J. Catal.* **73** (1982), 237.

45. P. J. Berlowitz, and D. W. Goodman, *J. Catal.* **108** (1987), 364.

46. E. Shustorovich, and A. T. Bell, *Surf. Sci.*, in press.

47. I. E. Wachs, and R. J. Madix, *J. Catal.* **53** (1978), 208.

48. K. Bange, D. E. Girder, T. E. Madey, and J. K. Sass, *Surf. Sci.* **137** (1984), 38.

49. E. M. Stuve, S. W. Jorgensen, and R. J. Madix, *Surf. Sci.* **146** (1984), 179.

50. R. A. Hadden, H. D. Vandervell, K. C. Waughy and G. Webb, in *Proceedings, 9th International Congress on Catalysis* (M. J. Philips, and M. Ternan, Eds.) Vol. 4, p. 1835, Chemical Institute of Canada, Ottawa, 1988.

51. C. T. Campbell, and K. A. Daube, *J. Catal.* **104** (1987), 109.

52. C. M. A. M. Mesters, T. J. Vink, O. L. J. Gijzeman, and J. W. Geus, *Surf. Sci.* **135** (1983), 428.

53. J. N. Russell, Jr., S. M. Gates, and J. T. Yates, Jr., *Surf. Sci.* **163** (1985), 516.

54. W. Ho, *J. Phys. Chem.* **91** (1987), 766.

55. (a) G. C. Chinchen, P. J. Denny, J. R. Jennings, M. S. Spencer, and K. C. Waugh, *Appl. Catal.* **36** (1988), 1. (b) M. Bowker, R. A. Hadden, H. Houghton, J. N. K. Hyland, and K. C. Waugh, *J. Catal.* **109** (1988), 263. (c) M. E. Fakley, J. R. Jennings, and M. S. Spencer, *J. Catal.* **118** (1989), 483.

56. J. A. Brown Bourzutschky, N. Homs, and A. T. Bell, *J. Catal.* **124** (1990), 73.

57. C. T. Au, and M. W. Roberts, *Chem. Phys. Lett.* **74** (1980), 472.

58. G. Wedler, and M. Ruhmann, *Surf. Sci.* **121** (1982), 464.

59. F. H. P. M. Habraken, E. Ph. Kieffer, and G. A. Bootsma, *Surf. Sci.* **83** (1979), 45.

60. (a) R. B. Anderson, *The Fischer–Tropsch Synthesis*, Academic Press, New York, 1984. (b) V. Ponec, *Catal. Rev. Sci. Eng.* **18** (1978), 15. (c) P. Biloen, and W. H. M. Sachtler, *Adv. Catal.* **30** (1981), 165. (d) A. T. Bell, *Catal. Rev. Sci. Eng.* **23** (1981), 203. (e) C. K. Rofer-DePoorter, *Chem. Rev.* **81** (1981), 447.

61. E. Shustorovich, *Catal. Lett.* **7** (1990), 107.

62. E. Shustorovich, and A. T. Bell, *Surf. Sci.* **248** (1991), 359.

63. (a) E. Shustorovich, and A. T. Bell, *J. Catal.* **113** (1988), 341. (b) A. T. Bell, and E. Shustorovich, *J. Catal.* **121** (1990), 1.

64. T. J. Vink, O. L. Gijzeman, and J. W. Geus, *Surf. Sci.* **150** (1985), 14.

65. (a) L. Whitman, L. J. Richter, B. A. Gurney, J. S. Villarrubia, and W. Ho, *J. Chem. Phys.* **90** (1989), 2050. (b) E. Umbach, and D. Menzel, *Surf. Sci.* **135** (1983), 199.

66. H. J. Krebs, H. P. Bonzel, and Gafner, *Surf. Sci.* **88** (1979), 269.

67. (a) D. W. Goodman, *Acc. Chem. Res.* **17** (1984), 194. (b) D. W. Goodman, and J. E. Houston, *Science* **236** (1987), 403.

68. (a) F. M. Hoffmann, and J. L. Robbins, *J. Electron Spectrosc. Relat. Phenom.* **45** (1987), 421. (b) F. M. Hoffmann, and J. L. Robbins, in *Proceedings, 9th International Congress on Catalysis, Calgary, 1988* (M. J. Phillips, and M. Ternan, Eds.), Vol. 3, p. 1144; Vol. 5, p. 373, Chemical Institute of Canada, Ottawa, 1988.
69. X. D. Q. Li, T. Radojicic, and R. Vanselow, *Surf. Sci.* **225** (1990), L29.
70. R. C. Brady, and R. Pettit, *J. Am. Chem. Soc.* **102** (1980), 6181.
71. R. C. Brady, and R. Pettit, *J. Am. Chem. Soc.* **103** (1981), 1287.
72. (a) E. Shustorovich, and A. T. Bell, *Surf. Sci.* **200** (1988), 492. (b) A. T. Bell, and E. Shustorovich, *Surf. Sci.* **235** (1990), 343.
73. F. Zaera, *J. Am. Chem. Soc.* **111** (1989), 4240.
74. (a) E. M. Stuve, and R. J. Madix, *J. Phys. Chem.* **89** (1985), 105. (b) M. A. Henderson, G. E. Mitchell, and J. M. White, *Surf. Sci.* **203** (1988), 378.
75. (a) T. B. Beebe, Jr., and J. T. Yates, Jr., *J. Phys. Chem.* **91** (1987), 254. (b) X.-Y. Zhu, and J. M. White, *Surf. Sci.* **214** (1989), 240.
76. (a) K. M. Ogle, J. R. Creighton, S. Akhter, and J. M. White, *Surf. Sci.* **169** (1986), 246. (b) C. M. Greenlief, P. L. Radloff, and J. M. White, *Surf. Sci.* **191** (1987), 93.
77. P.-K. Wang, C. P. Slichter, and J. H. Sinfelt, *J. Phys. Chem.* **89** (1985), 3606.
78. P.-K. Wang, J.-P. Ansermet, S. L. Rudaz, Zh. Wang, S. Shore, C. P. Slichter, and J. H. Sinfelt, *Science* **234** (1986), 35.
79. F. Zaera, (a) *Surf. Sci.* **219** (1989), 453. (b) *J. Am. Chem. Soc.* **111** (1989), 8744.
80. H. Ibach, and S. Lehwald, *J. Vac. Sci. Technol.* **15** (1978), 407.
81. P.-K. Wang, C. P. Slichter, and J. H. Sinflet, *Phys. Rev. Lett.* **53** (1984), 82.
82. N. R. Avery, *Langmuir* **4** (1988), 448..
83. E. A. Carter, and B. E. Koel, *Surf. Sci.* **226** (1990), 339.
84. W. Erley, A. M. Baro, and H. Ibach, *Surf. Sci.* **120** (1982), 273.
85. U. Seip, M.-C. Tsai, J. Küppers, and G. Ertl, *Surf. Sci.* **147** (1984), 65.
86. X.-Y. Zhu, M. E. Castro, S. Akhter, J. M. White, and J. E. Houston, *Surf. Sci.* **207** (1988), 1.
87. X.-Y. Zhu, M. E. Castro, S. Akhter, J. M. White, and J. E. Houston, *J. Vac. Sci. Technol. A* **7** (1989), 1991.
88. S. Lehwald, and H. Ibach, *Surf. Sci.* **120** (1982), 273.
89. J. A. Stroscio, S. R. Bare, and W. Ho, *Surf. Sci.* **148** (1984), 499.
90. J. E. Parmeter, M. M. Hills, and W. H. Weinberg, *J. Am. Chem. Soc.* **108** (1986), 3563; **109** (1987), 72.
91. P. Jacob, A. Cassuto, and D. Menzel, *Surf. Sci.* **187** (1987), 407.
92. S. J. Lombardo, and A. T. Bell, *Surf. Sci.* **206** (1988), 101.
93. S. J. Lombardo, and A. T. Bell, *Surf. Sci.* **224** (1989), 451.
94. S. J. Lombardo, and A. T. Bell, *Surf. Sci.* **245** (1991), 213.
95. F. Zaera, E. B. Kollin, and J. L. Gland, *Surf. Sci.* **166** (1986), L149.
96. J. A. Prybyla, P. J. Estrup, and Y. J. Chabal, *J. Vac. Sci. Technol. A* **5** (1987), 791.
97. J. A. Prybyla, P. J. Estrup, S. C. Ying, Y. J. Chabal, and S. B. Christmann, *Phys. Rev. Lett.* **58** (1987), 1877.
98. P. J. Estrup, *J. Vac. Sci. Technol.* **16** (1979), 635.
99. T. E. Felter, R. A. Barker, and P. J. Estrup, *Phys. Rev. Lett.* **38** (1977), 1138.

100. S. Semancik, and P. J. Estrup, *J. Vac. Sci. Technol.* **18** (1981), 541.
101. R. A. Baker, S. Semancik, and P. J. Estrup, *Surf. Sci.* **94** (1980), L162.
102. H. R. Han, and L. D. Schmidt, *J. Phys. Chem.* **75** (1971), 227.
103. P. J. Estrup, unpublished results, Department of Chemistry, Brown University, Providence, R.I.
104. J. M. White, *J. Phys. Chem.* **87** (1983), 915.
105. T. Yamada, T. Onishi, and K. Tamaru, *Surf. Sci.* **133** (1983), 533.
106. T. Yamada, T. Onishi, and K. Tamaru, *Surf. Sci.* **157** (1985), L389.

Index